Traffic Management
Planning, Operations, and Control

John E. Tyworth
The Pennsylvania State University

Joseph L. Cavinato
The Pennsylvania State University

C. John Langley, Jr.
The University of Tennessee

WAVELAND
PRESS, INC.
Prospect Heights, Illinois

Dedicated to
J. Richard and Dorothea Titsworth
El Cerrito, California

Mary Ann Cavinato
State College, Pennsylvania

Claremont J. and Anne G. Langley
Troy, Ohio

**Selected topical questions from past AST & L examinations
are reprinted with the express permission of the American
Society of Transportation and Logistics, Inc.**

For information about this book, write or call:

> Waveland Press, Inc.
> P.O. Box 400
> Prospect Heights, Illinois 60070
> (708) 634-0081

Preface

*T*his book is about the purchase and use of freight transportation services in the business enterprise—what we refer to as the traffic function. The traditional ways of managing this function have changed critically since 1980. Deregulation and other regulatory reforms have opened the transportation industry to new services, practices, and freedoms. In the aftermath of these changes, progressive companies quickly began to appreciate how aggressive and innovative traffic management could increase profits. Not surprisingly, companies now seek individuals who have the skills and knowledge to conduct sophisticated planning, operations, and control in the traffic area, people who can use this ability to exploit the opportunities available in a freer transportation market.

The purpose of this book is to offer a complete and modern introduction to traffic management. We believe this book will be of value to both business students and traffic professionals already active in the field.

In this book we abandon the customary technical-legal approach to traffic management and instead emphasize how to plan, organize, implement, and control traffic activities. The traffic function is presented in relation to the total business enterprise; we investigate the implications that decision making in the traffic area has for other relevant logistical, marketing, production, and purchasing activities.

This book covers the latest topics critical to effective traffic management in a deregulated environment—contracting, shipment costing, negotiating, antitrust, electronic data interchange, transportation planning, information system support, and control. It also takes new approaches to several traditional topics. The discussion of transportation pricing emphasizes issues, concepts, and planning implications instead of technical procedures. In addition, because of its importance to strategic transportation planning, shipment consolidation is treated as a major subject and not as a by-product of the traditional "terminal and accessorial services" topic.

The book is intended to have a strong practical orientation, which is accomplished by the use of checklists, step-by-step procedures, and other reference material found in end-of-chapter appendices, vignettes summarizing actual company experiences, and numerous practical examples. The book also includes selected topical questions taken from past American Society of Transportation and Logistics (AST&L) exams used in its professional certification program.

We have organized our chapters into three parts. Part I examines the evolution and nature of traffic management and the changing external business environment. It establishes a foundation for the discussion of more specialized subjects in Parts II and III. In Part II, our attention shifts to the internal decision environment of the traffic organization. Topics focus on elements of traffic planning and operations supportive of the overall strategic direction of the firm. Part III investigates information system support, management control, and organizational issues related to the traffic function.

We wish to acknowledge our gratitude and indebtedness to our former logistics and transportation teachers and mentors: Professor Emeritus Roy J. Sampson and the late Professor Harold K. Strom, University of Oregon; Professors Robert D. Pashek, John J. Coyle, John S. Spychalski, and Joseph L. Carroll, The Pennsylvania State University; and Professors Emeritus Joseph L. Frye and Frank L. Hendrix, the University of Tennessee.

Many individuals were helpful during the preparation of this manuscript. We especially want to thank Professor Thomas M. Corsi, University of Maryland, and Walter L. Weart, Road & Sea Transportation Services, Inc.; Professors John C. Sypchalski, John W. Bagby, and Kalyan Chatterjee, The Pennsylvania State University; Professor John Ozment, University of Arkansas; Professors John D. Palmer and Vernon C. Seguin, James Madison University; and Professor Gayton E. Germane, Stanford University; for their most helpful suggestions and comments.

We are grateful to Richard Haupt, Ford Motor Company; Neil S. Llewellyn, Interstate Commerce Commission; Michael V. Wells, Hershey Chocolate Company; Stanley Hoffman, Union Carbide Corporation; and John Kreitner, General

Electric Company, as well as the numerous companies and trade organizations, that generously supplied helpful materials and information.

Special appreciation is extended to Marianne Miller for her extensive and thoughtful editing support. Warm thanks also go to Judy Sartore, LuAnn Jaworski, and Janette Burns for their help in typing portions of the manuscript and to the hundreds of students who pretested chapter drafts.

Finally, we extend a special note of thanks to our wives, Stephanie, Mary, and Anne for their patience, support, and encouragement, and to our children, Mike and Cindy, Janet and Josh, and Sarah and Mercer.

University Park, Pennsylvania   J. E. T.
University Park, Pennsylvania   J. L. C.
Knoxville, Tennessee   C. J. L.

Contents

part one | *Traffic Management in a Changing Business Environment* 1

chapter 1 **Traffic Management in the Firm** 3

Introduction 4
Evolution of Traffic Management 6
Overview of the Traffic Management Framework 14
Summary 21
Selected Topical Questions from Past AST&L Examinations 22

chapter 2 **Transportation Services and Selection** 25

Introduction 26
Legal Influences on Transportation 27
Legal Forms of Carriage 28

Changing Legal Influences 33
Carrier Firms and Entities 41
Mode Selection 43
Carrier Selection 47
Summary 52
Selected Topical Questions from Past
AST&L Examinations 53

chapter 3 **Transportation Pricing in Transition** 57

Introduction 58
Rate Regulation of Common Carriage 60
Rate Regulation of Contract Carriage 69
Adapting to the Evolving Pricing Environment 71
Summary 75
Selected Topical Questions from Past
AST&L Examinations 75

chapter 4 **Tariff Pricing Systems** 79

Introduction 80
Class Rate Framework 82
Commodity Rate Framework 93
Information Technology in Pricing 94
Summary 102
Selected Topical Questions from Past
AST&L Examinations 104
Appendix 4.1   Procedures for Looking Up
a Freight Rate 106

chapter 5 **Freight Rates** 115

Introduction 116
Route-Related Rates 117
Distance-Related Rates 126
Volume-Related Rates 128
Miscellaneous Rates 134

Summary 136
Selected Topical Questions from Past
AST&L Examinations 137

| part two | *Planning and Operations* 141 |

chapter 6 **Transportation Planning** 143

Introduction 144
General Planning Framework 145
Developing a Strategic Business Plan
for the Traffic Area 148
Summary 160
Selected Topical Questions from Past
AST&L Examinations 161

chapter 7 **Making the Shipment** 163

Introduction 164
Documentation 164
Terms of Sale 175
Using the Carrier to Secure Payment
for the Goods 180
Rate Retrieval and Freight Payment Systems 182
Carrier Credit Arrangements 185
Summary 187
Selected Topical Questions from Past
AST&L Examinations 188
Appendix 7.1 Bill of Lading Preparation 190

chapter 8 **Contracting for
Transportation Services** 193

Introduction 194
The Contracting Cycle 195
Potential Pitfalls in Contracting 209

Summary 211

Selected Topical Questions from Past
AST&L Examinations 213

Appendix 8.1   Checklist for Contracts Between
Shippers and Railroads 215

Appendix 8.2   Background Research
for Contracting 229

chapter 9 **Costing and Negotiating** 235

Introduction 236

Shipment Costing 237

Shipment Costing as a Tool for
Traffic Management 250

Negotiating Rates and Services 252

Summary 257

Selected Topical Questions from Past
AST&L Examinations 259

chapter 10 **Shipment Consolidation
and Load Planning** 263

Introduction 264

General Forms of Shipment Consolidation 265

Planning Framework 268

Summary 290

Selected Topical Questions from Past
AST&L Examinations 290

chapter 11 **Accessorial Services and Charges** 293

Introduction 294

Accessorial Linehaul Services 295

Accessorial Terminal Services 301

Summary 314

Selected Topical Questions from Past
AST&L Examinations 315

chapter 12 **Carrier Liability for Loss and Damage** 317

Introduction 318
Liability Regimes 319
Summary 331
Selected Topical Questions from Past AST&L Examinations 333

chapter 13 **Claims Management** 335

Introduction 336
Making the Claim for Loss, Damage, or Delay 336
Controlling Freight Loss and Damage Claims 342
Other Claims Actions 346
Summary 350
Selected Topical Questions from Past AST&L Examinations 351

chapter 14 **Fleet Management** 355

Introduction 356
Private Motor Carriage 357
Private Rail Operations 375
Summary 376
Selected Topical Questions from Past AST&L Examinations 376

chapter 15 **International Traffic Management** 379

Introduction 380
Export-Import Transportation Services 381
Export-Import Facilitators 386
Pricing and Cost Factors 389
Method of Payment 393

Documentation   396

Marine Insurance   398

Export-Import Traffic Issues   399

Summary   401

Selected Topical Questions from Past
AST&L Examinations   401

chapter 16   **Social Regulation Affecting
Traffic Management**   403

Introduction   404

Hazardous Materials Transportation   404

Truck and Driver Highway Safety   412

Other Sources of Social Regulations   412

Summary   413

Selected Topical Questions from Past
AST&L Examinations   414

Appendix 16.1   Catalogue of Social Regulations
Relating to Traffic Management   415

Appendix 16.2   Hazardous
Material Definitions   419

part three   *Control*   423

chapter 17   **System Support and Control**   425

Introduction   426

System Support and Control Framework   426

Information System Requirements   428

Key Techniques   431

Computer Usage Patterns   449

Summary   452

Selected Topical Questions from Past
AST&L Examinations   453

Appendix 17.1   Measuring and Reporting Specific
Benefits Resulting from Transportation Cost
Control Management   456

chapter 18 **Organizing for Traffic Management** 459

Introduction 460
Modern Traffic Management Orientation 461
Elements of Good Organizational Design 464
Major Organizational Influences 465
Organizational Issues and Alternatives 471
Measuring Organizational Effectiveness 479
Organizational Growth in the Traffic Area 480
The Issue of Titles 481
Summary 482
Selected Topical Questions from Past
AST&L Examinations 482

Selected Bibliography 485

Author Index 503

Subject Index 506

part one

Traffic Management in a Changing Business Environment

*T*raffic management involves the purchase and use of freight transportation services for manufacturing and merchandising firms, as well as for governmental, educational, hospital, and other similar non-profit institutions. During the past several decades, many laws, regulations, and institutional practices of carriers and their ratemaking organizations have confined traffic management to a tightly regulated set of rate and service options. As a result, the traffic function assumed a technical-legal orientation. Much time and effort was spent pursuing regulatory matters before the federal and state agencies that regulate transportation.

By contrast, the 1980s have ushered in a challenging new era of traffic management. The marketplace for transportation services has undergone dramatic changes as a result of federal regulatory reforms and deregulation. New services, practices, and freedoms have emerged. Pricing is flexible, and traffic managers can negotiate easily for special services. These changes coupled with rapid advances in information processing technology have created new directions for the conduct of traffic management. It is now seen as an activity to manage intensively in the search for innovative ways to contribute to the organization's goals.

The new business environment has led one major firm, the General Electric Company, to change the title of its traffic executives from *traffic manager* to *transportation buyer*. This simple change, which affected several hundred executives, characterizes a broader pattern of recognition by industry that traffic managers must (1) identify the transportation needs of the firm, (2) translate these needs into plans for the selection and acquisition of carrier services, (3) actually purchase and use the services, and (4) apply controls to assure goals are met.

Unlike in the past, the traffic function is no longer hostage to a relatively inflexible array of transportation rates and services. Today's manager identifies, influences, and buys what is needed to make a maximum contribution to the organization. Companies require executives with the skills and knowledge necessary for sophisticated management of traffic planning, operations, and control.

Part One consists of five chapters that investigate this new business environment and creates a broad conceptual foundation for the presentation of more specialized topics in Parts Two and Three. The material is introduced primarily from the perspective of the profit-oriented user of freight transportation services. When appropriate, the discussion shows how traffic decisions relate to logistical, marketing, and production activities.

Specifically, Chapter 1 examines the evolution of the traffic function and the changes that have forged the current orientation. It introduces the current reasons for the growing recognition of the importance of traffic management by top management. This chapter also presents an overview of the decision framework for traffic management.

Chapter 2 introduces the concepts and issues related to the selection of transportation services. The first section of the chapter presents the legal forms and types of carriage to the student new to the field. This section also serves as a refresher for the more experienced reader. The second section provides an overview of mode and carrier selection criteria and highlights the nature and potential complexity of the transportation purchase decision.

The next three chapters concentrate on the changing environment for transportation pricing. Chapter 3 focuses on the changing scene for rate regulation and on the emerging antitrust issues of interest to traffic management. Chapter 4 investigates the key concepts embodied in current tariff pricing systems. It also discusses the problems and promise of using information technology to automate the rate search. In addition, issues related to this technology are integrated into relevant topics throughout the book. The traditional manual procedures for locating rates in tariffs are contained in the appendix to Chapter 4. Finally, Chapter 5 surveys major rate forms and examines their implications for traffic management.

chapter 1

Traffic Management in the Firm

chapter objectives

After reading this chapter, you will understand:
The role of traffic management in the firm.
The evolution of traffic from a reactive to an aggressively managed function.
The planning, operational, and control elements of traffic management.
The kinds of decisions and responsibilities that the traffic function assumes.

chapter outline

Introduction   *Summary*

Evolution of Traffic Management   *Questions*
Phase I: Limited
organizational integration
Phase II: A significant step forward
Phase III: Total integration

*Overview of the Traffic
Management Framework*
Planning
Operations
Control

Introduction

Transportation provides industry with the vital links between supply sources, production and distribution facilities, and markets that altogether define logistical networks. The larger firms in wholesaling, retailing, or manufacturing, require extensive transportation operations. Retailers such as Sears, J. C. Penney, and K Mart average more than twenty thousand shipments per week through networks that comprise thousands of stores and tens of thousands of suppliers. Likewise, manufacturers such as Ford, Dow Chemical, General Electric, General Mills, and Weyerhaeuser have extensive logistical structures over which thousands of different products — raw materials, component parts, supplies, and finished goods — move to points of production, storage, or consumption.

As a vital activity, transportation can significantly affect the firm's profitability. Transportation costs, expressed as a percentage of the sales dollar, generally range from about 6 percent (for high-value manufactured goods such as machine tools, plastics, and dry grocery products) to perhaps as much as 60 percent (for low-value bulk commodities such as coal or sand and gravel). These costs translate into billions of dollars each year for a company like General Motors and into hundreds of millions of dollars for large firms in food, forest, petrochemical, metallurgical, and other distribution-intensive industries. For the nation, the annual freight bill exceeds $240 billion and represents about 8 percent of gross national product.[1]

Transportation service, moreover, can have a critical impact on the firm's production and marketing efforts. A product has value only if it is available to buyers when and where they need it. Fast transportation, for example, may save the sale of an out-of-stock item. The reliable delivery of undamaged goods offers customers an additional purchase incentive because reliable transportation service permits buyers to coordinate operations with scheduled arrivals, which can reduce costly inventories. Thus, transportation costs and services affect a host of business decisions related to prices, products, markets, facility locations, sources of supply, and, ultimately, profitability.

We define *traffic management* as the planning, implementation, and control of the purchase and use of freight transportation services for the purpose of achieving organizational objectives. By *traffic managers*, we mean all persons involved with these activities. As defined, traffic management represents a key component of business logistics.

Business logistics, which has received extensive coverage in the management literature, is concerned with the cost, as well as the value, added to goods by making them available when and where they are needed. Besides transportation, other basic logistical activities include inventory, order processing, warehousing, materials handling, purchasing, packaging, and customer service. Management's challenge is to integrate traffic activities with other logistical, production, and marketing functions to obtain the maximum contribution for the overall system rather than for any single function.

Although the traffic manager now plays a key role in that challenge, the traditional view of the traffic manager as a highly specialized technician found in the ranks of middle management lingers on, indeed, and with some merit. The status of the traffic function in different organizations can vary significantly. In some small businesses, the person having this title may be a shipping clerk with some supervisory responsibilities. In large firms, this person may carry the title of director, vice-president, or general traffic manager and may be a member of the senior management team. In the General Foods Corporation, which is fairly representative of many of the larger firms, a general traffic manager heads the traffic function and has a multimillion dollar operating budget, exclusive of freight expenditures.

During the 1980s, the importance of traffic management and the need to manage this function aggressively have received increasing recognition from top management. In the first section of this chapter, we find out why this change has taken place. The second section completes our introduction; it outlines the decision-making framework for traffic management and maps the subject matter for the rest of the book.

Evolution of Traffic Management

The total cost concept forms the foundation of logistics and modern traffic management. Rather than treat each cost component separately, the concept is to trade off costs among all activities to minimize total costs. The rationale lies in the interrelated nature of transportation and other logistical, marketing, and production operations. Decisions made in one area affect costs in other areas. Thus, implementation of the total cost concept is likely to be far more effective when the various activities are organized as an integrated system under the control of a single manager.

Although the business community recognized traffic management as a separate occupation in the latter part of the nineteenth century, it was not until the total cost concept received widespread recognition in the 1950s that modern traffic management gained a solid foothold. Before 1950, the better managed firms, of course, were aware of the importance of balancing transportation costs against warehousing, inventory, or other logistical costs.[2] Nonetheless, distribution activities were organizationally fragmented. The growing recognition of the total cost concept in the 1950s gave impetus to the logic of an integrated organizational structure and the single manager. That recognition was primarily the product of two trends: (1) advancing computer technology, which gave managers the means to analyze complicated tradeoffs, and (2) rising domestic and foreign competition, which highlighted the need to control costs. These developments initiated a growing awareness that distribution remained a fertile area for cost reductions.

Since the 1950s, the logistics management discipline has taken on new dimensions as it has adapted to the changing business environment. Likewise, traffic management has grown and has experienced a changing management orientation. Evidence suggests that the logistics organization generally has undergone three phases of development.[3] The number of firms making the transition to the second and third phases has accelerated since the late 1970s. The three phases of development roughly coincide with the decades as follows: 1950s–1960s, 1970s, and 1980s. But in reality, changes in organizational and management orientation are part of a dynamic and continuous evolutionary process. We want to look at that process (see **Exhibit 1.1**) primarily from the perspective of the traffic manager.

Phase I: Limited Organizational Integration

During the first phase of development, the integration of logistical activities under a single manager generally was limited to finished goods transportation and warehousing. Top management saw logistics generally and transportation specifically as passive functions necessary for the support of marketing and production. The business environment, as well as the marketing practices and the

■■■■■■ *Exhibit 1.1* **Evolution of modern traffic management in the logistics organization**

Time Period	Organizational Phase	Traffic Management Orientation	Business Environment
	Activities fragmented Functional management	*Reactive posture*	*Less uncertainty*
1950s– 1960s	I—Limited integration Outbound transportation Private transportation Field warehousing Administration and systems planning	Cost control Procedural Technical	Relatively stable Rising competition Dispersion of industry Comprehensive transportation regulation
1970s	II—Step forward Inbound transportation Inventory Order processing Plant warehousing Customer service	Cost and service Assurance of supply Inbound and outbound Coordination and planning	Energy shocks Supply disruptions Double-digit inflation Lagging productivity Some deregulation
1980s	III—Total integration Sourcing/purchasing Raw materials/work-in-progress inventory Production planning Sales forecasting Distribution engineering	Integrated corporate strategic planning Aggressively influencing rates and services Analytical, costing, and negotiating skills Traffic management as purchasing function	Recession and growth Computer and telecommunication advances Transportation deregulation
	Activities integrated	*Proactive posture*	*More uncertainty*

strong cost-control orientation that prevailed during the 1950s and 1960s, formed that viewpoint.

Relatively stable economic conditions, rising competition, and dispersion of population and industry identify the general features of that business environment. Competitive marketing practices, such as product differentiation and scrambled merchandising, generated costly distribution support requirements by increasing the number of products and retail outlets. The total cost concept received top management attention because it offered a promising way to contain those distribution costs.[4]

Traffic management—reactive and technical. In Phase I, traffic management followed a reactive, technical approach that focused mainly on daily, monthly, or yearly operational tasks. Given marketing service requirements, the traffic manager purchased transportation services with an eye toward cost tradeoffs. Besides tradeoff analysis, technical expertise, especially in the area of rate regulation, played an important role in transportation cost control. Independent government regulatory agencies kept a tight rein on the transportation industry and generally fostered inflexible but predictable rates and services over which shippers had limited influence.

In that environment, knowledge of the legal duties of shippers and carriers was essential because these two groups maintained adversarial relationships. Whenever proposed or existing rates or services adversely affected the firm, the traffic manager typically sought remedy through the regulatory agencies. Furthermore, the tariff expert (like the tax accountant who wades through the swamp of Internal Revenue Service regulations to limit the firm's tax liability) could generate considerable savings through clever applications of complicated tariff rules and rates. Other priority areas for cost reduction, such as claims administration and freight bill audits for overcharges and duplicate payments, required a good deal of technical tariff and paralegal experience.

Service recognition. During the late 1960s, the service dimension gained more of top management's attention. More recognition was given to the premise that faster and more reliable delivery often makes products more valuable to the buyer. Therefore more expensive logistical service can actually increase profits. In other words, fast and reliable deliveries can create additional sales that offset the added expenditures needed to purchase the premium delivery service. As more firms recognized and adopted this view, the traffic-logistics management orientation began to shift from passive support of marketing or production to active participation in corporate strategic planning. That shift grew throughout the 1970s to fruition in the 1980s.

Phase II: A Significant Step Forward

In the second phase of development, organizational integration advanced significantly. Firms began to move order processing and plant warehousing—activities typically found in the marketing and manufacturing functions, respectively—under a single manager. More important, logistics management shed its primary focus on operational activities to adopt a broader planning and control orientation that gave the single manager responsibility for both inbound and outbound flows. As part of the Phase II development, top management created more senior management-level slots for logisticians. The new slots represented an important step forward. They helped to achieve greater coordination with marketing and production and to allow a direct voice in corporate strategic planning.

Technological and economic events essentially explain this transition. The new generation of more sophisticated computer software that firms installed to support order processing, measure transportation performance, and model logistics systems both highlighted the importance of logistics service and gave management more effective tools for planning and controlling that service. Nonetheless, the volatile business environment of the 1970s mainly propelled the maturation of the systems focus. The energy crisis created shortages, supply disruptions, and more rapidly rising costs of transportation. These events caused firms to reconsider past tradeoff calculations and to expand the planning focus to the inbound side of logistics to assure continuous supplies. Many firms began shipping in volume quantities to use the less energy-intensive, cheaper forms of transportation. In effect, tradeoffs favored reduced transport costs with increased warehousing and inventory costs.

During the latter part of the decade, *stagflation* joined the business lexicon to describe high inflation coupled with stagnating growth and productivity. Rapidly rising interest rates curbed the availability of capital and made capital-intensive assets, such as warehouses and inventory, more expensive. Once again, managers had to reconsider the basic tradeoff calculus. This time, the escalating cost of capital and energy forced firms to concentrate their efforts on ways to improve the productivity of capital-intensive assets such as inventories and warehouses. For many, these efforts led to materials requirements planning. Such planning helped management to secure the timely delivery of the raw materials and component parts needed for work in progress (as well as the supply of spare parts, repair equipment, and machinery required for production). The timely deliveries permitted companies to reduce supplies without imposing additional risks of production stoppages. Furthermore, the firms that had originally developed their logistical networks on the assumption of cheap fuel and low interest rates faced the task of redesigning their networks to reflect the fundamental changes in the costs of production and distribution.

Traffic management — broader horizons. In the wake of those events, traffic managers often found their responsibilities had expanded to include inbound transportation and greater planning coordination with purchasing, production, and marketing. Yet throughout most of the 1970s, the traffic manager still had relatively little influence over transportation rates and services, primarily because the transportation industry remained subject to comprehensive regulatory control. As a result, most traffic managers retained a technical orientation. Not surprisingly, other executives continued to view traffic managers, even during the second phase, as specialists rather than as key members of the management team.[5] The challenge remained for traffic managers to become managers first and technical experts second; to become concerned with the whole logistics system rather than with transportation itself; and to devote more attention to the needs of the market and far less to government, regulation, and tradition.

Phase III: Total Integration

In the third phase of organizational development, some firms have created what Bowersox refers to as the "materials logistics" organization.[6] This structure represents an attempt to integrate the total logistics process. Most Phase III firms are currently found in the electronic, food, pharmaceutical, and chemical and allied products industries.[7] **Exhibit 1.2** describes how the logistics organization of one major firm, the Hershey Chocolate Company, has evolved to Phase III.

Exhibit 1.2 **Hershey Chocolate Company at forefront of logistics integration**

Hershey Foods Corporation was like most other firms during the late 1960s in that the various operations related to logistics were organizationally fragmented. The finance department was responsible for order processing and billing. Field warehousing reported directly to the chairman of the board, whereas traffic, production planning, and inventory control reported to the secretary. Meanwhile, purchasing reported to the president. As a result, the goals of the department managers primarily focused on cost saving measures within each area of responsibility. Savings in one department often ended up costing the company money in another.

Hershey recognized the problem and, with the help of a leading consulting firm, decided to form a totally integrated logistics organization. In the early 1970s, the traffic, field warehousing, and production planning and inventory control functions, as well as distribution planning, were united under a director of distribution, as shown below.

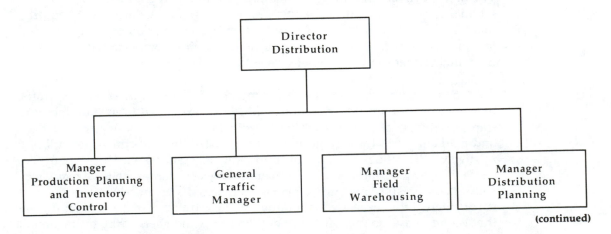

(continued)

Exhibit 1.2 (continued)

Earl J. Spangler, a manufacturing executive, was selected to head this new organization. Two years later, Spangler was named vice-president of manufacturing in addition to his distribution responsibilities. This dual role lasted about three years and, in retrospect, created a climate in which stock availability and order fulfillment were viewed as equally important to high-quality, efficient production. This transitional period also created an environment in which production planning became the facilitator between sales and marketing, and manufacturing.

In 1976, Spangler was appointed president of the newly formed Hershey Chocolate Company, a division of Hershey Foods Corporation. Over the next few years, the distribution organization took over responsibility for order entry, order processing, and billing from the finance department and created a customer service function, some of which had previously been performed in sales. The head of distribution was elevated to vice-president at that point. In 1981, the purchasing function was assigned to distribution, and the overall function was retitled logistics.

By 1984, the logistics organization had evolved to that shown below. Operations planning is responsible for production planning, inventory control, capacity analysis, material requirements planning, as well as logistics planning and logistics systems development. Field operations is responsible for inventory deployment and field inventory control, public field warehouses, and company-owned distribution centers. Purchasing has traditional procurement responsibilities along with management of production material inventories. The transportation department is organized into production materials and finished goods transportation groups. The distribution services unit enters and processes local less-than-truckload orders and all truckload orders, and has responsibility for customer service and billing.

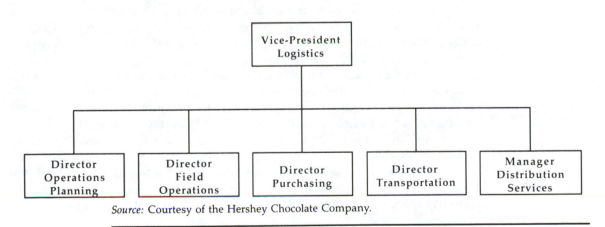

Source: Courtesy of the Hershey Chocolate Company.

In these firms, the single manager generally has responsibility for the following five additional activities: (1) distribution engineering (setting and monitoring productivity standards), (2) sourcing (selecting responsible sources of supply), (3) production planning, (4) raw materials and work-in-progress inventory, and

(5) sales forecasting. Except for distribution engineering, these functions generally have a greater impact on inventory assets than on transportation. From the traffic perspective, the inclusion of sourcing and raw materials and work-in-progress inventory contributes to the total integration of transportation planning and operations into all material flows.

Phase III developments, like Phase I and Phase II, represent an adaptation to the changing external environment. As already indicated, during the 1970s, the logistics service dimension, the energy crisis, and the capital crunch helped establish the strategic role of logistics. Near the end of that decade, the emerging economic, regulatory, and technological conditions created even more uncertainty and again changed the relative costs of capital, energy, labor, and information.

Stagflation was followed by recession and growth. Government policy promoted a freer market environment. And the tempo of technological change increased rapidly. Altogether, these changes created new opportunities, but more uncertainty, for the traffic manager. Shippers have considerably more influence over rates and services now than in the past. Furthermore, telecommunications and computer technology have given the manager powerful new tools to coordinate operations and to consider complex relationships far more comprehensively, cheaper, and faster than was thought possible a mere decade ago.

Firms continue to search for ways to reduce capital-intensive inventory and warehouse assets or to make those assets more productive. As part of this effort, management now gives considerable attention to *just-in-time deliveries*, a concept applied with great success by the Japanese. Suppliers schedule deliveries to arrive at the moment buyers need the goods (see **Exhibit 1.3**). This task requires the application of reliable transportation, sourcing design (the selection and location of suppliers), and information technology, such as direct electronic links to suppliers through which daily or weekly requirements are transmitted.

To summarize briefly, the changing external environment forced many firms to realign manufacturing (or merchandising) and distribution operations to conform to the conditions of the 1980s. This process made top management more aware that centralizing and integrating logistical functions with manufacturing and marketing activities at top corporate levels offered a highly effective method to accomplish the task and to exploit opportunities in this new environment. Phase III organizations represent tangible evidence of that development.

Traffic management — A new era. For traffic management, the 1980s have ushered in a challenging new era. Federal legislation, subsequent deregulation, and regulatory reforms have restructured the transportation industries. Coupled with technological advances in information processing, this restructuring has radically altered the decision environment of the traffic manager. New forms of carriage, practices, and freedoms have emerged. Inflexible rate structures have given way to more of an open market environment that places much greater emphasis on negotiation and bargaining. At the same time, the costs and the

■■■■■■■■ *Exhibit 1.3* **Direct computer links help Ford keep Firestone's tires rolling**

Firestone Tire and Rubber Company and Ford Motor Company have established direct computer links between Ford's Louisville truck plant and Firestone's Akron headquarters to coordinate production requirements, calibrate shipment scheduling for just-in-time deliveries, and cut inventory levels without jeopardizing either firm's plant operations. Each night, the two firms exchange important information (see diagram below). From Louisville, Ford computers send to Firestone's main computers the types and numbers of tires required for truck production for each of the next three days. In return, Firestone's computer sends the daily shipment status of tires in transit. In addition, each workday around 7 A.M., a Firestone inventory specialist contacts his or her counterpart at Ford to discuss and determine the number of tires needed to keep Ford's production line going. Firestone then schedules a shipment from either a tire factory or a distribution center to arrive at the truck plant the following morning.

Tires are now supplied about three days ahead of actual use instead of one week. This allows Ford to stock the truck plant with about 50 percent fewer tires. Likewise, with timely production planning information, Firestone has been able to reduce safety stocks for unexpected contingencies and to exploit shipment planning opportunities.

Tire Requirements — Next Three Days

```
 ┌─────────────┐                                              ┌─────────────┐
 │    Ford     │ ∿∿∿∿∿∿∿∿ (Direct computer links) ∿∿∿→        │  Firestone  │
 │ truck plant │ ←∿∿∿∿ (Information about tires in transit)∿∿ │ headquarters│
 │Louisville,KY│                                              │  Akron, OH  │
 └─────────────┘                                              └─────────────┘
```

Next-day delivery

Production and shipment
planning information

```
        ┌─────────────┐
        │  Firestone  │
        │   plant or  │
        │  warehouse  │
        └─────────────┘
```

Source: Adapted from Leon R. Brodeur, President and Chief Operating Officer, The Firestone Tire & Rubber Company, "Streamlining Tire Distribution." Reprinted, with permission, from the 1983–84 Presidential Issue of *Handling and Shipping Management*, Copyright © 1983, Penton/ IPC, Inc.

quality of similar transportation services, as well as the long-term financial viability of carriers, show more variability.

These changes have made top management sensitive to the importance of transportation costs and service and the need to capitalize on new opportunities. Altogether, they have created new directions for the conduct of traffic management. The traffic function is now seen as an activity to be managed intensively in the search for innovative ways to contribute to the profitability of the firm. Rates and services are no longer an inflexible constraint; they are an element of strategic planning. Traffic managers have the opportunity to custom tailor the transportation link to fit the optimal network design, rather than rely on off-the-rack carrier services. This change, probably more than any other, has led firms to treat traffic as a key element of logistics and corporate strategic planning — an outcome that has given the traffic professional greater visibility.

In a changing environment, companies must have executives with the skills and knowledge necessary for effective management of transportation planning, operations, and control. Successful managers will need solid analytical and interpersonal skills, besides knowledge of traffic and logistics principles and concepts. As aptly stated elsewhere, "power resides in the hands of those who understand how to use information, how to find it, organize it, and make it useful."[8] New, more powerful, and relatively inexpensive computer-supported management tools (for example, electronic spreadsheet, operations research, forecasting, statistical, shipment costing, and routing software for the microcomputer) assault the marketplace with increasing frequency. Those tools linked to information stored in mainframe computers (data bases) and electronic data interchange facilities (direct computer-to-computer communication) offer opportunities that will challenge the most able traffic manager's analytical and creative energy.

The new environment allows managers to extend that energy to negotiations and bargaining with carriers. Knowledge of contracting techniques and good interpersonal skills will greatly enhance the likelihood of successful transactions. These skills are also essential for the development of effective coordination and working relationships with other managers and staff, especially at top corporate levels.

Overview of the Traffic Management Framework

An overview of the decision-making framework for traffic management is best introduced schematically in a flowchart. All the basic relationships among the key elements that constitute the firm's internal and the external environments can be seen at a glance. As shown in **Exhibit 1.4**, traffic planning, operations, and control are an integral part of logistical and corporate strategies. The schematic also shows traffic management in relation to the transportation industry, the regulatory and legal institutions, and the other basic factors that also make up the firm's external environment. Note that bidirectional arrows simply imply a dynamic, interactive process in which influence flows both ways.

Planning

In this overview, the discussion of transportation planning focuses on objectives, selection, acquisition methods, and resources.

Exhibit 1.4 Overview of traffic management decision-making environment

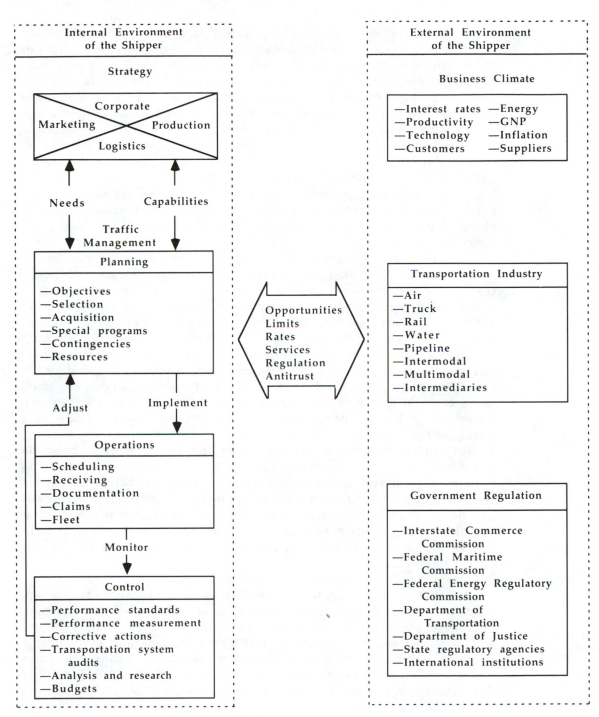

Internal Environment of the Shipper

Strategy

Corporate
Marketing — Production
Logistics

Needs   Capabilities

Traffic Management

Planning
—Objectives
—Selection
—Acquisition
—Special programs
—Contingencies
—Resources

Adjust   Implement

Operations
—Scheduling
—Receiving
—Documentation
—Claims
—Fleet

Monitor

Control
—Performance standards
—Performance measurement
—Corrective actions
—Transportation system audits
—Analysis and research
—Budgets

Opportunities
Limits
Rates
Services
Regulation
Antitrust

External Environment of the Shipper

Business Climate
—Interest rates   —Energy
—Productivity   —GNP
—Technology   —Inflation
—Customers   —Suppliers

Transportation Industry
—Air
—Truck
—Rail
—Water
—Pipeline
—Intermodal
—Multimodal
—Intermediaries

Government Regulation
—Interstate Commerce Commission
—Federal Maritime Commission
—Federal Energy Regulatory Commission
—Department of Transportation
—Department of Justice
—State regulatory agencies
—International institutions

1. Objectives. In more advanced phases of organization development, the traffic manager participates in the development of logistics and corporate goals. The interactive process of coordinating logistical objectives with marketing, operations (manufacturing or merchandising), and overall corporate goals, as well as creating strategic plans to attain those objectives, determines traffic management's needs. These needs are defined in terms of products, purchases, supply sources, facilities, flows, competitive posture, and service levels. The capabilities of traffic management include elements such as the amount of managerial expertise and talent, the nature of the organization, the level of information system support, and the size of the traffic department's budget. Together, needs and capabilities shape the traffic management objectives and plans that apply to various management levels (corporate, regional, group, division, department, or plant) and guide operations to maximize their contribution to the firm.

Making a maximum contribution to the firm requires the traffic manager not only to assure the availability of required transportation services and control costs but to anticipate and aggressively pursue new opportunities. In what is nearly an open-market situation, the prudent manager must assume that the availability of carrier services will not remain constant and must plan accordingly. Developing cooperative (rather than adversarial) relationships through long-term contracts, in which the firm has an interest in the well-being of the carrier, is one way the traffic manager might actively contribute. Further, because transportation affects the overall quality of customer service, the traffic manager might also contribute by using the firm's leverage (say, from desirable volume traffic) to negotiate higher quality transport service at the same or even a reduced cost.[9]

Although assurance of supply and cost control remain important goals, the key word is *contribution*. As stressed earlier, the total cost concept guides the traffic manager to evaluate the impact of transportation purchases on other cost centers and on customer service. In this effort, the traffic manager must move beyond the traditional reactive posture. Like the industrial purchasing executive, the traffic manager must plan carefully and aggressively seek ways to influence the price-service mix in a way that makes a greater contribution to overall corporate goals.

2. Selection. Once objectives have been established, the traffic manager needs to consider all applicable transportation alternatives in the development of strategic plans. The planning horizon extends both to inbound, outbound, and interfacility transportation and to related services. Plans must address trends in energy, technology, and other elements that influence the business and transportation environment. Further, planning needs to consider probable carrier actions and strategies related to activities such as mergers, abandonments, and multimodal operations.

The choice of transportation can get quite complex. A multitude of different combinations of modes, carriers, and types of equipment create a vast array of services that have different price and performance attributes. Fortunately, in most instances, relatively few combinations are appropriate for a given product and market.

3. Acquisition methods. The method of acquisition is closely related to the selection process, as well as to corporate purchasing (see **Exhibit 1.5**). The traffic manager can use tariffs, contracts, lease-buy agreements, or intermediaries (agents, brokers, or cooperatives) to acquire transportation services. Acquisition planning also involves the selective use of terms of sale for control of both acquisition and selection.

As suggested earlier, one of the most interesting and creative aspects of the acquisition process involves opportunities available to negotiate new alternatives with the carriers. New alternatives are not confined to linehaul transportation (over-the-road movement) but may involve demurrage, credit, documentation, communications, switching, weighing, and other accessorial or special services. In addition, the acquisition process includes special programs for hazardous materials transportation or government traffic, and contingency planning for events such as labor strikes, energy shortages, product recalls, or material shortages.

4. Resources. To accomplish its mission, the traffic function requires resources in the form of information, personnel, organizational structures, facilities, and funding. Information, of course, is essential to planning and decision making. The more volatile pricing environment and the more rapid pace of change in the transportation industry make access to timely and accurate information increasingly important. Integrated information system support also helps achieve greater coordination among activity centers.

Practitioners frequently note that traffic is data rich but information poor. Although traffic operations generate abundant data from bills of lading, freight bills, arrival notices, claim forms, and other documents, resources are needed to organize and integrate that data into information that supports planning and operations. For example, information systems, capable of creating freight flow profiles (shipment size and frequency by product type, mode of transportation, and carrier for origin-destination pairs) represent key resources to support planning and control.

Besides information, the nature of the organization and the qualities of the staff represent important resources. As already indicated, there appears to be a trend among many of the top industrial firms toward the formation of a materials-logistics organization, a structure that integrates all inbound and outbound logistical activities and places them under a single manager. (The traffic

■■■■■■ *Exhibit 1.5* **Traffic management as a corporate purchasing process**

Task	Purchasing	Traffic
Quality policy	Define the quality of goods required by the firm. Does not always mean highest quality, but quality level in light of what is needed.	Determine what type and quality of carrier services will satisfy the movement needs of the company's goods. Many times this is initially established by corporate customer service standards and translated into movement terms by traffic managers.
Price policy specifications	Determine appropriate prices to pay for the quality of product or services to be obtained. Represent targets in negotiations.	Establish ranges of prices appropriate when negotiating for carrier services. Provide input to budgeting process.
Sourcing	Continuous process of finding, using, and developing firms from which to buy. A wide range of vendors will often assure competitive selling to the firm. Ethical buying practices included here.	Develop contracts and relationships with carriers that will assure the firm of transportation service in future.
Make–buy	Evaluation of whether to buy product from the outside or produce it in-house	Evaluate alternative methods of purchasing transportation service, including tariffs, contracts, and intermediaries in relation to what private carriage would cost and the service configuration it would provide the firm. Includes analysis of providing some of the services typically handled by carriers.
Quality-control standards	Determine what is to be measured in the quality area.	Defining parameters of carrier services and prices to have tangible measures of transportation supportive of the firm's customer service goals
Inspection	The means for testing and determining measures of products and services obtained	Measure process of transit times, pickup reliability, rates charged, and loss and damage experience.
Standards	Active participation in the development of budgets and performance measures that superiors will use to judge departmental performance	Determine appropriate measures of traffic department performance. Develop management information systems in traffic area.
Value	Evaluation of present products and processes with intent of reducing costs, improving quality, or creating other means of improvement.	Includes evaluation and testing of new services; methods of loading or bracing; other forms of packaging; rating; bill payment; and any interfacing inventory, warehousing, or other functional activities.

organization, as well as issues involving centralization versus decentralization and line versus staff structures, is discussed in Chapter 18.)

Facilities and equipment, and the budget to support planning and operations, identify key resources. The plants, distribution centers, and other facilities that define the logistical network represent a crucial dimension of traffic management. Perhaps less obvious, but of great importance, computer and telecommunication equipment offers exciting new ways to manage traffic. Although start-up costs (equipment, administration, training, and so on) may be high, the potential gains in productivity are tremendous. For example, automated rate retrieval and paperless billing or claims transactions are two significant areas of opportunity.

Operations

To the casual observer, many elements of operations appear to take place with little or no effort. Yet without careful management supervision, it is probable that the firm unnecessarily will experience higher inventories and product loss or damage, as well as aggravated relations with customers and other departments. This section of the chapter introduces the major operating tasks, responsibilities, and terminology, which have been organized into five groups:

1. Scheduling. Following planning and policy guidelines, traffic personnel at the various locations must schedule transportation services to mesh with production and marketing operations. This task requires advance coordination with carriers to assure that the proper equipment (capacity and design) arrives for loading in suitable, that is, usable and clean condition. More flexible policies, of course, will allow some discretion in the selection process to take advantage of local conditions. Other responsibilities include screening potential new carriers that show an interest in bidding for the firm's traffic and entering the schedule-order data into the corporate information system.

2. Making and receiving shipments. Making and receiving shipments involves tactical planning and control of shipments and supervision of loading or unloading freight. Tactical planning and control requires supervision of tasks such as

- Taking advantage of shipment consolidation opportunities
- Arranging pallet or container load configurations
- Tracing and monitoring the progress of shipments
- Changing the routing to expedite shipments or to divert shipments to new destinations
- Ordering special services such as refrigeration or exclusive use of the vehicle

Loading and unloading responsibilities include supervision of penalty charges (detention or demurrage) for detaining carrier's equipment after the expiration of free time, weighing and inspection procedures, and industrial switching activities. In addition, the traffic function is responsible for proper packing, marking, blocking, bracing, and stowing of freight in accordance with carrier rules and regulations. These latter tasks, of course, are especially important for hazardous material shipments.

3. Preparing and processing shipping documents. Shipments produce a great deal of paperwork, especially in international trade. Traffic personnel are responsible for the proper use, control, and administration of shipping documents (such as bills of lading, freight bills, arrival notices, and delivery receipts). They are also responsible for supporting paperwork (for example, purchase orders, sales invoices, and requisitions). In the preparation of the bill of lading, traffic personnel frequently prerate shipments. That is, they extract rates from computer-supported data bases or look up rates in tariffs and then annotate the correct rate on each bill of lading. The traffic function must also install administrative controls, such as procedures for postshipment audits of freight bills, to minimize or eliminate overcharges and duplicate payments.

4. Claims prevention and processing. Claims prevention and settlement identify another important part of traffic operations. Prevention is essential to good cost-control management. Experts estimate that shippers and carriers collectively have the capacity to control about one third of the billions of dollars lost annually. Preventing loss or damage of freight, moreover, is essential to good logistics service. Damaged freight clearly ruins whatever benefit timely or fast delivery service might have on the customer.

When loss or damage does occur, the traffic function initiates or administers the claims recovery actions that include informal settlements, arbitration proceedings, and formal litigation. The different levels for land, water, and air common carriage make the settlement of loss and damage claims a challenging task. In addition, the decision to purchase insurance to protect against freight loss or damage is an important collateral responsibility, especially in the management of export-import traffic.

5. Supervising fleets. Fleet supervision encompasses privately owned and assigned vehicles or equipment. Many shippers run their own private trucking operations. In addition, some companies manage private and assigned fleets of railroad freight cars. Private fleets comprise cars that shippers own or lease. Assigned cars belong to railway companies but are assigned to the exclusive use of large shippers primarily in the auto, food, petrochemical, metallurgical, and forest industries. Ford and General Motors, for example, each have combined fleets of bilevel and trilevel automobile "rack" cars and "parts" boxcars that exceed twenty thousand vehicles. These fleets alone generate more than $900 million in freight revenues each year. Fleet management responsibilities also may involve supervision of intermodal transportation assets such as trailers, shipping containers, or other equipment with similar functions.

Companies may want to organize private fleet operations as a separate activity in the logistics organization. Traffic management can then view such transportation as another price-service option without developing a vested interest in perpetuating it by having daily responsibility for its operations.

Control

Effective management of the planning and operations processes requires a control system. A good control system consists of the following elements: (1) a set of standards to measure progress toward organizational goals, (2) a system of measurements to compare actual performance against standards, and (3) a mechanism for correcting deviations from the standards. As part of the control process, firms may conduct annual reviews (transportation system audits) of all aspects of planning and operations, as well as analyses and research of new alternatives and services.

Summary

Transportation touches virtually all commercial activity. For the industrial or institutional user, it is the vital link in the logistics network over which products flow from sources to end users. Transportation costs and services can have a significant impact on production and marketing activities, and, ultimately, on profitability.

Traffic management is the professional effort to manage the purchase and use of transportation services to contribute to the goals of the organization. It involves the planning and control of transportation services as part of an interrelated system that encompasses logistical, marketing, and production activities.

The total cost concept forms the foundation of modern traffic management. Since the 1950s, when the total cost concept gained widespread recognition, logistics and traffic organizations have experienced three phases of development in response to the changing external environment. Owing to these developments, the general organizational structure and management orientation has changed. These developments explain why so many firms now recognize traffic management as an area to be managed intensively in the search for ways to contribute to organizational goals. They also explain why good traffic managers need solid analytical and interpersonal skills.

The decision-making framework for traffic management consists of (1) planning the selection and acquisition of transportation services, (2) managing day-to-day operations, and (3) controlling plans and operations.

Traffic planning is an integral part of corporate and logistics strategy and comprises the following basic elements: objectives, selection, acquisition methods, and resources. Besides cost control and assurance of supply, the fundamental objectives of traffic management include the aggressive search for ways to influence the price-service mix to make the maximum contribution to organizational goals. With corporate and traffic goals in hand, the planner must consider the alternative transportation services and acquisition methods available in a changing industry-regulatory environment. The availability of resources—information, personnel, facilities, and funds—both permits and limits effective traffic management.

Traffic plans are implemented through day-to-day operations, which include (1) scheduling, (2) making and receiving shipments, (3) preparing and processing shipping documents, (4) claims prevention and processing, and (5) supervising fleets. A good traffic control system comprises a set of standards to measure progress, a way to measure actual performance, and a mechanism for correcting substandard performance.

Selected Topical Questions from Past AST&L Examinations

1. Throughout the years, the total cost criterion has been recommended to determine the most appropriate distribution system configuration. Recently, however, the alternative of maximizing contribution to profit (rather than simply minimizing cost) has received substantial attention and support. For what specific reasons might management choose the criterion of maximizing the contribution to profit rather than minimizing cost? (Spring 1981)
2. It has recently been said that transportation has evolved into essentially a purchasing function, the "purchasing of a service." Do you agree or disagree? Explain fully. (Spring 1985)

Notes

1. *Transportation in America*, 3d ed. (Washington, D.C.: Transportation Policy Associates, March 1985), p. 4.
2. See G. Lloyd Wilson, *Industrial Traffic Management Part 1* (Chicago: Traffic Service Corp., 1935–1936), p. 16; Roy J. Sampson, Martin T. Farris, and David L. Shrock, *Domestic Transportation: Practice, Theory, and Policy*, 5th ed. (Boston; Houghton Mifflin, 1985), p. 290.
3. Jack W. Farrell, ed., "Organization study: Distribution departments gain ground," *Traffic Management* 20 no. 9, (September 1981), pp. 42–50; see also Douglas M. Lambert, James F. Robeson, and James R. Stock, "An appraisal of the integrated physical distribution

management concept,'' *International Journal of Physical Distribution and Materials Management* 9 no. 1 (1978), p. 84; Wendell Stewart, ''Traffic and transportation managements' role in the evolving distribution organization,'' in *American Society of Traffic and Transportation 36th Annual Meeting, August 12–14,* 1981 pp. 11–44; Donald J. Bowersox, ''Emerging from the recession: The role of logistical management,'' *Journal of Business Logistics* 4 no. 1 (1983), pp. 21–34.

4. For additional discussion, see Bowersox, ''Emerging from the recession,'' pp. 21–34; see also *Logistics Management,* 2d ed. (New York: Macmillan, 1978), pp. 4–6.

5. Harold Koontz, ''Management and transportation,'' *Transportation Journal* 6 no. 4 (Summer 1966), p. 22; see also Alan J. Stenger, ''Organizing and managing tomorrow's traffic management activities,'' *32nd Annual Meeting of the American Society of Traffic and Transportation* (1977), pp. 53–67.

6. Bowersox, ''Emerging from the recession,'' pp. 21–34.

7. Jack W. Farrell, ed., ''Organization study,'' *Traffic Management,* pp. 42–50; see also Earl J. Spangler, ''Full scope activities through logistics,'' *Handling and Shipping Management* (Presidential Issue 1982–1983), pp. 30–38.

8. Presentation by Peter F. Eder, ''Report on joint conference — Part one: Changing concepts of transportation,'' *Traffic Quarterly* 27 no. 2 (April 1983), pp. 166–192.

9. See Masao Nishi, ''Measuring the transportation manager's contribution to company profits,'' *Handling and Shipping Management* (Presidential Issue 1983–1984), pp. 83–88; Patrick Gallagher, ed., ''Corporate transportation finds its role at FMC,'' *Handling and Shipping Management* (May 1982), pp. 85–92; Masao Nishi and Patrick Gallagher, ''A new focus for transportation management: Contribution,'' *Journal of Business Logistics* 5 no. 2 (1984), pp. 19–29.

chapter | 2

Transportation Services and Selection

chapter objectives

After reading this chapter, you will understand:

The sources of law that influence the transportation community.

The legal forms of carriage and types of services found in the transportation industry.

The changing legal influences on transportation.

The cost and service elements involved in the purchase of transportation.

chapter outline

Introduction

Legal Influences on Transportation

Legal Forms of Carriage
Common carriage
Contract carriage
Private carriage
Exempt carriage

Changing Legal Influences
Motor carrier industry
Rail carrier industry
Air freight industry
Surface freight forwarder industry

Carrier Firms and Entities

Mode Selection

Carrier Selection
Screening carriers
New directions

Summary

Questions

Introduction

The field of traffic management is closely linked with the legal and regulatory environment. Historically, the transportation industry has been subject to regulation and other legal influences in the United States, and the various legal forms of carriage have evolved under these constraints. The deregulatory movement of the 1970s and 1980s relaxed many long-standing restrictions governing new carriers and gave carriers more freedom to expand the scope of operations or to introduce completely new services. Today, shippers and receivers have wide ranges of services for their selection. In addition, new forms of services can easily be obtained.

This chapter presents an overview of the sources of law that influence the transportation field. It reviews the traditional legal forms of carriage and introduces the different kinds of companies and legal entities that populate the transportation industry. The chapter concludes with a discussion of the cost and service elements to consider when purchasing freight transportation.

Legal Influences on Transportation

The transportation industry has a long tradition of economic and safety regulation. Economic regulation affects transportation pricing, entry into and exit from the transportation industry, financial elements, and service standards. Safety regulation concentrates on practices such as training requirements, vehicle operations, and maintenance standards.

As shown in **Exhibit 2.1**, seven major sources of law influence carrier and shipper activities. The *transportation laws* that Congress passes and the president signs into law are found in the *United States Code* (U.S.C.), a compilation of all federal laws in force. This legal source is broken down into various "titles" or "chapters." Title 49, for example, contains the legislative provisions that regulate domestic rail, motor, and water carriers. Title 14 addresses aviation matters, and laws governing international water transportation are a part of Title 46.

Another source of legal influence comes from *transportation regulations*. Many points of law are not spelled out in detail in the United States Code. Instead, the laws empower specific agencies created by Congress to fill in the details and administer many provisions of the laws. Such administrative regulations are the specific rules created by the agencies like the Interstate Commerce Commission

Exhibit 2.1 Sources of legal influences in traffic management

Transportation Laws
Congress enacts, and president signs bill making it law. Law appears in United States Code. Domestic surface laws are found in Title 49 U.S.C. Revised Interstate Commerce Act is part of 49 U.S.C.

Transportation Regulations
Laws passed and found in U.S.C. allow regulatory agencies to fill in the gaps in laws passed. Agencies develop administrative regulations. When put into force, these are found in weekly publication *Federal Register* and updated yearly in *Code of Federal Regulations.*

Contract Law
State laws passed and adopted into Uniform Commercial Code dealing with contract law and law of claims

Court Decisions
Decisions of courts in interpreting transportation laws

Agency Decisions
Decisions of regulatory agencies in matters under their jurisdiction

State Laws
Similar process to federal laws

State Agency Decisions
Can be similar to federal level

(ICC). These regulations are published in the *Federal Register* as they are created. The ICC publishes its own register of actions, and this is called the *ICC Register*. These regulations are updated annually in a publication known as the *Code of Federal Regulations* (CFR). For example, the ICC's rules are found in Title 49 of the CFR.

Contract law also influences carrier-shipper relations. Although many principles of contract law stem from statutory law and administrative regulations, others arise from the federal *Uniform Commercial Code* (UCC), especially Articles 2 and 7, which address issues in contracting and carrier liability. The Uniform Commercial Code was a model law developed for individual states to adopt in order to bring about some conformity of contract law among the states. It is a form of law that was not written with transportation in mind, since transportation was governed by the Interstate Commerce Act. Today, there are some difficult areas of interpretation open in the application of some parts of the UCC to transportation practices.

Court decisions made in cases involving transportation laws and regulations also help define the law. Such decisions establish boundaries and fine points that guide future actions.

Agency decisions involve cases in which the federal regulatory agencies address issues of concern to the transportation community and establish rules or provide interpretations of laws or regulations. These decisions have the same influence as do court case decisions. Examples of these decisions are found in ICC reports.

State laws related to transportation affect carriers and shippers. Likewise, *state agency decisions* about transportation-related issues influence the transportation community. These two sources, however, primarily pertain to intrastate matters.

No single source of legal influence predominates. Traffic managers need to be aware of all the sources, as well as of the changes taking place in each.

Legal Forms of Carriage

Classifying carriers according to legal form comes from the economic regulation of transportation. The four legal forms of carriage are common, contract, private, and exempt. These legal forms generally characterize the nature of services performed and indicate the amount of economic regulation applied to carriers.

Common Carriage

A *common carrier* holds itself out to serve the general public for compensation. Independent regulatory agencies like the ICC previously protected common carriers from excessive transport competition by making it difficult for potential or existing carriers to obtain new or expanded operating authority. In other words, this policy created legal barriers to entry for all legal forms of carriage. In exchange for this protection, common carriers were required to serve all shippers on reasonable request. In addition, regulation emphasized reasonable and non-discriminatory rates, as well as a high degree of safety and service. Since 1980, the transportation community has seen a fundamental change in regulatory policy. Deregulation and regulatory reforms have reversed that emphasis. Common carriers have new rate and service freedoms, but the legal barriers to participation in all forms of carriage have been largely dismantled. Generally speaking, the current policy is for competition in the marketplace to replace much of the past economic regulation as the means of providing reasonable rates and services. (The next section of this chapter elaborates further on recent changes in transportation regulation.)

The recent changes in regulatory philosophy notwithstanding, the common carrier's obligation to serve all shippers on reasonable request never was absolute. Traditional legal defenses for refusing service include

- No special equipment to haul a particular commodity
- No operating ability or authority to serve the origin, destination, or route in between
- Insufficient equipment
- Improper packaging and loading that pose danger to equipment and crew
- Goods of extreme value that are specifically exempted from the requirement to carry them
- Embargoes

Historically, the transportation community has viewed common carriers as the backbone of the transportation industry in the United States. These carriers are capable of handling freight of all types to and from any community in the nation. The vast network of routes and terminals is giving shippers service unparalleled in any other country. Examples of common carriers include Conrail, Santa Fe Railway, Roadway Express, Yellow Freight, Consolidated Freightways, Averitt Express, United Airlines, all the prime passenger carriers, and Dravo Bargeline. In 1985, nearly all of the more than five hundred railroads existed as common carriers, as did approximately forty thousand motor carriers in terms of outstanding ICC certificates.

Contract Carriage

Contract carriage is a legal form of for-hire transportation that provides specialized services to relatively few shippers or receivers. Contract carriage involves specific shipper-carrier arrangements; it is not a service held out to the general public as is common carriage.

Contract carriers in the motor and water modes of domestic transportation are separate legal entities that require permits for operating authority. ICC restrictions on the scope of motor contract operating authority limited this form of carriage until the late 1970s. For example, dual rights, whereby one carrier possesses both common and contract carriage operating rights, were prohibited with a few specific exceptions. In addition, the ICC had a policy of limiting carriers to eight separate shipper contracts. These two policies have been dropped. Dual rights and more than eight contracts are now commonplace.

Although virtually all railroads serve as common carriers, the Revised Interstate Commerce Act (RICA) states that they may "enter into contracts with purchasers of rail services to provide specified services under specified rates and conditions."[1] Railway companies do not need permits to perform contract services and, therefore, do not operate as separate legal entities called contract carriers.

The RICA permits railway carriers to engage in contract carriage as long as they do not tie up more than 40 percent of their equipment by type in this form of service. Once carriers exceed the 40 percent level, they might begin to falter in their normal common carrier obligations. Nonetheless, the ICC may use its powers to exempt traffic or services from economic regulation. With this exemption, carriers may establish any legal relationship with the shipping public, and the application of the 40 percent rule is set aside. Contract carriage is used extensively for both regulated and exempt traffic— specifically, for bulk commodities such as chemicals, minerals, coal, and grain; forest products; foodstuffs; and metals (see **Exhibit 2.2**).

The relaxation of restrictions on railway contract carriage has greatly boosted this form of service. Before 1978, there were no pure contracts in the rail area except for limited unit-train services. By contrast, a survey taken in late summer 1983 indicated that Conrail received 26 percent of its revenue from contracts, the Santa Fe Railway was up to 40 percent, and the Chessie System was so busy negotiating contracts that it did not know for sure how much revenue was earned through them.[2] In December 1981, 768 contracts were on file at the ICC. By December 1985, 33,500 new contracts had been filed (see **Exhibit 2.3**).[3]

Private Carriage

Private carriage is not subject to economic regulation. It is subject to noneconomic regulations covering hazardous goods movement, employee safety, vehicle safety, and other social regulations established by government agencies such as the Department of Transportation (DOT).

Historically, the ICC has applied the "primary business test" to determine whether motor carriage is private or for hire. Using this test, the ICC examined a proprietary trucking operation to see if the shipper used it in the furtherance of a primary form of business (other than transportation) and if the shipper had actual control over the transportation. Until 1978, a strict interpretation of this

■■■■■■ *Exhibit 2.2* **Railway contracts by major commodity groups**

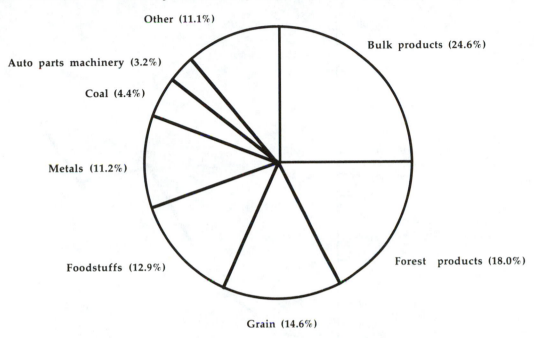

Other (11.1%)

Bulk products (24.6%)

Auto parts machinery (3.2%)

Coal (4.4%)

Metals (11.2%)

Foodstuffs (12.9%)

Forest products (18.0%)

Grain (14.6%)

Source: "Report on Railroad Contract Rates Authorized by Section 208 of the Staggers Rail Act of 1980." (Washington, D.C.: ICC Office of Transportation Analysis, March 13, 1984), p. 8.

test meant that private carriage essentially was confined to not-for-hire transportation provided by manufacturing or merchandising firms to haul their own goods in their own vehicles with their own employees. An example is a General Electric truck that moves light bulbs from a GE plant to a customer.

Currently, the concept of private carriage permits participation in for-hire services as follows:

1. Toto for-hire service. In the Toto case of 1978, the ICC decided that private trucks may offer for-hire service to second parties in certain situations, which include filling empty backhauls or promoting other efficiencies such as better equipment utilization during off-peak periods (see Chapter 14).[4]

2. Compensated intercorporate hauling (CIH). The Motor Carrier Act of 1980 opened the way for shippers to consolidate private fleet operations into wholly-owned subsidiary trucking companies that offer for-hire service to members of a "corporate family." This family includes "the parent corporation and all subsidiaries in which the parent corporation owns directly or indirectly a 100 per-

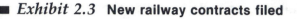

Exhibit 2.3 **New railway contracts filed**

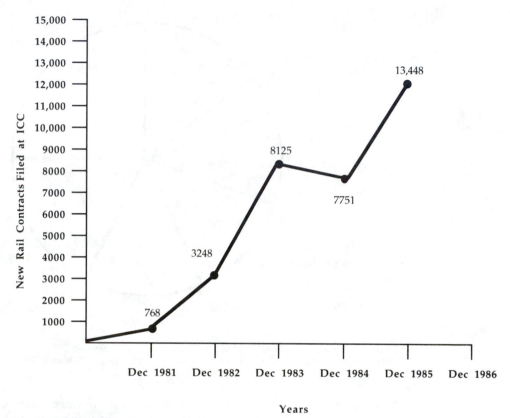

Source: Interstate Commerce Commission, Contract Advisory Service.

cent interest.'' The ICC has limited jurisdiction over such compensated inter-corporate hauling by requiring an intercorporate hauling certificate before initiating such activities.[5]

   3. Trip leasing. Shippers that have private truck operations may lease their equipment with drivers to other for-hire carriers for individual trips. These shippers may also use owner-operators and other second parties that provide both the vehicle and the driver.

Exempt Carriage

Exempt carriage refers to for-hire transportation not subject to economic regulation by regulatory agencies. Exempt carriers have no obligation to serve and no requirement to meet certain rate standards to which common carriers must adhere. Motor carriers hauling fresh vegetable and fish products, as well as live-stock, have long been exempt from ICC controls. The ICC now has expanded powers to exempt rail traffic and services from regulation and has used this power to exempt perishables traffic, piggyback service, and some boxcar movements.[6] Shippers negotiate directly with carriers for exempt carriage service. They also secure this service through brokers.

As this discussion indicates, the legal regulations governing a form of carriage define the kind of service it provides. In the past, carriers were either common, contract, private, or exempt. The current national policy, however, permits single carriers to provide one or more of these legal forms of service. Therefore it is more appropriate to look at these legal forms as kinds of "carriage" by a transportation company rather than as kinds of "carriers" or transportation companies.

Changing Legal Influences

Federal economic regulation has been administered by independent regulatory agencies such as the Interstate Commerce Commission, Federal Maritime Commission, and until recently, the Civil Aeronautics Board, which is now defunct. The realm of economic regulation of the domestic transportation industry grew from passage of the Act to Regulate Commerce in 1887 (for rail), the Motor Carrier Act of 1935 (for motor), the Civil Aeronautics Act of 1938 (for airlines), the Transportation Act of 1940 (for domestic water carriers), and the Freight Forwarder Act of 1942. This body of legislation prevailed until the late 1970s when political pressure for deregulation took hold. The ICC responded by taking administrative moves to relax its regulation. More important, however, Congress passed into law the Motor Carrier Act of 1980 and the Staggers Rail Act of 1980. This legislation gave tremendous impetus to deregulation and regulatory reform. And more recently, Congress enacted the Surface Freight Forwarder Deregulation Act of 1986.

Motor Carrier Industry

The Motor Carrier Act of 1980 and the ICC's subsequent implementation of its provisions greatly altered marketing practices and operations in the motor carrier industry. The transportation policy statement found in the act states that competition and market forces are to guide development of the industry in the future, regulatory influences are to be reduced, and shipper needs are to dictate the qualities and levels of services.

Exhibit 2.4 highlights important changes produced by the Motor Carrier Act. In the area of entry regulation, it is relatively easy for motor carriers to obtain new or expanded operating authority today. Carriers are permitted to

Exhibit 2.4 Changes in motor carrier regulations

Item	Old Law	New Law	Implications for Traffic
National Transportation Policy Section 10101	ICC to preserve competition among carriers and modes	ICC to promote services that meet needs of public; allow price-quality options; strive for productive use of resources	Opens way to creativity in developing service that meets needs of shipper; different services can be provided by same carrier.
Common Carrier Operating Rights Section 10922	Old law protected existing carriers against new competition. Applicant had to prove need for new service, how existing carriers not serving, and that new service would not harm existing carriers. Also had to prove experience in trucking and sound financial background Existing carriers could easily protest and prevent new service.	Prove shipper need and financial fitness. Burden of proof on protestants, not on applicant as in old system	Much easier to obtain new carrier for needed service; cost and effort of obtaining new service much less on traffic department
Contract Carrier Operating Rights Section 10923	Same as with common carriers Must show (1) number of shippers served, (2) nature of service, (3) effect of new service on existing carriers, (4) effect on applicant of denying permit, and (5) changing character of shipper's service patterns Could not serve more than eight different shippers (Rule of 8) Could not possess both common and contract rights at same time (dual authority)	Must show same items as in old law, but ICC reviews them in much more liberal light Rule of 8 no longer applies. Dual authority OK	Much easier to obtain service specifically tailored to firm's needs; examples: ■ assigned vehicles ■ scheduled service ■ special equipment ■ volume rate movements ■ services that include more than movement

Exempt Transportation Sections 10521, 10524, and 10526	Services not subject to economic regulation of ICC; include private carriage, moves within one state, agricultural product moves, and others	Expanded list made shippers' association operation more viable.	More services available Buying of service is more flexible, subject to fast negotiation, and greater use of brokers. More difficult to ascertain prevailing rates and amout of service available Courts must be used to settle disputes. ICC was easier.
Private Transportation Section 10524	Shipper must control and be responsible for the vehicle. Could not handle any other freight than firm's Difficult to handle freight for firms within same corporate parent ownership Could not lease driver and vehicle from same place; difficult to use owner-operator	More liberal rules apply. Can obtain common or contract operating rights to handle freight for other firms on backhaul (*Toto* decision) Can haul for corporate affiliates Leasing rules relaxed	Simplified ability to handle other movements of freight Eliminated many operating rules in private carriage Can recoup expenses by handling freight for others
Rate Filing Sections 10761, 10762, and others	Rates must be filed for thirty days before effective Other carriers and shippers could easily protest new rate and hold up effective date.	Rates can come into effect on shorter notice. More difficult for others to protest	Carrier being used can more easily place a new and favorable rate into effect for the firm. In tight carrier supply periods, shipper has little recourse in fighting new rate.
Rate Flexibility Section 10708	Did not exist	Carriers can raise or lower rates by amount up to that of Producers' Price Index and not be subject to protest.	Same as above

conduct both common and contract carriage simultaneously. The act also expands the number of services and the types of traffic exempt from ICC regulation. In addition, private carriage is permitted without many of the previous restrictions for ownership, use of vehicles, and operations.

The Motor Carrier Act has drastically changed industry pricing practices. Rates published in tariffs filed with the ICC can go into effect in less than the thirty days required in the past. These rates have considerable freedom from ICC interference. It is also extremely difficult for a protesting shipper or competing carrier to block new rates from going into effect. Further, collective rate-making by rate bureau organizations is restricted. For example, carriers have not been permitted to use these organizations for single-line ratemaking since 1984. Single-line ratemaking by motor carriers allows pricing flexibility in that other carriers in the bureau could no longer protest or vote on such rates. The advent of increased motor carrier supply and competitive pressures caused many carriers to publish their own rate structures in the late 1970s and early 1980s. Independent rate action is but one additional tool available to carriers to depart from rate bureau structures where negotiations and competitive pressures lead them to do so.

To a large extent, the Motor Carrier Act of 1980 has attained its goals. More carriage service is available than in the past; rates are more flexible now; and the industry has shifted to a buyers market, at least temporarily, where traffic managers can obtain desired services and competitive prices. It is difficult to ascertain to what degree the Motor Carrier Act of 1980, the ICC's own relaxation of administrative rules in the motor carrier area, or the recession in the early 1980s was responsible for many of the changes in the field. In any event, the buying of motor carrier service is vastly different today from what it was just a decade ago.

Rail Carrier Industry

Whereas the Motor Carrier Act was designed to bring about service flexibility and competitive rates, the Staggers Rail Act of 1980 was passed mainly to assist the railway industry out of financial problems and into a period of profitability. Fears of nationalization, stemming from the railway company bankruptcies of the 1970s, were behind this movement.

This act contains both favorable and unfavorable points for traffic managers (see **Exhibit 2.5**). Increased competition and reduced regulatory restrictions are goals of this act, but the major thrust is to provide rail carriers with new pricing and operating freedoms that will help them become financially healthy enterprises.

One very powerful provision of the Staggers Rail Act allows the ICC to exempt certain railway traffic and services from economic regulation. For example, the ICC has used this provision to exempt piggyback (trailer- or container-on-

Exhibit 2.5 Changes in railroad law

Item	Old Law	New Law	Implications for Traffic
Rail Transportation Policy Section 10101a	ICC to preserve competition among carriers and modes	Free-market forces to dictate competition Minimize regulatory control over carriers. Reduce regulatory burden to entry and exit. Shift toward individual carrier pricing and away from collective pricing Prohibit predatory pricing.	More freedom to obtain specific rates and services from individual carriers
ICC Exemption Power Section 10505	No rail services were exempt from regulation.	Section 10505 allows ICC to unregulate rail carrier services.	ICC has unregulated virtually all agricultural and food products, except grain; also piggyback traffic Shipper can have sudden rate and service shifts result from ICC actions here; some positive and others negative.
Abandonments Sections in Chapter 109	Railroad had to prove service was no longer needed by public. Easy for shippers to stall or prevent abandonment	Shipping community has burden to prove why carrier should continue. Carrier can abandon if not profitable enough. Shippers and states to take over some lines that railroads seek to drop	Possibility of railway line loss of service is greater. Can have large business impact on shipping firm Gets traffic manager involved in rail retention activities
Contracts Section 10713	No rail rate contracts allowed, only unit trains	Rail rate and service can easily be negotiated and obtained.	Greater rate and service flexibility Firm can suddenly lose business if competitor obtains favorable contract to market via other railroad. **(continued)**

Exhibit 2.5 (continued)

Item	Old Law	New Law	Implications for Traffic
Minimum Rates Section 10701a	In issues between two modes, minimum lawful rate was subject to dispute.	Minimum lawful rate is "going concern value," at or near variable cost.	Railroads have freedom to seek traffic in easier manner than before through lowering of rate.
Maximum Rates Section 10709	A very difficult level to determine	Defined by certain percentages over variable cost, and through definition of "market dominance"	Still very difficult to build a case that a rail rate is too high
		Defines levels at which rates can be protested, and/or suspended, as well as which party has burden of proof	None of this applies to exempt traffic . . . almost no recourse.
Rate Freedom Section 10707a	None	No challenge to inflation-adjusted rates.	Less ability to block rail rate increases than in past
		Establishes "jurisdictional threshold" for market dominance, revenue adequacy criterion for assessing reasonableness	

Provision			
Surcharges — Light Density Lines Section 10705a	None	Railroad can apply rate surcharge to regular rate to eliminate loss from serving branch line.	No recourse other than to stop using service
Surcharges — Joint Lines Section 10705a	None	One railroad can apply rate surcharge if its division in a joint-line move does not cover variable costs.	Either pay new total charges, use another lower cost route, or stop using service.
Released Value Rates Section 10730	ICC applied strict rules to application. Only few products allowed to have released value.	Greatly relaxed for application in broader trade of products	Carriers easily able to apply via contract rates
Rates on Recyclable Commodities Section 10731	None	ICC can impose maximum rates over variable costs on movements involving recyclable products.	Shippers of scrap benefit
General Rate Increases Sections 10706 and 10712	Allowed by carriers in collective rate-making bodies	Not permitted except for costs affecting entire industry in pure inflationary manner	Pure cost increases go into effect with no ability to protest or prevent. All other rate increases are to be implemented by individual carriers only; can enhance negotiating power of individual shipper.

flatcar) service and some perishable goods traffic and to attempt to deregulate boxcar and export coal traffic.

Carriers may abandon certain railway lines without going through the lengthy procedures required in the past, where the ICC grants exemption from the regulatory procedures.[7] To retain such lines, many shippers must pay higher rates, or they must help establish locally operated short-line carriers, perhaps with financial assistance from state and local governments. This situation has forced many shippers to rethink plant location decisions.

Contracts between railroads and shippers are sanctioned by the law. As stated previously, this form of pricing has grown tremendously. These contracts are useful in reducing total shipping costs and obtaining guaranteed services, as well as many other service elements.

Railway companies enjoy new ratemaking flexibility today. The standard for minimum rates is the "going concern value" of the carrier, which is set at or calculated close to variable costs. At the other extreme, the act placed new limits on the ICC's ability to set reasonable maximum rates.

In addition, carriers may apply surcharges on light-density lines and joint-line traffic if the revenue for such operations is insufficient. Generally speaking, the use of surcharges has increased the cost of shipping via many branch lines. Surcharges also reduce the number of viable routes available between two points. Not surprisingly, many shippers have started to divert their branch-line traffic to other modes.

Air Freight Industry

Air freight became unregulated in 1977, and the passenger airline industry followed in 1978. Complete route and pricing freedom exist today in the airline industry. The shakeout resulting from deregulation in rail and motor modes represents a major uncertainty for the future. The air freight industry is three years ahead in this experience. A large number of new firms, especially courier and package express firms, came into existence following air deregulation. This is an entirely new industry in that documents traditionally moved by the U.S. Postal Service.

Air freight was the first area that traffic managers engaged in corporatewide negotiation and contracting. Many issues had to be resolved that dealt with rate structures, discounts, billing and paying procedures, and obligations and claims. This became a common activity in the early 1980s when such negotiations and contract arrangements were gaining popularity with motor and rail firms.

Surface Freight Forwarder Industry

The Surface Freight Forwarder Deregulation Act of 1986 ended entry, ratemaking, ownership limitations, securities liability, and other federal regulations affecting freight forwarders. This change creates another new dimension in the transportation industry and more service options for the traffic manager. The next section of the chapter describes the nature of freight forwarders.

Carrier Firms and Entities

Many different kinds of carrier firms and entities provide services to shippers. *Railway-based transportation companies* offer common, contract, and exempt transportation through single-car, multiple-car, and unit-train services. In addition, hybrid services such as trailer-on-flatcar (TOFC) or container-on-flatcar (COFC) are offered (see **Exhibit 2.6**). Furthermore, some of these companies have established trucking subsidiaries to move piggyback trailers to destinations not reached by the rail lines.

Trucking companies offer common, contract, or exempt truckload (TL) and less-than-truckload (LTL) services. Truckload operations are limited to the movement of single volume shipments, but many offer stopoff services. These carriers concentrate on commodity groups such as building materials, automobiles, and bulk liquids. Some trucking firms confine operations to TL movements in special-purpose equipment such as tank and auto-rack trailers.

Less-than-truckload services involve the pickup and delivery of many small shipments. The carrier assembles these shipments into large consolidated loads at its terminals, moves them in volume to destination terminals, disassembles the load or "breaks bulk," then delivers the individual shipments.

Domestic inland waterway companies run common, contract, and exempt barge operations. Barge shipments usually require very large "barge-size" lots. Domestic water transportation also includes ship movements between U.S. coastal ports and on the Great Lakes.

Air service can be obtained directly from airline companies or from air freight forwarders. Movement is by the package, pallet load of packages, or container. All passenger carriers move freight, and several all-freight airlines are available as well.

Some transportation companies specialize in *small package express and courier service*. United Parcel Service (UPS), Roadway Package Express, Federal Express, American Delivery Systems, and some of the airlines offer this type of service. Bus companies, as well as regional and local "package" trucking companies, also provide small-shipment movement.

Pipeline companies offer common carriage of bulk liquids, principally oil, gas, and chemicals. Pipelines, though, are sometimes used to transport other commodities such as woodchips and coal slurry. Some petrochemical firms have contracted with pipeline companies to use the idle small diameter lines made obsolete by large diameter pipe for underground storage of bulk liquids.

Freight forwarders are carriers that generally handle small shipments. They collect small shipments from many shippers in an area, consolidate the ship-

███████ *Exhibit 2.6* **Railroad intermodal service**

Plan I: Railroads carry trailer or container loads of freight for motor common carriers as a flat charge per unit. The railroad has no direct contact with the shipper and merely substitutes trailer, or container, on flatcar transportation in place of the motor common carriers' transportation by truck. The motor common carrier solicits and bills customers at truck rates, takes trailers or containers to, and picks them up from, TOFC terminals.

Plan II: This is a total railroad door-to-door transportation service. The railroad furnishes all transportation equipment; delivers the trailers or containers to shipper's loading dock and, in some cases, helps load them; transports them to its loading ramp; loads and transports them to consignee's unloading dock and sometime helps unload the units.

Plan II 1/4:   Same as Plan II, except railroad either picks up or delivers the trailer or container, but not both.

Plan II 1/2:   Provides for the shipper and receiver to perform pickup and delivery of railroad-owned trailers or containers with railroad performing ramping and deramping and linehaul service between ramps.

Plan III: This is a railroad transportation service restricted to ramp operations. The shipper loads his or her own or leased trailer or container and provides for its delivery to loading ramp. The railroad loads the trailer or container on the flatcar at origin piggyback terminal, transports it to destination piggyback terminal, and grounds it.

Plan III—Minibridge: Containerized shipments having origin or destination from or to an international foreign port and moving by rail to or from originating or final destination U.S. ports. This rail-water service is instead of an all-water move.

Plan III—Landbridge: Containerized shipments moving by water into U.S. ports for transport by rail to another port for further transport by water to destination.

Plan IV: Rail carrier handling of loaded and empty privately owned or leased trailers and flatcars between specific points or areas under charges published in regular rail tariffs. These are flat per car charges to compensate the rail carriers for service in moving the loaded or empty trailers on flatcars and not for the commodity being shipped.

Plan V: Railroads and motor common carriers participate in a joint rail-truck transportation service at a published single-factor joint through rate. Each participating carrier gets a division of the through rate.

Source: Association of American Railroads, *Railroad Ten-Year Trends* Vol. 2 (Washington, D.C.: Economics and Finance Department, 1985), table III-C-14.

ments by destination to obtain volume rates, use transportation offered by third parties (frequently railway companies) to ship the consolidated lots to their own terminals, disassemble the lots, and deliver the goods to various consignees served by the destination terminal. The freight forwarder's service may be faster than LTL service and often is less costly for the shipper. For longhaul movement, forwarders use the railroad's piggyback service.

Shippers' associations are not-for-profit cooperative entities that operate in the same manner as freight forwarders, except these associations serve only member shipper-receiver firms. In addition to forming consolidated loads to take advantage of quantity discounts, shipper associations offer efficient scheduling and handling of small shipments. Any surplus funds earned above operating costs for a year are redistributed back to the members according to a preset formula. Currently, there are about 150 viable shippers' associations, more than twice the number that existed in 1983.

Shipper's agents act as retailers of unregulated piggyback services. They typically contract with railway companies for large volumes of TOFC/COFC service at discount rates and then resell portions of this service to shippers at higher, but competitive, rates. Unlike shipper's associations, shipper's agents are profit-making entities and do not concentrate on shipment consolidation. Agent services include shipping arrangements, expertise, single bills, and credit.[8]

Special property brokers coordinate transportation arrangements for shippers, receivers, and carriers, as well as find shipments of exempt traffic for carriers and vehicle owner-operators. Working for a commission, the broker arranges the service, and often checks vehicle condition, insurance, and driver qualifications. This transportation group used to be limited to the exempt fruit, vegetable, and animal movement area. Today, there are about four thousand brokers licensed to handle regulated freight; although most are small operations, this sector will no doubt grow.

Motor carrier brokers must obtain a license from the ICC to operate. Before 1978, these brokers played a minor role in the motor carrier industry because the ICC imposed many restrictions on their operations. Today, they are permitted to go beyond making arrangements and actually control the shipments, including entering into agreements with contract carriers. Some "marriage brokers," for example, match up shipments going to common destinations and negotiate rates for consolidated loads that are less than the rates that shippers might negotiate individually. Many brokers now offer services such as storing, routing, billing, and tracing. They are becoming more popular as more products and forms of traffic become exempt from economic regulation.[9]

Nevertheless, shippers will need to screen special property brokers carefully to make sure they are reputable and financially sound.[10] This effort, coupled with a written agreement clearly establishing the agency relationship for the collection of freight charges, will assure that the shipper does not end up paying the broker and then being held liable to pay the carrier, if the broker defaults. The Transportation Brokers Conference of America, located in Oak Forest, Illinois, maintains a list of property brokers that shippers can use as a source for screening potential brokers.

Mode Selection

The mode selection decision focuses on the choice of railway, highway, waterway, pipeline, or air transportation service. This choice is an important part of strategic transportation planning. Different modes can have a significant impact on logistical system design elements such as the layout and location of plants

Exhibit 2.7 **Average revenue per ton-mile of modes of freight transportation**

	Cents/Ton-Mile
Air	50.20
Truck (class I)	
General LTL	22.16
Specialized TL	9.55
Rail (class I)	3.09
Oil pipeline	1.27e
Barge	0.82

e = estimated
Source: *Transportation in America*, 4th ed., April 1986 Supplement, p. 11.

and warehouses and the type of material handling systems needed. In practice, the mode selection decision involves a limited set of alternatives, typically air–truck, truck–rail, rail–water, water–pipelines, or rail–pipeline.

Several basic elements affect mode choice. Clearly, shippers consider *cost* a major determinant. By using average revenue per ton-mile as a proxy for cost, it is possible to get an approximation of costs for different mode choices. As shown in **Exhibit 2.7**, the average revenue per ton-mile proxy shows a wide range of values among the modes of freight transportation. Minimizing transportation costs, though, is not the objective of the traffic function. As stressed in the previous chapter, the primary goal of traffic management *is to contribute to the profitability of the firm*. This task is accomplished by assessing the impact of transportation cost and performance on "total costs" and on customer service.

Accessibility is an important service element affecting the mode decision. This element includes access to loading and unloading facilities, proximity to the mode, and any storage facilities that would also have to be employed.

Speed is another modal choice consideration. Long-distance movement is fastest by air, followed by motor, then by rail and water. Speed is important with high-value goods where an investment in inventory is an important factor to the seller or buyer. On the other hand, a slow mode can provide the advantage of also being a warehouse in transit, thereby reducing storage charges in warehouses at the origin or destination.

Capability can mean many different things to shippers. For some, it might mean the ability to serve many or all customers with the same mode. Motor has this advantage. For others, it might mean the capability to handle high, wide, and heavy shipments (see **Exhibit 2.8**).

Traffic characteristics and market requirements will determine the importance of modal elements. For example, a shipper that must get a large volume shipment of gas from Texas to New York will probably find pipeline the most

Exhibit 2.8 **Mode capability**

Barge	15 Barge tow	Jumbo hopper car	100 unit train (grain)	Large semi
1500 ton	22,500 ton	100 ton	10,000 ton	25 ton
52,500 bushels	787,500 bushels	3500 bushels	350,000 bushels	875 bushels
453,600 gallons	6,804,000 gallons	30,240 gallons	3,024,000 gallons	7560 gallons

Cargo Capacity

 = =

1 Barge   **15 Jumbo hoppers**   **60 Trucks**

 =  =

1 Tow   **2 $\frac{1}{4}$ Unit trains**   **900 Trucks**

Source: Paul F. VanWicklin, "Opportunity Rides the Rivers." Reprinted by permission, from May issue of *Handling & Shipment Management*, copyright © 1982, Penton/IPC, Inc.

practical and efficient mode of transportation. On the other hand, a shipper of important documents will likely find courier or air services to be fastest and safest. The type of equipment (see **Exhibit 2.9**) required also will influence the mode of transportation.

Exhibit 2.9 **Overview of transportation equipment**

Name	Variations	Comments
Rail Equipment Boxcar	Plain, 40', 50', and 60'	Capacities from 40 to 100 tons
	Specially equipped; above sizes; insulated, refrigerated, or with bracing and cushioning features	Cubic size from 3300 to 6000
		More expensive
		Carriers assign or select shippers for use of these cars.

(continued)

Exhibit 2.9 (continued)

Name	Variations	Comments
Plain hopper	Open top loading, bottom unloading 40 and 100 ton capacities	Prime coal and other bulk commodity car when weather protection not important
Covered hopper	Hatch loading on top, bottom unloading Covered for weather protection Sizes from 1900 to 5700 cubic feet	Used for bulk limestone, grain, cereal, flour, and other bulk movements Small cubic foot size cars used for heavy density products; large cars for light density goods like plastic pellets and cake mixes
Plain flat	40' to 80'	Used for machinery and other large goods
Equipped flat	Usually auto-rack cars	Used for auto movement
Gondola	A flatcar with sides Same sizes	Used for products that can be loaded efficiently only from top crane Used for pipe, steel, machinery, and some very dense ores
Piggyback	Truck trailer on a special flatcar Container on same car	Trailers loaded by being driven onto car or lifted by crane Trailer lifted from wheel chasis by crane; moved without wheels
Tank car	Varying sizes	Railroads will not provide; shipper-receiver must own or lease and provide for move.
Motor Equipment		
Van	Highway trailer from 24' to 65'	Size limited by state and federal highway restrictions; most are 45' and 55' Can be hauled in double configuration by single tractor—called ''twin''
Flat bed	Flat body; can have protection from stake and canvas side and top	Good for steel hauling and any other goods not needing weather protection and that can be loaded only from top or side via crane or other special equipment
Straight body	Truck and body as same unit; usually up to 20' and 24'	Common in city pickup and delivery operations
Tank and bulk	Special equipment for particular goods	

(continued)

Exhibit 2.9 (continued)

Name	Variations	Comments
Dump	Used for ores, dirt, stone, gravel, etc.	
Barges		
Standard barge	195' by 35'	Standard river freight carrier
	Holds 1500 tons and almost 50,000 cubic feet	Most firms charge for minimum of 600 tons if shipment is less.
	Covered barge for weather protection	
Tank and chemical barge	Special designs for bulk liquid products	
Air		
Pallet — igloo	Packages on pallet; shaped like igloo to conform to shape of main plane floor and body	
Container	Varying sizes; many specific to particular plane	

Carrier Selection

In evaluating carriers, there are several important factors to consider, including

- Rates
- Willingness to negotiate
- Transit time or speed
- Reliability of transit time and pickup
- Ability to handle single haul
- Breadth of points served
- Equipment availability
- Loss and damage experience
- Promptness of settling claims
- Computer interface on rates, billing, and tracing
- Quality of carrier sales force
- Financial stability
- Information services
- Flexibility to meet unexpected needs
- Insurance coverage

Many surveys of traffic managers, taken during the 1960s and 1970s, pointed to consistent and reliable service as the most important carrier selection factor, with price often being of secondary importance.[11] These surveys were taken during an era when prices were fixed by rate bureaus. Now that rates may be negotiated, and each carrier prices its services individually rather than through rate bureaus, many shippers have become more sensitive to price in the selection process.[12]

As the discussion that follows will illustrate, the evaluation of mode or carrier alternatives can be accomplished in an analytical manner. The key elements that determine total logistical costs related to the transportation of goods include:

Key elements	Example	
	Irregular route TL carrier	Intermodel TOFC service
Product shipped:		
Annual volume (lb)	1,000,000	1,000,000
Unit weight (lb)	1	1
Unit value ($)	1.00	1.00
Transportation cost and service elements:		
Shipment Size (lb)	40,000	40,000
Cost ($/cwt):		
Linehaul freight rate	$4.50	$2.80
Other:		
Pickup and delivery	$0.00	$1.25
Loading and unloading	$0.00	$0.00
Loss and damage	$0.01	$0.01
Total	$0.01	$1.26
Service:		
Transit time (days)	4	10
Percentage on-time	90%	60%
Credit period (days)	7	30
Other logistical cost elements:		
Carrying cost factor:		
In-transit inventory	10%	10%
Base and safety stock	30%	30%
Cost per order	$25	$25
Safety stock (units)	10,000	50,000
Packaging ($/cwt)	$1.00	$1.25
Cost of money	10%	10%

With information like this in hand, a traffic manager can compute the total annual cost of each method of shipping as follows:

1. Direct shipping cost. This cost is usually the most visible single cost in the entire set of cost elements. Direct shipping costs include the linehaul freight rate applicable to the origin–destination movement, any additional pickup, delivery, loading, or unloading charges, and the average amount of loss and damage expected during the year. To calculate direct shipping costs, it is necessary to compute the annual linehaul cost and then add that amount to any other applicable shipping charges. The calculation of linehaul cost is as follows:

$$\text{Cost per shipment (\$/cwt)} \times \text{Number of shipments per year} \qquad \text{(1a)}$$

In our example, the annual linehaul cost of the truckload (TL) service is the $1,800 per shipment (40,000 lb × $4.50/cwt) times 25 shipments (1,000,000 lb annual volume/40,000 lb per shipment) or $45,000. (Note that if the $4.50 rate required a minimum weight of 50,000 lbs., the cost per shipment would be 40,000 lbs. shipped as 50,000 lb × 4.50/cwt, or $2,250.) The calculations for trailer-on-flatcar (TOFC) service produce an annual linehaul cost of $28,000.

Any other applicable shipping costs are determined as follows:

$$\text{Annual volume (in cwt)} \times \text{Other charges (\$/cwt)} \qquad \text{(1b)}$$

The TL service involves another $100 (10,000 × $.01) in direct shipping cost, whereas the TOFC alternative produces an additional $12,600 (10,000 × $1.26).

2. In-transit inventory. Goods in transit represent funds tied up that could be earning money invested elsewhere or that had to be borrowed from another source and cannot be paid off until payment is received from the customer. The cost of money or opportunity cost usually represents a good estimate of the carrying cost factor for inventory-in-transit. The method of calculating in-transit inventory costs is:

$$\text{Transit time/365} \times \text{Annual volume (in \$)} \times \text{Carrying cost factor} \qquad \text{(2)}$$

which, for our example, equals 4/365 × $1,000,000 × 0.10 or $1,096 for the TL service and equals 10/365 × $1,000,000 × 0.10 or $2,740 for the TOFC service. The shorter the transit time, the less costly this element is to the firm. Similarly, the lower the value of the product, as well as the carrying cost factor, the less costly this element will be.

3. Inventory carrying cost at destination. The size of the shipment (the order quantity) affects base–inventory costs at the destination facility. Base stock is the amount ordered during the replenishment process. The average amount of base stock on hand during the year is one-half of the order quantity. Handling, storage, insurance, taxes, depreciation, obsolescence, and interest typically constitute the carrying cost elements for base stock. For purposes of computation, logisticians express such elements as a percentage of the dollar value of the product. The computation then proceeds as follows:

$$\text{Average base stock (\$)} \times \text{Carrying cost factor} \qquad \text{(3)}$$

which is $40,000/2 × 0.30 = $6,000 for both transportation service alternatives.

4. *Delay cost.* When a shipment arrives late, a company may not have enough inventory on hand to serve customers or to meet production schedules. Safety stock is held to cover such delays and short-range variation in demand. The carrier's on-time performance can have a significant impact on the amount of safety stock held. The cost of holding this additional inventory is calculated as:

$$\text{Safety stock (\$)} \times \text{Carrying cost factor} \tag{4}$$

In the example, this cost is \$3,000 (\$10,000 × 0.30) for the TL option and \$15,000 (\$50,000 × 0.30) for the container option.

5. *Packaging cost.* Certain transportation services require more rugged packaging to protect the goods against rough handling, contamination, or other sorts of freight damage. This cost is determined as follows:

$$\text{Packaging cost (per cwt)} \times \text{Annual volume (in cwt)} \tag{5}$$

which is \$10,000 for TL transportation and \$12,500 for the TOFC alternative.

6. *Ordering cost.* There are certain costs associated with the task of placing or processing an order for goods. This task may involve activities such as vendor and carrier selection, document preparation, order transmission, and production set-up arrangements like retooling. The total annual cost of ordering is:

$$\text{Cost per order} \times \text{Number of orders per year} \tag{6}$$

In our example, the cost per order is \$25. The number of orders per year is 25 (1,000,000 lbs./40,000 lbs. per order), which produces the same total annual cost of \$625 for both service alternatives. Nonetheless, the higher minimum weights that are typically attached to freight rate discounts may require larger order sizes that generate fewer orders per year. As a result, shipping and order costs decline, while base-inventory costs increase.

7. *Credit terms.* These terms limit the opportunity of the company to make short-term investments with the money owed the carrier for freight charges. The calculation of credit-term savings is:

$$\text{Annual freight charges} \times \text{Cost of money} \times \text{Credit period}/365 \tag{7}$$

In our example, the seven-day credit period allowed for TL service permits the shipper to earn \$86.49 (\$45,100 × 0.10 × 7/365), while the thirty-day period for TOFC service produces savings of \$333.70 (\$40,600 × 0.10 × 30/365). Such earnings represent a negative cost because they reduce the total cost of shipping.

8. *Total logistical costs.* As the calculations that follow show, the sum of cost elements 1 through 7 in our example equals the total logistical costs arising from each transportation mode or service option.

Cost element	Truckload service	TOFC service
1. Direct shipping cost		
(a) Linehaul	$45,000	$28,000
(b) Other charges	100	12,600
2. In-transit inventory	1,096	2,740
3. Inventory carrying cost		
at destination	6,000	6,000
4. Delay cost	3,000	15,000
5. Packaging cost	10,000	12,500
6. Ordering cost	625	625
7. Credit terms	−86	−334
Total logistical costs	$65,735	$77,131

The irregular-route TL carrier that charges a linehaul rate of $4.50/cwt will cost the firm a total of $65,735 per year. On the other hand, the TOFC service that offers a rate of only $2.80/cwt will actually cost the firm $77,131. The value of this approach is that changes in transportation cost and service elements can be tested in terms of their impact on the total cost.

Screening Carriers

Before undertaking a total cost analysis, traffic managers often eliminate the carriers that do not meet minimum performance requirements for some or all of the selection criteria already listed.[13] This screening action is especially important to assure that carriers are properly insured. Regulated motor carriers are required by law to maintain minimum levels of insurance. Despite the efforts of the ICC and the Department of Transportation, as well as state enforcement agencies, the number of uninsured and underinsured carriers has been growing. In recent years, escalating insurance premiums for vehicles, drivers, and cargo have increased costs for all and have cast severe doubt on continuance of operations for many carriers. Some trucking companies have initiated surcharges on the regular rate to cover this increased cost.

The insurance issue is significant to shippers, because they are exposing themselves to a serious risk if they use uninsured or underinsured carriers. One authority notes, for example, that a shipper might be held ''collaterally'' liable by virtue of employing the uninsured carrier, and recommends that shippers ''routinely oblige the carriers serving them to submit copies of their insurance certificates.''[14]

New Directions

Carrier selection includes approaches today that are completely new to traffic management. For example, it is common to find shippers seeking bids from carriers for all, or major parts, of their movements. Where a firm might have employed five hundred carriers each year in the late 1970s, it is not uncommon today to find those same shippers concentrating their traffic with fewer than a dozen carriers. Furthermore, as firms come to realize that effective traffic management actually contributes to the profitability of the company, they are engaging in long-range planning for the selection and acquisition of transportation services. This planning entails the development of projections of future transportation costs, service availabilities, and technologies. It also requires a forecast of regulatory changes and an assessment of future carrier strategies.

Summary

Seven major sources of law influence shipper and carrier activities:

1. Federal *transportation laws* found in the United States Code, especially Title 49
2. *Transportation regulations* promulgated by independent regulatory agencies such as the ICC
3. *Contract law* stemming from statutory law, administrative regulations, and the *Uniform Commercial Code*
4. *Court decisions* made in cases involving transportation laws and agency regulations
5. Federal *agency decisions*, which establish rules and interpret transportation laws
6. *State laws* affecting intrastate transportation activities
7. *State agency* decisions, which interpret state laws and establish rules for intrastate transportation matters

Four major legal forms of carriage have evolved in the legal and regulatory environment of transportation. A for-hire carrier performing *common carriage* holds itself out to serve the general public. These carriers are capable of handling freight of all types just about anywhere in the nation. *Contract carriage* is a specialized form of for-hire transportation arranged by contract with specific shippers. *Private carriage*, which is not subject to economic regulation, occurs when shippers haul their own goods in their own vehicles with their own employees. Today, it also embraces for-hire services including Toto operations, compensated intercorporate hauling, and trip leasing. *Exempt carriage* refers to for-hire transportation not subject to economic regulation by regulatory agencies.

The Motor Carrier Act of 1980, the Staggers Rail Act of 1980, and the Surface Freight Forwarder Deregulation Act of 1986 have dramatically changed the legal and regulatory environment. As a result, the structure of the transportation industry is changing. Restrictions on a carrier's ability to offer common and contract carriage have been relaxed. Carriers have freedom to enter and exit as well as to tailor services for specific shippers. Intermodalism is evolving to offer many services not in existence a decade ago. Contract carriage is becoming nearly as popular as common carriage. Opportunities for private carriage have expanded. Likewise, the list of commodities and transportation services eligible for exempt carriage has grown. In addition, carrier pricing today is largely free from ICC regulation.

The selection and acquisition of transportation services is the primary responsibility of the traffic manager. Many different kinds of carrier firms and entities provide services to shippers. Transportation companies in each of the five modes of transportation (rail, truck, water, air, and pipeline) offer various common, contract, or exempt carriage services. Some transportation companies specialize in small package express or courier service. Others, including freight forwarders, shipper's associations, shipper's agents, and special property brokers perform as intermediaries. Agents and brokers have assumed a more important role today than in the past.

Mode selection is an important part of strategic transportation planning. Major determinants of mode choice include cost, accessibility, speed, and capability. Carrier selection requires an evaluation of carrier service elements as well as freight rates. Traffic managers need to assess the impact of both transportation cost and service levels on total logistical costs and ultimately on the profitability of the firm.

The changing legal and regulatory environment and the changing nature of the transportation industry have important implications for traffic managers. These include a nonuniformity of carrier obligations and claims processes, as well as the risk of depending on a carrier to an increasing degree only to have the carrier withdraw service through bankruptcy. On the positive side is the constant need to review new and evolving services with an eye toward considering them for the firm. The entire transportation buying activity is turning from one of complying with regulations to one of making business decisions.

Selected Topical Questions from Past AST&L Examinations

1. Explain carefully the distinctions between common and contract carriage. How is the status of exempt carriage changing? (Fall 1979)
2. Many shippers have justified the use of a higher cost mode or carrier on the basis of total cost. Explain. (Fall 1979)
3. The director of transportation for Western Electric recently published a formula that his company uses in evaluating truck versus air movements. The formula considers the rate, the time in transit, and the firm's cost of capital. In some instances, a lower rate carrier is actually the higher total cost option. How can this be so? (Spring 1981)
4. The continuance of high interest rates is motivating many firms to adjust their modal choices to use higher priced transportation. Explain. (Fall 1981)

5. You are the traffic manager for a major consumer product manufacturer. Your firm is ready to begin a daily stock transfer movement from the Philadelphia, Pennsylvania, plant to a warehouse in Los Angeles, California. Daily tonnage for this move is projected as being 120,000 pounds. You have received the following rate proposals:

Carrier	Cost	Delivery
Countryside Contract Carrier, Inc.	$2900	4th A.M.
Coastal Piggyback Service, Inc.	$2500	7th A.M.
SoCal Agricultural Cooperative	$2750	6th A.M.
Railroad (boxcar rate)	$2200	9th A.M.
Irregular Route Common Carrier, Inc.	$2700	6th A.M.

You know that the contract carrier uses company driver teams; the ag coop and irregular rate common carriers use owner-operators; and the piggyback rate is door-to-door.
  (a) What criteria would you use in evaluating which carrier(s) to use for this move? Explain.
  (b) Whom would you give the traffic to? Explain. (Fall 1982)
6. Brokers are emerging as deregulation continues.
  (a) What is the legal definition of a broker?
  (b) What role do they play for a shipper and a carrier?
  (c) How are they different from a freight forwarder?
  (d) What cautions should a shipper be aware of when using them? (Fall 1984)
7. You are asked to give a talk to general business students about the field of transportation and freight movement in particular. One of them asks, "Just what does it mean when a firm's movement of, say, widgets that were traditionally subject to ICC regulations becomes exempt in the eyes of transportation law?" How would you answer? (Fall 1984)
8. Describe three different intermediaries found in U.S. transport. What fundamental or useful services do intermediaries provide? (Spring 1985)

Notes

1. 49 U.S.C. 10713.
2. Dennis M. Gawlik, "Railroad contract rates: Their managerial and organizational implications for rate-making," unpublished masters thesis (University Park, Pennsylvania: The Pennsylvania State University, 1982), chap. 1.
3. Interstate Commerce Commission, *1984 Annual Report* (Washington, D.C.: U.S. Government Printing Office, April 1985), p. 33.
4. *Toto Purchasing and Supply Company, Inc.,* 128 MCC 873 (1978); see also *Ex Parte Number MC-118, Grant of Operating Authority to an Applicant Who Intends to Use It Primarily as an Incident to the Carriage of Its Own Goods and Its Own Nontransportation Business,* 43 Fed. Reg. 55051

(1978); *Mercury Motor Express, Inc.* v. *United States of America and Interstate Commerce Commission,* Number 79–10542, and *East Texas Motor Freight System and Regular Common Carrier Conference of American Trucking Associations* v. *ICC and USDA,* Number 79–1121, June 18, 1981.

5. 49 U.S.C. 10524.

6. Thomas M. Corsi, Curtis M. Grimm, and Robert Lundy, ''ICC exemptions of rail services: Summary and evaluation,'' *Proceedings—Twenty-Sixth Annual Meeting Transportation Research Forum* 26 (1985), pp. 86–92.

7. 49 CFR 1152.50; Ex Parte No. 274 (Sub-No. 8A), ''Exemption of out of service lines (discontinuance of service and trackage rights)'' (not printed), served April 20, 1984.

8. See Terrence A. Brown, ''Shippers' agents and the marketing of rail intermodal service,'' *Transportation Journal* 23 no. 3 (Spring 1984), pp. 44–52.

9. For further reading about the role of brokers, see Michael R. Crum, ''The expanded role of motor freight brokers in the wake of regulatory reform,'' *Transportation Journal* 24 no. 4 (Summer 1985), pp. 5–15.

10. For further reading on this issue, see John Martell, ''Broker double liability,'' *The Private Carrier,* (September 1985), pp. 30–33; Hank Urycki, ''The pitfalls of working with property brokers,'' *The Private Carrier* (September 1985), pp. 33–35; Ronald N. Cobert and Robert L. Cope, ''How to avoid paying twice for the same shipment when using a broker,'' *Traffic World* (January 1985), pp. 72–75.

11. See Roger E. Jerman, Ronald D. Anderson, and James A. Constantin, ''How traffic managers select carriers,'' *Distribution Worldwide* (September 1978), pp. 21–24; James R. Stock and Bernard J. LaLonde, ''The transportation mode decision revisited,'' *Transportation Journal* 17 no. 2 (Winter 1977), pp. 51–57.

12. See, e.g., Lisa H. Harrington, ''Carrier-selection criteria change with the times,'' *Traffic Management* (September 1983), p. 60; Edward R. Bruning and Peter M. Lynagh, ''Carrier evaluation in physical distribution management,'' *Journal of Business Logistics* 5 no. 2 (1984), p. 40; cf. Gwendolyn H. Baker, ''The carrier elimination decision: Implications for motor carrier marketing,'' *Transportation Journal* 24 no. 1 (Fall 1984), pp. 25, 28.

13. See, e.g., Baker, ''The carrier elimination decision,'' pp. 20–29.

14. Colin Barrett, ''Insurance woes put shippers at peril, too,'' *Distribution* (February 1986), p. 70.

chapter | 3

Transportation Pricing in Transition

chapter objectives

After reading this chapter, you will understand:
Why the transportation industry now makes extensive use of three different methods of pricing.
The changing nature, function, and legal status of tariffs and rate bureaus.
The significance of key regulatory concepts of interest to shippers considering protest or complaint actions.
The significance of antitrust law to traffic management.

chapter outline ■

Introduction

Rate Regulation
of Common Carriage
Tariffs
Rate bureaus
Protests and complaints
Maximum railway rates
Minimum railway rates
Nonrailway rates

Rate Regulation
of Contract Carriage

Railway contracts
Nonrailway contracts

Adapting to the Evolving
Pricing Environment
Management of pricing information
Antitrust

Summary

Questions

■ *Introduction*

The changing character of rate regulation and the growth and development of computer and telecommunication technology have greatly altered the transportation pricing environment. Generally speaking, the structure of freight rates remained relatively stable from the end of World War II until 1980. During this period, common carriers collectively fixed rates through their rate bureaus. Federal regulation, moreover, greatly limited competition from contract, exempt, and private carriage, which meant that the published rates of common carriers predominated.

Regulatory changes, especially since 1980, have created a more dynamic pricing environment. Although tariffs are still important, rates for rapidly expanding contract and exempt transportation services now form important threads of the current pricing fabric. On one hand, shippers face an onslaught of independently filed, and less stable, published tariff rates. On the other, new opportunities to negotiate contract and exempt rates have come into play. The Burlington Northern Railroad, for example, showed the extent to which the pricing environment has changed by indicating that 90 percent of its pricing is now outside the realm of rate bureaus.[1]

Thus, three methods of pricing are predominant today: published tariffs, negotiated contracts, and unregulated open-market transactions. As shown in **Exhibit 3.1**, the type of carriage (common, contract, or exempt) determines the pricing method (published tariffs, negotiated contracts, or open-market transactions) used by freight carriers. Regulated common carriers rely on a complex scheme of class and commodity rates published in tariffs and filed with the Interstate Commerce Commission (ICC). Contract carriers negotiate price-service agreements with individual shippers, including agents, shipper associations, brokers, and freight forwarders. Exempt carriers, of course, are not subject to rate regulation. They rely on price circulars or daily price quotations typically given over the phone in a spot-market atmosphere. Exempt carriers also enter into contracts with shippers.

This chapter examines the changing nature of rate regulation for common and contract carriage and discusses the implications for adapting to the evolving pricing environment.

Exhibit 3.1 **Overview of transportation pricing methods**

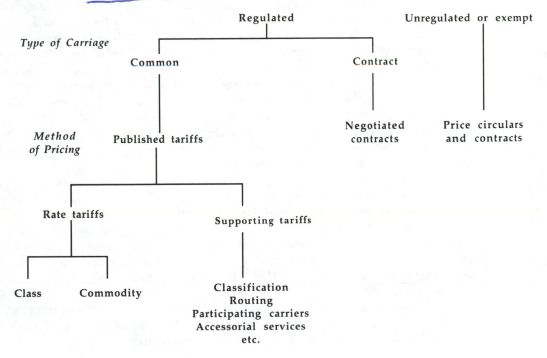

Rate Regulation of Common Carriage

Tariffs

The domestic transportation companies that provide ICC-regulated common carriage services have the legal duty to publish, file, and make available for public inspection the rates applicable to all of their routes. These common carriers use an elaborate system of tariffs to implement this duty. A *tariff* is a transportation industry publication that contains rates, rules, and other useful information needed to purchase common carrier services.

This method of pricing generally requires the use of two groups of tariffs. The first group consists of rate tariffs, which contain the rates for the linehaul (door-to-door or terminal-to-terminal) shipments. The rate search, though, often leads to the use of other tariffs that make up the second group. These tariffs provide supporting information about the correct application of rates, routes, and charges for accessorial services like stopoffs or refrigeration. For example, the classification tariff (examined in Chapter 4) does not contain rates for different classes of freight; its purpose is to assign articles to these classes. Other tariffs give information about routing options, participating carriers, mileage, demurrage, detention, and many other accessorial or special services. **Exhibit 3.2** lists the different categories of tariffs and shows the standard coding system used by the ICC to identify them.

Tariffs are official documents; they have the "force and effect of statutory law."[2] This legal status is the basis of the *filed-rate doctrine*, which states that the charges based on filed tariff rates must be assessed and collected, regardless of any separately negotiated agreements. In the wake of deregulation, however, ICC enforcement of this long-standing doctrine has been relatively lax. Today, it appears that tariff-filing requirements are ignored almost as much as they are observed. This practice has forced shippers to be alert for "phantom discounts," which refer to verbal and even written discounts to shippers that are not subsequently published in tariff form. The problem is that some motor carriers later ended up in bankruptcy courts, and their trustees sued thousands of shippers to collect undercharges based on legally filed tariff rates.

Such phantom discounts led the National Industrial Transportation League (NITL) to petition the commission to declare valid all rates negotiated in good faith by shippers and carriers, regardless of whether the rates were actually filed. After investigating the issue, the ICC denied the NITL request and adopted the policy that it could construe such undercharge collection actions as unreasonable practices.[3] If such actions are found unreasonable, the shipper may find relief from paying the filed tariff rate. Thus, this policy opens a small loophole in the filed-rate doctrine.

The Revised Interstate Commerce Act (RICA) gives the ICC the authority to create specific tariff regulations. The primary objective of such regulation is standardization. As **Exhibit 3.3** illustrates, these regulations address virtually all

Exhibit 3.2 **Tariff designation codes**

Item	Length (characters)	Description		
ICC designation	3	"ICC" prefix		
Carrier/agent code	2–4	Unique code authorized by ICC		
Tariff number	1–4	**First digit**	**Tariff**	**Type***
Rail		1	Class rates	R
		2	Class exception	R
		3	General commodity	R
		4	Specific commodity	R
		5	Routing guides	S
		6	Governing tariffs	S
		8–9	Miscellaneous (accessorial and special service)	S
Motor		1	Governing publications	S
		2	Commodity	R
		3	Combined class and commodity	R
		4	Commodity column	R
		5	Class	R
		6	Miscellaneous tariffs	S
		7	Import-export	S
		9	Emergency movements	R
Tariffs other than above		Codes assigned by carriers or agents		
Reissues	1	Next letter in alphabet; first issue begins with "A"		

*R = rate tariff, S = supporting tariff

aspects of tariffs, including their form, arrangement, and printing. Whenever carriers haul regulated traffic, they must comply with the commission's tariff regulations.

The ICC tries to balance the need for tariff uniformity and strict compliance with the need for pricing flexibility. Flexibility is desirable because it allows carriers to experiment with tariffs to make them as simple and clear as possible to attract business. In 1984, with this goal in mind, the ICC's Bureau of Traffic revised and consolidated its rules concerning the format and content of tariffs. Four separate sets of rules (for rail, motor, water, and freight forwarder services)

■■■■■■■ *Exhibit 3.3* **Index to ICC tariff regulations (49 CFR 1312)**

Section	Title	Page Number
1312.1	Introduction and General Provisions	1–5
.2	Applications for Special Tariff Authority	5–5a
.3	Size of Paper and Where to Send Tariff Publications and Other Documents	5a–6
.4	Filing Tariffs	6 –11
.5	Posting Requirements	11–16
.6	Furnishing Copies of Tariff Publications	16–17
.7	Form and Printing	17–19
.8	ICC Tariff Designations	19–24
.9	Codes for Identification of Places	24
.10	Powers of Attorney, Concurrences, Transfer of Agent	24–27
.11	Tariffs Issued by Joint Agents	27–28
.12	Title Page of Original Tariffs	28–31
.13	Contents of Tariffs	32–40
.14	Statement of Rates and Fares	40–46
.15	Routing	46–48
.16	Sectional Tariffs	48
.17	Amendments	48–64
.18	Supplements	64–68
.19	Cancellation of Tariffs and Transfer of Provisions	68–69
.20	Transfer of Operations— Change in Name and Control	69–75
.21	Suspended Matter	75–81
.22	Rates and Other Provisions Prescribed by the Commission	81
.23	Expiration Dates	82
.24	Index of Tariffs	82–83
.25	Tariffs Listing Carrier's Operating Authority	83–84
.26	Participating Carrier Tariffs	84–86
.27	Rate Basis Tariffs	86–87
.28	Classification, Exceptions, Rules, Dangerous Articles, and Station List Tariffs	87–89
.29	Accessorial, Terminal, and Other Service	89–92
.30	Released Rate Provisions	92
.31	Distance Rates	92–94
.32	Commodity Rates Determined by the Use of Rate Base Numbers	94

(continued)

Exhibit 3.3 (continued)

Section	Title	Page Number
.33	Application of Rates from or to Intermediate Points	94–95
.34	Class Rates from or to Unnamed Points	95
.35	Continuous Service Rates	95–96
.36	Time-Volume Rates	96–98
.37	Seasonal Water Rates	98
.38	Export and Import Traffic and Joint Rates with Ocean Carriers	98–100
.39	Substitution of Service	100–102
.40	Miscellaneous Provisions Which May Be Filed on Less Than Statutory Notice	102–106
.41	Claims Rules	106
.42	Contracts and Contract Summaries	106–110

were condensed into a single set, and the new rules set forth basic guidelines rather than precise requirements.[4] As part of this initiative, the ICC has authorized the publication of zip-code tariffs, which facilitate computer-supported rate retrieval.

Carriers publish tariffs in three ways. First, each carrier may choose to create its own tariff. Second, carriers may choose instead to authorize regional rate bureaus such as the Southern Freight Association (for rail) and the Middlewest Rate Bureau (for motor) to publish rates on their behalf in "agency" tariffs. Third, since agency tariffs generally have only regional application, two or more rate bureaus often establish a joint-agency agreement to publish interregional tariffs. Shippers can determine if a carrier is a party to a tariff by checking the section of the tariff titled *Participating Carriers* or by reviewing separate participating carrier tariffs.

Rail carriers must give twenty days public notice for new or increased rates, one day for rate reductions, and forty-five days for surcharges or joint-rate cancellations. Not until the required number of days passes after filing does a rate become effective. The RICA also requires nonrailway carriers to give thirty days public notice, but the ICC has used its authority to reduce the thirty-day notice period for independent motor carrier rate filings. The current rules permit rate reductions and new rates to become effective on one day's notice and rate increases to become effective on seven day's notice. The two main goals are (1) to make it possible for motor carriers to respond quickly to new market demands and (2) to reduce the numerous special permission applications for short notice authorization.[5]

Rate Bureaus

Carriers create nonprofit rate bureaus to perform the following functions:

■ Collectively set rates within the limits of ICC-approved bylaw agreements
■ Prepare, publish, file, and distribute tariffs and supplements
■ Formulate uniform freight classifications and standard rules for packaging, marking, loading, unloading, billing, and many other aspects of transportation
■ Offer data collection services
■ Process general rate increases and independent actions

Railroad and motor carrier rate bureaus are organized by mode and by territory, although the ICC recently has given several bureaus nationwide authority. Currently, three major bureaus serve the railway industry, whereas about a dozen major bureaus operate in the trucking industry. Water carriers have the Waterways Freight Bureau and the Intercoastal Steamship Freight Association. The Airline Tariff Publishing Company publishes rates for participating air carriers.

Membership in rate bureaus is open but voluntary. Most major rail and trucking firms are members of at least one rate bureau. The larger carriers, which have interregional operations, belong to several different bureaus. Members, however, have the unrestricted right to pursue independent actions.

The regulatory climate since 1980 has significantly altered the historical legal status and role of rate bureaus. In 1948, the Reed-Bulwinkle Act amended the Interstate Commerce Act to grant carriers immunity from antitrust when they collectively set rates through rate bureaus, but it also made these organizations subject to ICC oversight. That is, members could fix prices (including charges for facilities and equipment, allowances, credit, and classification) but only within the confines set by the ICC-approved bylaws of the bureau. After 1960, the ICC marginally restricted the scope of immunity allowed in the bylaws. The real breakdown in collective ratemaking came with the passage, in 1980, of the Staggers Rail Act and the Motor Carrier Act. These acts codified the previous ICC restrictions and added new ones.

Now the ICC will approve rate bureau agreements only when there is a specific finding that national transportation policy considerations justify the antitrust exemption. The current goal of bureau regulation is to foster price competition through independent actions in open markets. In pursuing this goal, the ICC has severely limited the conduct of permissible collective actions. Generally, the ICC will not authorize a rate bureau to operate unless its bylaws meet the following conditions:

1. Single-line rates. Bureau members can no longer vote on or discuss rates applicable to the lines of a single carrier.

2. *Joint-line rates.* Bureau members are prohibited from discussing or participating in joint-rate agreements unless they practically participate in the route. A joint rate applies to a through movement over the lines of two or more carriers (see Chapter 5). In 1980, the ICC established the "direct-connector" doctrine to determine the meaning of "practically participate" in the railway industry.[6] This doctrine confines joint-rate agreements to the directly connecting rail carriers involved in a specific movement of a specific commodity. For motor carriers, joint-rate immunity does not extend to shipments weighing less than 1000 lb.

3. *Independent actions.* Rate bureaus may not interfere with the independent actions of its members. A member that initiates an independent action also has the right to decide if it should be docketed and publicized.[7] Additionally, bureau staff cannot initiate any rate proposals.

4. *Broad tariff changes.* Motor carriers may discuss and vote on general (across-the-board) rate increases and other broad tariff changes. The discussion and voting, however, must not involve individual markets and single-line rates. To gain immunity in these actions, railway companies must meet the direct-connector standard.

5. *Commodity classification and rules.* Bureau members may discuss and vote on commodity classifications and rules that have general application to all rates.

6. *Contract and exempt rates.* Antitrust immunity does not extend to contract or deregulated or exempt transportation. Further, from the perspective of antitrust enforcement, contract transportation embraces the so-called Section 10721 rates that carriers quote to government agencies (see Chapter 5).

Besides these basic conditions, rate bureaus must publicize meeting schedules, including the main details of the agenda. Meetings must be open to the public, and a railway rate bureau must make recordings and transcripts of its meetings. Additionally, bureaus must dispose of proposals within 120 days.

Collective ratemaking, therefore, no longer represents the predominant form of pricing in the transportation industry. It appears increasingly likely, moreover, that Congress will strip away all antitrust immunity from rate bureaus, although this issue is still controversial.[8] Rate bureaus now concentrate on compiling, publishing, and distributing rate information, electronically or in printed form, in competition with emerging independent third-party enterprises. The Western Trunkline Tariff Bureau, for example, was given permission by the Justice Department in 1983 to act as an electronic rate clearinghouse for exempt rate information. In 1984, a nonbureau enterprise, DSI RAIL, a unit of Distribution Sciences, Inc., initiated a National Railway Ratemaking and Rate Retrieval system. This system will allow participating railroads to establish through rates electronically, while preserving the confidentiality of rates that is necessary to avoid antitrust violations.

Protests and Complaints

Shippers can challenge a rate by submitting a protest or a complaint to the ICC. Generally speaking, a shipper files a protest to prevent a proposed rate or service from going into effect, but a complaint registers dissatisfaction with an existing rate. To settle such disputes, the ICC can (1) investigate the reasonableness of proposed rates, rules, or practices to gather information, (2) suspend the effective date of the proposed change and investigate the rate, (3) declare an existing rate unreasonable and establish a new rate that it considers reasonable.

Nevertheless, the Staggers Rail Act and the Motor Carrier Act have greatly curtailed the ICC's powers to investigate or suspend tariff rates, impose maximum or minimum controls to make rates reasonable, or remove unreasonable discrimination. The commission's interpretation and implementation of these acts, moreover, has limited the scope of rate regulation even further. As the following discussion will demonstrate, it is extremely difficult for shippers to obtain regulatory remedy for rates considered too high or too low.

Maximum Railway Rates

The RICA limits the ICC's jurisdiction over railway rates to circumstances in which a railway company is market dominant—meaning that the carrier has an effective monopoly over a shipper's traffic. Congress formally established the market dominance concept in the Railroad Revitalization and Regulatory Reform Act of 1976 to restrict the ICC's maximum rate-setting power to situations in which competition could not assure reasonable rates and services. In 1980, Congress used the Staggers Rail Act to expand the rate and service freedoms of the railway industry.

The RICA, as amended by these two acts, defines market dominance as "an absence of effective competition from other carriers or modes of transportation for the transportation to which a rate applies."[9] Nevertheless, in 1981, the ICC stretched the concept of "effective competition" to include indirect forms of competition either from the availability of substitute products or from the shipper's ability to purchase the same product at alternative locations.[10] The initial guidelines for rate disputes required a railway company to identify the source of competition (transportation, product, or geographic). To establish ICC jurisdiction, shippers then had to show that this competition was not effective. The effect of these rules was to increase the amount of justification needed to prove market dominance. In 1985, though, the National Industrial Transportation League negotiated an agreement with representatives of the railway industry for a more equitable distribution of the burden of proof. Under the proposed rules, which have been recommended to the ICC for approval, railroads have the burden of proving the existence of effective product or geographic competition.

Even with the adoption of these rules, the law still expressly allows carriers to escape ICC jurisdiction by showing that a proposed rate does not exceed a certain "jurisdictional threshold." The RICA specifically defines this jurisdictional threshold as the "Cost Recovery Percentage" (CRP). This measure is supposed to indicate the level at which the railway industry would earn enough revenue to cover total costs if all the traffic that generated revenue-to-variable-cost ratios above the CRP were to produce ratios only equal to the CRP. The ICC is responsible for determining the CRP each year. As shown in **Exhibit 3.4**, the RICA does not permit the jurisdictional threshold to fall below 170 percent or to exceed 180 percent.

For a proposed rate, this ratio consists of the revenue generated from the move compared to the variable cost of providing the service. If the ratio does not exceed the jurisdictional threshold, the RICA instructs the ICC to find that the carrier does not have market dominance.[11] By contrast, a ratio that does exceed the threshold creates no presumption about market dominance or reasonableness.

The first major test in rate protests or complaints is whether market dominance prevails so that the ICC has jurisdiction to assess reasonableness. Once

Exhibit 3.4 **Jurisdictional threshold to assess reasonableness of rail rates**

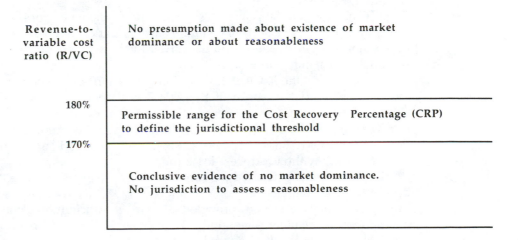

Revenue-to-variable cost ratio (R/VC)

No presumption made about existence of market dominance or about reasonableness

180%

Permissible range for the Cost Recovery Percentage (CRP) to define the jurisdictional threshold

170%

Conclusive evidence of no market dominance. No jurisdiction to assess reasonableness

October 1, 1984 and beyond

this test has been passed, a second major roadblock appears— *revenue adequacy.* To determine the reasonableness of proposed rates, the ICC must consider the adequacy of carrier revenues. That is, along with the general goal of reasonable rates for transportation, the Staggers Rail Act emphasizes the need for railway companies to earn adequate revenues to meet the needs of interstate commerce. A carrier is "revenue adequate" if it has a rate of return on investment equal to the current cost of capital. At this writing, no revenue adequate Class I rail carrier has been found.[12]

Furthermore, railway companies may raise their rates quarterly to recover inflationary cost increases. The ICC uses the Association of American Railroads' cost recovery index to determine the authorized percentage increase that reflects rising railroad costs. A proposed rate increase that does not exceed this level cannot be found to exceed a reasonable maximum.[13]

Minimum Railway Rates

All proposed rates must contribute to the "going value of the concern." This condition is met when the proposed rate produces revenues greater than the directly variable costs of the service.[14] A rate that does not contribute to the going value of the concern is *presumed* unreasonable. But a rate that does contribute establishes *conclusive* evidence that it is not unreasonably low. The burden of proof rests with the protestant to show initially that the proposed rate is below the variable cost floor.

Nonrailway Rates

The ICC has the general power to suspend and investigate proposed new rates, rules, and practices of nonrail common carriers under its jurisdiction for up to seven months. The burden is on the carrier to show that proposed changes are reasonable or nondiscriminatory.

Nevertheless, the RICA defines a Zone of Ratemaking Freedom (ZORF) for motor carriers. The purpose of the ZORF is to permit market-oriented rate changes free from ICC interference. The base rate, defined as the rate in effect one year before the effective date of the proposed rate, represents the initial floor of the ZORF. Base rates are automatically adjusted for inflation, which is determined by the Producers Price Index.

The RICA initially created the zone's ceiling at 10 percent above the adjusted base rate. The commission may expand the ZORF by five percentage points during any one-year period, provided (1) there is sufficient actual and potential competition to regulate rates, and (2) the public will benefit from further rate flexibility. In 1986 the ICC adopted a rule to make a 5 percent increase in the

ZORF automatic each year, unless the commission takes action to the contrary.[15] On October 23, 1986, the ZORF was 20 percent. It now appears that the ZORF will increase by another 5 percent each year thereafter. Although zone rate-making must represent independent actions, rate bureaus may still publish rates established within this framework.

Shippers also may make formal complaints about existing rates, rules, and practices. The commission has authority to investigate on its own initiative or on complaint by the shipper. The burden of proof that rates are unreasonable or discriminatory rests with the party making the complaint. Finally, the general standards for minimum rate regulation, like the guidelines for rail carriers, prohibit rates below variable costs.

Rate Regulation of Contract Carriage

Railway Contracts

In 1982, the ICC formally established rules for filing and adjudication of railroad transportation contracts.[16] These rules require rail carriers entering into contracts with purchasers of transportation services to file the original and one copy of the written contract accompanied by three contract summaries. At this writing, the commission is still developing rules for disclosure of contract information to the public. Currently, details of the contract remain confidential, and only summaries, which contain a brief general description of the essential elements of the contract, are made public. Regulations require additional detail for contracts involving the transportation of agricultural, forest, and paper products or the domestic ports. In addition, the ICC may release some confidential contract information when certain conditions associated with a complaint are met.[17]

Although carriers must file contracts with the ICC, these agreements may apply immediately to shipments moving after negotiations have been concluded, if the parties agree. The ICC stipulates that contracts must be filed within a reasonable period after shipments begin. In addition, the ICC may institute a proceeding to review a contract on its own initiative or after a complaint has been made. Only shippers and ports (a port authority or a government body with responsibility for port facilities) have the legal standing to file complaints against proposed rail contracts. Shippers include consignors and consignees, although the ICC may recognize freight forwarders and brokers as shippers in certain cases.

A shipper may lodge a complaint if it will be individually harmed because the proposed contract unduly impairs the rail carrier's performance as a common carrier. The ICC will examine the impairment issue case by case. Car supply, though, is a critical element in this judgment.

In addition, shippers may challenge the contracts that involve the transportation of agricultural, forest, or paper products on the following grounds:

1. Individual harm will result because the rail carrier(s) engaged in unreasonable discrimination. Here "unreasonable discrimination" means the railroad has refused to enter into a contract with the shipper for transportation of the same commodity under similar conditions.
2. Individual harm will result because the proposed contract constitutes a destructive competitive practice. Generally speaking, carriers engage in such practices when they set prices at very low levels (below variable costs) to drive out competitors.

A port may file a complaint only on the grounds that it will experience harm because the proposed contract will allow unreasonable discrimination against that port. In this situation, "unreasonable discrimination" refers to the use of devices such as rebates, special rates, and drawbacks to receive different compensation from shippers for essentially the same service.[18]

By statute, the ICC must decide whether to institute a proceeding to review contracts within thirty days of the filing date. Thus, the ICC has reserved a minimum of twelve days to decide whether to institute an investigation, allow an appeal, and decide that appeal. Shippers and ports have only eighteen days after a carrier files the contract to register a complaint.

Unchallenged contracts gain approval automatically thirty days after filing. If the ICC institutes a proceeding to review a contract, the contract is approved

1. On the date the ICC renders a favorable decision, provided that at least thirty days have passed since the filing date
2. On the thirtieth day after filing the contract, if the ICC denies the complaint
3. On the sixtieth day after filing the contract, if the ICC fails to disapprove the contract

Once the contract has been approved, the shipper must pursue any subsequent legal actions through the courts, at which point the RICA exempts the contract from ICC regulation.

Nonrailway Contracts

Although the RICA requires nonrailway contract carriers to publish and file tariffs, formerly "schedules," containing new or reduced rates and to give thirty day's public notice, the ICC began to exempt individual contract motor carriers from tariff filing requirements in 1982. In 1983, the ICC went one step further and issued a blanket exemption from all contract tariff filing requirements (and is now considering the same action for water contract carriers).[19]

Motor contract carriers, in general, view this action as a positive initiative that will eliminate annual administrative costs of about $30,000 for a medium-sized operation.[20] Other interested parties, however, fear the elimination of filing requirements will lead to anticompetitive secret agreements between shippers and carriers.

Regulation of nonrailway contract pricing practices offers few legal remedies to the shipper. The ICC does have jurisdiction to assess the reasonableness of minimum rates of nonrailway contract carriers but only to protect common carrier service from predatory practices.

Adapting to the Evolving Pricing Environment

More than a decade ago, a leading marketing expert recommended that carriers should view themselves as transportation businesses rather than as railroads, airlines, truckers, and so forth and should operate to meet the needs of shippers at a profit.[21] Since 1980, new pricing freedoms coupled with increased competition have spurred many carriers to adopt the marketing concept to survive.[22] Carriers now offer expanded "product lines" that include many new rate and service offerings designed to meet shipper needs.

To adapt to this evolving pricing environment, traffic managers must develop new skills in contracting, costing, and negotiating. Furthermore, traffic managers must give special consideration to the proliferation of rates, as well as to the antitrust implications of various pricing arrangements. These special considerations are discussed next.

Management of Pricing Information

A freer market environment for transportation has greatly increased the volume of rates. Since 1980, the number of tariff filings has continued on a steep upward trend. Carriers now file more than one million tariffs a year or about ten thousand pages a day.[23] This avalanche of new and rapidly changing rates has increased the need for simplified pricing schemes, as it has for timely and accurate management information systems to help the traffic manager seize pricing opportunities.

Fortunately, advancing computer and telecommunication technology, along with new software applications, can greatly facilitate the process of adaptation. Nonetheless, tariff computerization and telecommunication technology still face obstacles. Converting 20 million tariff pages to electronic data bases is inefficient. Carriers and shippers must continue to search for ways such as zip code systems to simplify tariffs. In addition, the transportation community needs to make further progress in setting widely accepted standards for computer-to-computer communication (see Chapter 4). To meet the challenges of a competitive market environment, the transportation industry will have to apply the innovations in information technology to the publication and distribution of rates and related matter. And traffic managers will need to be more sophisticated in the management of pricing information systems.

Antitrust

Antitrust enforcement is a relatively new concern for traffic management. As previously discussed, the Railroad Revitalization and Regulatory Reform Act of 1976, the Motor Carrier Act of 1980, and the Staggers Rail Act of 1980 have greatly restricted the antitrust immunity of rate bureaus. Further, these acts confer no immunity to contract or exempt ratemaking. Shippers and carriers now are largely accountable for their conduct on the basis of antitrust laws. This point is especially significant because penalties for antitrust violations may include (1) fines up to $100,000 per individual and $1,000,000 per corporation, (2) imprisonment up to three years, and (3) treble private damages.

The traffic department will need to educate all employees engaged in purchasing transportation services about the basics of antitrust law. Legal experts recommend training programs that include elements such as compliance manuals, seminars, and periodic audits of arrangements with carriers.[24] This subject is complex; many volumes and learned treatises have been written about it. The discussion that follows, therefore, attempts to provide only an overview of antitrust law as it pertains to the field of transportation.

Three acts essentially constitute antitrust law: the Sherman Act of 1890, the Clayton Act of 1914, and the Federal Trade Commission Act of 1914. Nevertheless, legal scholars and enforcement officials have indicated that the Sherman Act contains the essence of the ''law of antitrust'' that applies to transportation.[25] Section 1 of the Sherman Act prohibits all restraints of trade. Section 2 makes it illegal for a business to monopolize or attempt to monopolize any segment of interstate commerce.

In the words of the chief of the antitrust division's transportation section:

> *The central purpose to be served by the antitrust laws is the preservation of free markets—markets characterized by intense competition—such that our national resources may be developed efficiently so as to yield the largest possible quantity and the richest possible variety of goods and services to our population.*[26]

In the purchase of transportation services, traffic managers essentially have to steer clear of Section 1 violations. These violations fall into two categories: per se and rule of reason. Per se violations are illegal regardless of any harm done and are prime targets of antitrust enforcement. They generally involve horizontal relationships, that is, agreements among shippers only or among carriers only. Certain vertical relationships, which involve agreements between shippers and carriers, also fall into this category.

The basic types of per se violations are as follows:

1. Price fixing. Any unreasonable interference by formal or informal agreement with the competitive pricing system of open markets is illegal. Shippers should avoid collective agreements with other shippers aimed at setting minimum rates, discounts, credit terms, or services. As noted previously, though, carriers have limited immunity to act jointly in certain rate matters. As a general rule, when participating in joint-rate discussions with carriers, shippers should not assume the same immunity extends to them.[27] The safest approach is to work closely with legal counsel having antitrust expertise in discussions with representatives of the carriers directly involved in the haul. Additionally, shippers should prevent such discussions from straying in the area of single-line rates or services.

The *Keogh Doctrine*, which has received a lot of attention in a recent Supreme Court case, is an interesting facet of antitrust law that relates to price fixing in the transportation industry.[28] The doctrine states that rates considered legal under the regulatory regime of the RICA cannot be illegal under antitrust laws. Consequently, common carriers operating under filed tariff provisions have a strong defense against suits brought by shippers alleging price-fixing schemes or other antitrust violations.

2. Division of markets. Agreements by shippers or carriers not to compete with one another in certain markets are illegal. This type of behavior, however, will most likely involve carriers. For example, two or more trucking companies might decide to divide up specific customers or routes, promising not to compete for a shipper's business. Shippers may also be liable for this type of trade, though, when they agree to allocate traffic among certain carriers.

3. Boycotts. Groups of shippers may not collectively refuse to deal with a carrier, say, to prevent the carrier from serving competitors or from offering rates to competitors at the same or lower levels. Further, as one legal observer has noted:

> *Even if a shipper thinks it has a good reason for advising other shippers to cease using a carrier, e.g., failure to pay loss and damage claims, the joint effort could be a violation of the antitrust laws if its effect is to foreclose the carrier from any substantial market.*[29]

4. Tying agreements. Tying agreements in the transportation industry involve vertical relationships between shippers (or receivers) and carriers. A tying agreement occurs, for example, when the shipper refuses to purchase a carrier's services unless the carrier buys products made by the shipper's firm. This type of pricing arrangement generally becomes a per se violation when (1) the seller of the tying product or service has enough economic power to impair competition in the market for the tied product and (2) a substantial amount of interstate commerce is affected.[30]

The second category of Section 1 violations involves a rule-of-reason analysis to determine whether an agreement reflects reasonable business behavior or harms competition. Agreements that fall into this category do not necessarily violate Section 1 of the Sherman Act unless there is economic proof of unreasonable damage to competition. In the transportation arena, these agreements generally involve contracts between shippers and carriers. The prime targets for analysis include the following kinds of agreements:

1. *Exclusive deals.* In the exclusive-deal arrangement, a carrier may agree to serve only one shipper. Conversely, a shipper may offer its traffic so long as the carrier agrees not to serve a competitor.

2. *Requirements contract.* The requirements contract, which is closely related to the exclusive deal, requires the shipper to purchase all its needs for certain transportation services from only one carrier.

3. *Joint bargaining.* In certain circumstances, shippers may organize to negotiate rates and services. The National Small Shipments Traffic Conference (NASSTRAC), for example, has received approval from the Justice Department to negotiate freight rates for its members. Likewise, shipper associations that pool traffic and get volume discounts from carriers generally do not face antitrust enforcement. Issues of concern to the Justice Department, however, include such things as a member's ability to act independently in the selection of carriers, the restriction of membership in order to place nonmembers at a competitive disadvantage, and the extent to which such organizations dominate a particular line of commerce.[31]

4. *Joint action among affiliates.* Many large firms with multiple divisions and subsidiaries have centralized the purchase of transportation services in a corporate traffic organization. This purchasing approach gives the corporate traffic manager additional negotiating leverage. It appears that such arrangements with affiliates will not violate antitrust law as long as the affiliates are partners, and not competitors, acting as a single economic unit.[32]

The Justice Department has established a Business Review Procedure to reduce uncertainty about antitrust enforcement.[33] Shippers contemplating joint actions may have proposed activities reviewed first. Commenting on the legal standing of this procedure, the deputy assistant attorney general of the antitrust division of the Justice Department has said:

> *While such a review is not formally binding, we have always honored our clearances and have not brought criminal enforcement actions against the conduct that was allowed,*

although we have reserved the right to seek civil, prospective relief if the facts or the relevant law changes.[34]

In addition, the Justice Department's review procedure represents prosecutorial intent of only one agency under one administration and is not binding on later administrations or the Federal Trade Commission or private plaintiffs. Shippers should also know that they will be required to divulge all relevant data during litigation; excepting special circumstances, such as a protective order for trade secrets, these data can be made public.

Summary

During the 1980s, the transportation community has witnessed fundamental changes in its pricing environment. Deregulation, regulatory reform, and information technology have been the principal catalysts in the transformation of the environment. No longer is stability the hallmark of freight rates. Currently, a dynamic situation exists in which traffic managers encounter an explosion of independently filed published tariff rates, as well as a host of opportunities to negotiate contract and exempt rates.

The type of carriage performed determines the method of pricing used by freight carriers. Common carriage requires published tariffs. Contract carriage relies on agreements with shippers, freight forwarders, agents, brokers, and shipper associations. Exempt carriage makes use of price circulars and contracts in an open-market setting.

In the 1980s, rate regulation of common and contract carriage has changed dramatically. Reform legislation and subsequent ICC interpretations have significantly altered the nature, function, legal status, and role of tariffs and rate bureaus. Concepts such as market dominance, revenue adequacy, inflationary cost recovery, zone of rate freedom, and going value of the concern make it difficult for today's shipper to challenge published tariff rates or services as unreasonable or discriminatory. Similarly, the regulation of contract carriage affords few opportunities for regulatory remedy.

To adapt to the evolving pricing environment, traffic managers need to develop new skills in contracting, costing, and negotiating. In addition, traffic managers must give special consideration to the management of pricing information and to the issue of antitrust enforcement.

Selected Topical Questions from Past AST&L Examinations

1. A manufacturer of heavy electrical equipment is faced with a 30 percent rail rate increase for movements of electrical machinery from River Front, Pennsylvania, to transcontinental destinations in California, Idaho, and Washington. As your

client, the manufacturer asks you to protest the rate increase on the grounds it is unreasonably high and will prevent movement of electrical machinery to major markets in transcontinental territory. Preliminary analysis reveals that all the traffic moves via rail at a rate level that gives a revenue yield of more than 60 percent of rail variable cost.

(a) In your protest and verified complaint to the Interstate Commerce Commission, describe the areas that you would specifically cover.

(b) When does protest have to be received by the commission to be considered timely filed? (Fall 1980)

2. The new Staggers Rail Act of 1980 incorporates a number of changes that will permit railroads greater pricing freedom and marketing opportunities.

(a) Describe three of these changes.

(b) Describe the changes that will restrict rate bureau handling of rail rates. (Fall 1980)

3. Discuss when a shipper and when a carrier has the burden of proof in a matter concerning the maximum reasonableness of rail rates under the Staggers Rail Act of 1980. (Fall 1981)

4. For the first time, the Interstate Commerce Act, pursuant to the Staggers Rail Act amendments, permits railroads to act as contract carriers. Complaints may be filed by interested parties against rail contracts. Specify the grounds on which these complaints may be lodged. (Spring 1982)

5. The Sherman Act of 1890 prohibits contracts, combinations, and conspiracy for the restraint of trade, monopolization, and attempts to monopolize. Although the language of the act is broad, certain practices are considered clearly illegal and are described as per se offenses. Both shippers and carriers can be found in violation. Identify some per se practices, and discuss what a shipper should avoid doing. (Fall 1982)

6. In a proceeding where the reasonableness of minimum railroad rates is at issue, what is the standard of reasonableness? (Fall 1982)

7. In the past, all rates were subject to public scrutiny in the form of tariff publishing, as well as protest and complaint processes. What processes are available today for a shipper or railroad to ascertain the term of a rail contract in effect between another competing shipper or railroad? (Fall 1984)

8. A responsible, well-intentioned member of the business community is considering a marketing arrangement whereby he and his competitors would be participants in the establishment of prices for merchandise sold by them. He seeks your advice, through review of the proposed arrangement, as to whether the conduct involved might lead to investigation and subsequent prosecution by the Justice Department for an antitrust law violation. Discuss the response that might be anticipated from the Justice Department and the procedure used. (Spring 1985)

9. The Staggers Act relaxed maximum railroad rate regulation and, in the process,

formally created two important concepts: market dominance and revenue adequacy.
(a) Briefly explain each concept.
(b) What roles do they play in maximum rate regulation?
(c) What implications do the two concepts, as currently applied, hold for shippers who plan to seek regulatory remedy for the rate changes that they consider unreasonably high? (Spring 1985)

Notes

1. "Carrier officers speak on strategies for success at Indiana conference," *Traffic World* (April 18, 1983), p. 45.
2. 49 U.S.C. 10761 and 10762; for a historical perspective of this issue, see G. Lloyd Wilson, *Industrial Traffic Management Part 1* (Chicago: Traffic Service Corp., 1935–1936); Kenneth U. Flood, Oliver G. Callson, and Sylvester J. Jablonski, *Transportation Management*, 4th ed. (Dubuque, Iowa: Wm. C. Brown, 1984), pp. 95–97.
3. Ex Parte No. MC-177. *National Industrial Transportation League—Petition to Institute Rulemaking on Negotiated Motor Common Carrier Rates*, October 29, 1986.
4. The new rules located at 49 CFR 1312 *(Revised Interstate Commerce Act)* replace those previously found in 49 CFR 1300–1310; see also No. 37321, "Revision of tariff regulations, all carriers" (not printed), served October 1, 1984.
5. Ex Parte No. MC-170, "Short notice effectiveness for motor carriers and freight forwarder rates" (not printed), served May 29, 1984.
6. 364 ICC 16 (1980).
7. 364 ICC 655 (1980).
8. For further discussion, see *Collective Ratemaking in the Trucking Industry, A Report to the President and to the Congress of the United States*, submitted by the Motor Carrier Ratemaking Study Commission, June 1, 1983; Grant M. Davis, "Evidential issues in collective ratemaking," *Traffic World* (May 9, 1983), pp. 35–40; Grant M. Davis, "The collective ratemaking issue: Circa 1984," *Transportation Practitioners' Journal* 52 no. 1 (Fall 1984), pp. 60–68; Elliot Bunce, "Special problems relating to collective ratemaking," *ICC Practitioners' Journal* 51 no. 6 (September 1984), pp. 583–590; Garland Chow and Richard Poist, "Rate bureau cost and benefits: The carrier perspective," *Proceedings—Twentieth Annual Meeting Transportation Research Forum* 20 no. 1 (October, 1979), pp. 432–439.
9. 49 U.S.C. 10709(a).
10. *Market Dominance Determinations and Considerations of Product Competition* 365 ICC 118 (July 18, 1981); *Western Coal Traffic League, et al.* v. ICC, 719 F.2d 772, (5th Cir. 1983); see also William B. Tye, "On the effectiveness of product and geographic competition in determining rail market dominance," *Transportation Journal* 24 no. 1 (Fall 1984), pp. 5–19.
11. 49 U.S.C. 10709(d)(2).
12. See, e.g., Ex Parte 393, Sub. 1, *Standards for Railroad Revenue Adequacy*, October 7, 1986.
13. 49 U.S.C. 10707(a)(2).
14. For a discussion of what makes up directly variable costs, see Interstate Commerce Commission, *Ex Parte 355 Cost Standards for Railroad Rates*, May 12, 1981, pp. 24–25.
15. Ex Parte MC-169, Sub. 1, *Automatic Expansion of ZORF for Motor Common Carriers of Property and Freight Forwarders*, September 22, 1986.
16. Interstate Commerce Commission, *Ex Parte 387 Railroad Transportation Contracts*, October 8, 1982. These rules are formally codified in 49 CFR 1039 and 49 CFR 1300.300–1300.399.
17. For additional information, see Interstate Commerce Commission, *Ex Parte 387 Railroad Transportation Contracts*, October 8, 1982; 49 CFR 1300.310.
18. See 49 U.S.C. 10741.

19. *Exemption of Motor Contract Carriers from Tariff Filing Requirements,* 133 MCC 150 (1983); Ex Parte MC-165, Sub. 2, served October 14, 1986.
20. See Francis J. Quinn, ed., "Contract carriers start to flex their muscles," *Traffic Management* (September 1983), pp. 84–87.
21. Theodore Levitt, "Marketing myopia," in *The Great Writings in Marketing,* Howard A. Thompson, ed. (Tulsa, Oklahoma: Pennwell, 1981), p. 24.
22. Virtually every modern marketing textbook establishes the "marketing concept" as the guiding philosophy for management. Stated briefly, the concept is to have the entire enterprise focus on satisfying the needs of customers at a profit.
23. Interstate Commerce Commission, *1985 Annual Report,* p. 71.
24. Steven J. Kalish, "Antitrust considerations for shippers," *Transportation Practitioners Journal* 52 no. 2 (Winter 1985), p. 196; see also John C. Bradley, "Antitrust compliance programs in the motor carrier industry— A primer on why and how," *ICC Practitioners' Journal* 49 no. 4 (May–June 1982), pp. 395–412.
25. See E. Stephen Heisley, "Antitrust implications for the future," presented to the Regional-National Educational Conference, American Society of Traffic and Transportation, Louisville, Ky. (May 1983), p. 1; Ronald G. Carr, "Railroad-shipper contracts under section 208 of the Staggers Rail Act of 1980: An antitrust perspective," *ICC Practitioners' Journal* 50 no. 1 (November–December, 1982), pp. 29–41.
26. Remarks by Elliot M. Seiden, Chief of the Transportation Section, Antitrust Division, before the spring meeting of the National Defense Transportation Association Military Traffic Management and Ports Committee, Crystal City, Virginia (April 19, 1983), p. 5.
27. This issue involves the concept of "derivative immunity." For further discussion, see Donald L. Flexner, "Potential problem areas for shippers and carriers under antitrust laws," *ICC Practitioners' Journal* 51 no. 6 (September 1984), pp. 572–575.
28. *Keogh* v. *Chicago & Northwestern Railway Co.,* 260 U.S. 156 (1922); reaffirmed in *Square D Co.* v. *Niagara Frontier Tariff Bureau, Inc.* 106 S.Ct. 1922 (1986); see also Donald L. Flexiner and Wm. Randolph Smith, "The Keogh Doctrine: Practical Implications after Square D," *Traffic World* (June 16, 1986), pp. 77–82.
29. Kalish, "Antitrust considerations for shippers," p. 189. Defaulters lists have been found legal, however, if used for self-protection by a trade association.
30. See *Fortner Enterprises, Inc.* v. *United States Steel Corp.,* 394 U.S. 495 (1969). For further discussion, see Carr, "Railroad-shipper contracts," pp. 32–35; Kalish, "Antitrust considerations for shippers," pp. 190–193.
31. For additional discussion, see Carr, "Railroad-shipper contracts," p. 39; Flexner, "Potential problem areas for shippers," pp. 575–577.
32. Kalish, "Antitrust considerations for shippers," pp. 194–195; see also, George W. Selby, Jr., "The nine commandments of lawful cooperation among competitors," *Traffic World* (March 31, 1986), pp. 87–91.
33. 28 CFR 50.6.
34. Carr, "Railroad-shipper contracts," p. 40.

chapter 4

Tariff Pricing Systems

chapter objectives

After reading this chapter, you will understand:

The changing nature, purpose, and design of class and commodity rate systems.

The role and key characteristics of the three major classification tariffs.

The underlying relationships between classification elements and transportation costs.

The role of the traffic manager in the classification process.

The problems and promise confronting the application of information technology to transportation pricing.

chapter outline

Introduction

Class Rate Framework
Commodity classification
Distance scales
Weight groups

Commodity Rate Framework

Information Technology in Pricing
Problem areas
Promising trends

Summary

Questions

Appendix 4.1—Procedures for Looking Up a Freight Rate

Introduction

Carriers must provide shippers with the information necessary to purchase transportation services. This information includes rates, packaging and marking requirements, routes, and services. Accomplishing this task is no small feat. There are millions of products with a vast array of different characteristics and billions of city-pair combinations. In addition, the weight of individual articles ranges from less than one pound to more than one million pounds. Further, carriers offer many different services such as refrigeration, exclusive use of the vehicle, and expedited movements. The result is an astronomical number of price-service combinations.

As already indicated in Chapter 3, the transportation industry accomplishes this formidable pricing task with published tariffs, negotiated contracts, and open-market arrangements. Class and commodity pricing systems define published tariff rates, and in varying degrees, comprise elements of contract and exempt ratemaking. Class rates are analogous to list prices and should therefore be the highest rate the carrier offers. A class rate can be found for virtually every commodity transported and for every shipping or receiving point throughout the nation. Class rates meet the common carrier's basic legal duty to publish reasonable and nondiscriminatory rates for all routes in which carriers participate. Historically, the ICC has considered class rates to represent maximum reasonable rates. During the 1980s, however, regulatory reform has nullified this traditional principle.[1]

By contrast, commodity rates represent reduced wholesale prices for selected commodity groups and for regional or local markets. Commodity rates normally apply instead of class rates applicable to the same shipment. About 20 percent of motor common carriage and 85 percent of railway common carriage moved at commodity rates during the 1960s and 1970s. The current situation does not present such a clear picture because pricing practices have changed dramatically. The trucking industry has adopted the widespread practice of offering discounts from class rates, which has diminished the importance of commodity rates. In the railway industry, the explosion of contract rates has similarly affected the importance of commodity rates.

This chapter presents the conceptual framework of tariff pricing methods — specifically, the nature, purpose, and design of class and commodity rate systems. For technical reference, the procedures for looking up a freight rate have been placed in the appendix to this chapter. By studying tariff pricing systems, we do not wish to convey the impression that traffic personnel spend hours searching through tariffs to find published freight rates. Modern traffic organizations, in fact, have little use for rate clerks. With attractive traffic to offer, today's traffic manager negotiates specific rates with carriers. Carrier representatives, especially from the trucking industry, often initiate competitive deals to obtain that traffic. These deals usually involve percentage discounts from published class rates. Some carriers even develop special tariffs to meet the needs of specific shippers. The General Electric Company was one of the first firms to seek a customized tariff.

In the current pricing environment, traffic managers make productive use of their time by concentrating on documenting the type and volume of traffic they have to offer carriers, by determining how that traffic affects carrier costs, and when necessary and practical, by using that knowledge to alter the nature and volume of traffic to make it attractive. Knowledge of the conceptual framework of tariff pricing supports these tasks and makes negotiating and costing processes more effective. In addition, this framework lays the foundation for understanding current issues related to the task of using computers and telecommunications technology to automate transportation pricing. The last section of this chapter presents these issues.

Class Rate Framework

When setting rates, carriers consider three fundamental shipping dimensions that affect both the costs and risks of transportation and the value of service to the shipper: commodity, distance, and weight. In the class rate system, tariffs define rates by scales that represent a systematic progression of these three dimensions.

Commodity Classification

Carriers use the classification process to assign commodities with similar transportation characteristics to relatively few groups or classes. This process greatly simplifies the task of ratemaking because it allows carriers to publish rates for only a few classes rather than for the millions of individual articles. A scale of numbers, called ratings, identifies each class. In practice, the terms *rating* and *class* are used interchangeably. The idea is to use (class) 100 as the benchmark or the initial rating in the product scale. Linehaul rates for this class of articles are established for progressive intervals of distance or weight. Other ratings in the product scale then define percentages of the class 100 rate. A class 70 article, for example, will move over the same route at 70 percent of the rate for a class 100 article. Carriers can use simple percentage computations like this to construct rates for classes other than 100, as well as for complete rate tables.

Unfortunately, the similarity of the terms *rating* and *rate* sometimes leads to confusion. The rating identifies the class to which an article is assigned for the purpose of developing class rates. The rate is the purchase price for linehaul freight transportation service.

Classification tariffs. Classification tariffs identify the classes to which commodities are assigned for the purpose of applying class rates. An alphabetical "Index to Articles" identifies the location of an article's rating and technical transportation description. These tariffs also contain standard rules for packaging, billing, loading, and related matters.

In domestic transportation, only the railway and trucking industries publish major classification tariffs (see **Exhibit 4.1**). Carriers in other modes of transportation generally participate in one or more of the three classification tariffs that follow:

1. Uniform Freight Classification (UFC). The railway industry publishes the UFC. This tariff has thirty-one classes that range from 13 to 400 and about 8000 descriptions. The UFC lists ratings by less-than-carload (LCL) and carload (CL) shipment sizes.

The Uniform Classification Committee (UCC), which will soon change its name to the National Railroad Freight Committee, performs the task of classi-

Exhibit 4.1 Major classification tariffs

Features	Uniform Freight Classification (UFC)	National Motor Freight Classification (NMFC)	Coordinated Freight Classifications (CFC)
Classes			
Number	31	23	9
Range	13–400	35–500	1–5 and 3, 2.5, 2, 1.5 × 1
Principal mode	Rail	Motor	Motor — New England
Classification criteria			
Cost-oriented	Density	Density	Density
	Stowability	Stowability	Liability
	Handling	Handling, including:	Value per pound
	Excessive weight	■ Excessive weight	
	Excessive length	■ Excessive length	
		■ Special care	
	Liability to damage	Liability, including:	
	Value per pound	■ Value per pound	
	(Susceptibility to:)	(Susceptibility to:)	
	Damage	■ Damage	
	Damage other freight	■ Damage other freight	
	Theft	■ Theft	
	Combustion or explosion	■ Combustion or explosion	
	Perishability	■ Perishability	
Demand-oriented*	Trade conditions		
	Ability to bear charges		
	Commodity competition		

*Removed from NMFC classification process in 1983 and to be removed from UFC process.

fying railroad freight and establishing uniform rules that address virtually all aspects of railway transportation. The UCC also works closely with the Standard Transportation Commodity Code Technical Committee that assigns standard transportation commodity codes (STCCs) to commodities. The Interstate Commerce Commission requires the railroad industry to assign STCCs to support broader federal government statistical reporting programs such as the *Census for Transportation.*

So far, however, shippers and nonrailroad carriers have found limited applications for the STCC. This coding system defines relatively broad commodity groups, which lack the precision necessary for the article descriptions that go on the bill of lading. Furthermore, the STCC does not adequately distinguish products by their transportation characteristics, so it has little value in the classification process.

2. *National Motor Freight Classification (NMFC).* The NMFC is the main classification tariff used by the trucking industry. Except for shipments within the New England Rate Bureau territory, which the Coordinated Freight Classification normally governs, it has nationwide application. The NMFC assigns twenty-three ratings that range from 35 to 500 and about ten thousand separately described articles. Classes do not go below 35 because motor carriers generally haul traffic having a relatively high value per pound. Such traffic increases the costs or risks of service, as well as the ability of the shipper to bear higher freight charges. In fact, some class rate tariffs provide rates only for class 50 and above.

The ratings may not represent exact percentages of the class 100 rate. In the Middle Atlantic Conference, for example, only shipments in the 2000–5000 lb weight group have ratings that follow the exact percentage relationship. Minor departures from the percentage formula occur for several reasons. Direct operating expenses comprise a high proportion of the motor carrier's costs and vary considerably by region and carrier.

Although the NMFC provides both less-than-truckload (LTL) and truckload (TL) ratings, the ICC rules now prohibit the two-tiered (LTL-TL) system. To comply with these rules, the NMFC has adopted a single-rating system for interstate shipments. The tariff still shows both LTL and TL ratings, as well as truckload minimum weights, since some intrastate motor carriage uses the dual system. However, its "Rule of General Application" states that LTL ratings apply to all sizes of interstate shipments, and truckload ratings and minimum weight provisions in the NMFC apply only as provided for in other tariffs of participating carriers.

The National Classification Committee (NCC) directs commodity classification. The NCC is an autonomous unit of the National Motor Freight Transportation Association, Inc. (NMFTA), which is the organization of motor common carriers that publishes the NMFC. Unlike the Uniform Classification Committee, the NCC does not help establish (and the NMFTA does not publish) STCCs in the NMFC because these codes are not viewed as sufficiently reflecting the differences in the transportation characteristics of the commodity. Otherwise, the NCC operates in a similar fashion.

3. Coordinated Freight Classification (CFC). Motor carriers operating in six New England states and in parts of New York and New Jersey participate in the CFC. The New England Motor Rate Bureau publishes this tariff, and practitioners usually refer to it as the "New England Classification."

The CFC has only nine classes (1 through 5 and 1.5, 2, 2.5, and 3 times class 1) and about 6000 descriptions. The tariff shows only one rating per article; it does not provide separate ratings for LTL and TL shipments. The density of an article essentially determines the rating. The purpose of this method of classification is to enable carriers to develop rate tables so that, regardless of the product's density, a vehicle load will produce about the same total revenue. The New England Rate Bureau, however, also considers value per pound as a secondary factor in the classification process. Like the NMFC, the CFC contains no STCC codes.

Exception ratings. Since operating conditions, traffic patterns, and intermodal competition vary by region, carriers may need to modify ratings found in the NMFC or the UFC. Some articles, for example, may move in much larger volumes within certain regions. Likewise, carriers may face stiffer competition in different territories. Exception ratings, which remove the application of the class rating, serve as a vehicle for regional adjustments to classification scales. As an aid to the shipper, ICC regulations require that descriptions of articles showing exception ratings use approximately the same wording shown for the corresponding classification tariff descriptions.

Classification criteria. Before 1983, the railway and trucking industries generally used fifteen criteria to assess the transportation characteristics of a commodity. As shown in Exhibit 4.1, three of these criteria (trade conditions, value of service, and competition) generally reflect the ability of shippers to bear freight charges. The other twelve classification elements identify the characteristics of a commodity that affect efficient transportation.

In 1983, after nearly five years of investigating the motor carrier classification system, the ICC substantially revised the traditional way motor carriers classify freight.[2] The new rules eliminated the ratemaking dimension of classification and modified cost-of-service criteria. And, in 1986, the commission ordered the railway industry to overhaul its commodity classification system to make it similar to the system now used by the trucking industry.

Specifically, the ICC made the following four changes related to ratemaking:

1. Demand-oriented elements. The ICC eliminated trade conditions, value of service, and competition with other commodities transported from the list of factors used for classification, stating:

> *The characteristics at issue are ratemaking or economic considerations and are not proper classification considerations. They do not involve the transportability of the commodity itself. These characteristics reflect the degree to which a commodity may bear transportation charges and have no significance as a transportation related quality of a commodity. Moreover, for classification purposes, transportation characteristics should be considered only where the characteristics of an article are largely the same in all territories and for all carriers. Trade conditions, value of service, and competition with other commodities may vary significantly from territory to territory and from carrier to carrier and are economic factors more properly to be considered by ratemakers.*[3]

In other words, the ICC considers the use of demand-oriented factors to be part of ratemaking, a function that should be undertaken individually rather than collectively through classification committees.

2. Differences in LTL and TL ratings. The ICC ruled that LTL and TL ratings may differ only to the extent they reflect changes in transportation characteristics. When such differences represent volume discounts, they indicate a ratemaking function and are outside the proper scope of classification.

3. Minimum weight factors. The minimum weight factor establishes the amount of weight that shippers must pay for to qualify for volume or truckload ratings. The ICC also considers decisions about minimum weights a ratemaking function. The National Classification Committee, however, may publish the amount of a commodity that can be loaded into various types of trailers.

4. Any quantity (AQ) ratings. For relatively few articles, the NMFC shows ''AQ'' instead of a minimum weight factor, which means the LTL rating applies to any amount of freight. Historically, AQ ratings have been assigned to low-density commodities that fill a truck without meeting the minimum truckload weight. In addition, market conditions or custom of the trade may prevent some commodities from moving in the volumes necessary to justify TL ratings. The ICC ruled that such AQ ratings are inappropriate, unless they are derived strictly from cost-of-service criteria.

With respect to cost-oriented criteria, the new rules condense the twelve traditional cost characteristics into four distinct factors: density, stowability, handling, and liability. In the future, the ICC will consider only these four factors in judging the reasonableness of motor carrier classification.

Role of the traffic manager. The traffic manager has several important responsibilities related to commodity classification. Perhaps the most important, however, is understanding the underlying relationships of classification factors. Classification principles extend well beyond the realm of published tariff rates and form an essential part of costing studies and rate negotiations applicable to common, contract, or exempt carriage. Thus, the discussion that follows briefly examines four cost-of-service factors in the ICC-designated order of importance:

1. Density. Essentially all classification begins with an evaluation of density, and it is of critical importance to shippers. Since shippers normally pay for the weight and space capacity of vehicles in dollars per unit weight, low-density (bulky) articles, which take up more space than high-density products to produce the same weight, will not generate as much revenue. In effect, bulky articles take up the space capacity of (or "cube out") the vehicle but fall short of the weight-carrying capacity. Thus, low density products get higher ratings.

The same article, however, may obtain a lower rating when the method of packaging creates a higher density configuration. For example, when items are "knocked down" (k.d.) or "nested," classification ratings are generally lower.

2. Stowability. This term refers to how the cubic dimensions of the freight affect the use of vehicle (freight car, trailer, or container) space.[4] Odd sizes and shapes, as well as excessive weight or length, do not stow well and waste space. Although stowability is similar to density, it is possible to have articles with the same density that "stow" differently. The size of the shipment also may affect stowability. For instance, the carrier might find that certain odd-shaped articles can be nested when shipped together in large lots but do not fit well with small shipments of other commodities.

3. Handling. Articles of excessive weight or length that require special handling (for example, the use of material handling equipment or heavy lift cranes) affect the efficient use of equipment and personnel because such freight often stows poorly. The configuration of the freight (loose, palletized, unitized, or containerized) will also affect handling costs.

4. Liability. This cost element comprises six dimensions that primarily affect the risk of damage (and, thus, the incidence of claim payments) and the level of special care and handling: (1) susceptibility to damage, (2) property damage to other freight, (3) perishability, (4) susceptibility to theft, (5) susceptibility to spontaneous combustion or explosion, and (6) value per pound.

Fragile articles are clearly more susceptible to damage than other commodities and generally require special care or handling. Commodities susceptible to damage, moreover, often adversely affect the load density.[5] For example, carriers may have to "top load" these articles to assure safe transportation. Therefore classification tariffs frequently show lower ratings for fragile articles shipped in sturdy containers. For example, some articles earn lower ratings because they are packaged "in containers, in boxes" rather than "in glasses, in boxes."

Articles such as paints and chemicals may contaminate other freight or even explode. The carrier has to take additional precautions in the care, handling, or routing of the shipment of these kinds of articles. Sometimes more expensive specialized equipment is necessary for safe transportation. Thus they take higher ratings.

Likewise, perishable products often require specialized heating and refrigeration equipment or special handling to expedite movement and will have ratings higher than other commodities.

Items susceptible to theft or pilferage, such as stereo appliances or televisions, impose additional transportation costs and have higher ratings. Articles susceptible to spontaneous combustion or explosion also have higher ratings. At least to some extent, carriers must isolate these articles or assume higher risks for the destruction of other freight. In addition, carriers must exercise special care in handling, and may have to provide specialized equipment.

High-value-per-pound commodities impose higher liability risks on the carrier and, therefore, increase expected claims payments or require more security. As a result, classification committees often tie the ratings of these commodities to the degree of protection afforded by the different types of packaging or product containers (for example, glass versus steel), the value per pound, or the stage of production (for example, rough versus finished).

Another important responsibility of the traffic manager is to negotiate changes in ratings, descriptions, or packaging requirements for the firm's product line whenever conditions warrant the effort. Successful negotiations will largely depend on the development of relevant supporting information about the product in question and on a good working knowledge of classification principles. These negotiations are especially appropriate for new products or new (or different) packaging materials like stretch wrap that may materially reduce damage. In this regard, the traffic department must educate its staff about the classification rules that permit experimental testing of new packaging.

A third important responsibility involving freight classification is to establish procedures to assure that bills of lading and other applicable documents show the correct classification descriptions. Shippers should not substitute the firm's product nomenclature or brand names (for example, "Pearl Beer") for the legally applicable tariff description ("beverages, alcoholic"). Using the correct classification description will minimize incorrect billings, especially when the customer pays the freight charges, and will help eliminate overcharge claims.

The traffic manager also must be prepared to respond to changes that significantly affect the company's transportation operations. The Uniform Classification Committee and the National Classification Committee publish their agendas and scheduled hearings in the *Traffic Bulletin*, and tariff subscribers may obtain similar information free of charge directly from the New England Classification Committee.

Distance Scales

Like the process of classification, the use of distance scales simplifies the pricing task. The transportation industry has created a uniform scale of mileage blocks (101–110 miles, 111–120 miles, and so on), and carriers publish rates by block rather than by mile. Determining the size of each block, however, is not an easy task. Costs can vary significantly even within relatively small blocks, especially for shorter hauls by truck.

Carriers also have to devise a practical method to determine the distance between city pairs. This task could easily become unmanageable if the actual distance between every possible pair of shipping points must be shown. For example, the National Motor Freight Transportation Association estimates there are 156 billion city pairs.[6]

Historically, the transportation industry has relied on the rate basis number (RBN) system to solve this problem. Rate basis numbers generally represent the distance over the shortest route between "base points," the principal shipping or receiving location (in terms of tonnage) within roughly a forty-mile square area. All communities within the defined area use the same base point to determine rate basis numbers.

Each class rate tariff has a section that lists RBNs for all shipping or receiving base points within the scope of the tariff. Other supporting tariffs, such as the *National Rate Basis Tariff* (published for the railroad industry) and "group guides" (used by motor carriers), identify the base points used by each community. For example, a shipper planning to make a shipment from Knobknoster, Missouri, to El Cerrito, California, would find that Knobknoster uses the Sedalia base point, and El Cerrito uses San Francisco. Those familiar with the task of looking up the distance between two cities in any commercial travel guide are already acquainted with the method for finding the rate basis number. One simply locates the base points (cities) as column and row headings and finds the column-row intersection to identify the RBN.

Class rates generally exhibit a tapering relationship with the distance. As shown in **Exhibit 4.2**, this relationship shows rachetlike step increases that become progressively smaller for longer distances. The rates originate at a fixed level above the origin because carriers must recoup dispatch, pickup, billing, and other terminal-related costs that do not vary with the length of the haul. As distance increases rates go up proportionately less because the fixed terminal-related costs are spread over more miles, and the carriers generally experience greater operating efficiency (for example, in fuel consumption and labor) for longer hauls.

Exhibit 4.2 **Tapering relationship between class rates and distance scale**

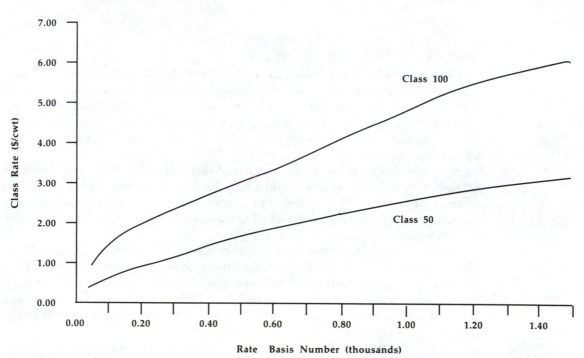

Although the RBN system has worked reasonably well for decades, carriers and shippers are now looking at other alternatives. During the early 1980s, for example, the transportation community witnessed the introduction of zip-code systems that greatly simplify the rate search process. In some of these systems, three-digit zip codes establish zones that replace base points (see **Exhibit 4.3**). These zones define a larger area than base points; thus, fewer zones define a given geographical area. The rate bureau that initiated zip-code rates estimates the new system will produce about a 75 percent reduction in the number of tariff pages required for the same pricing task.[7]

In addition, zip-code systems facilitate the use of customized computer printouts ("rate ponies") tailored to specific shipper traffic patterns. Rate ponies list class rates from a zip-code zone (perhaps encompassing a plant location) to all destination zones within the scope of the tariff. The shipper simply locates the three-digit zip code of a customer to find the rate.

The main disadvantage of this pricing innovation is inaccuracy. Population

Exhibit 4.3 Rate basis points versus zip code zones

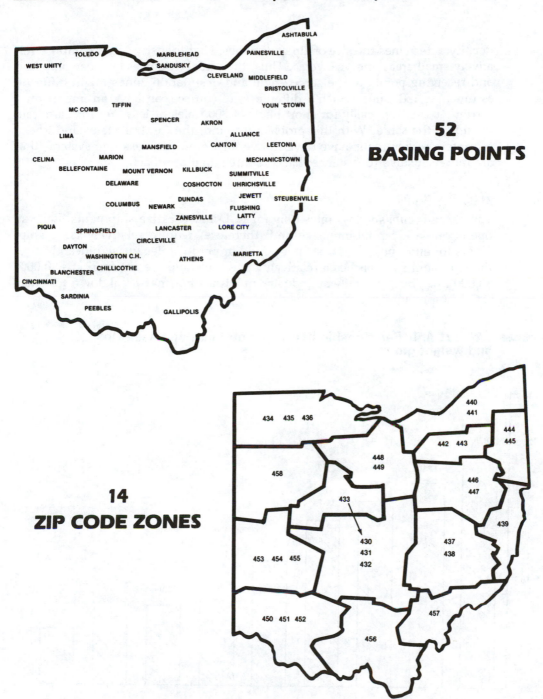

52 BASING POINTS

14 ZIP CODE ZONES

Source: Eastern Central Motor Carrier Association, ''Statement of evidence in justification of zip code tariff publications,'' Appendix 1 (submitted to the Interstate Commerce Commission in Ex Parte MC-82), effective January 3, 1983.

density determines the size of zip-code zones. High-density zones comprise relatively small areas and vice versa. Thus, the actual distances between shipping and receiving points in the same origin and destination zones might differ by as much as 150 miles.[8] Such a difference, of course, can have an impact on a carrier's costs, especially for short hauls of 300 miles or less. Yet the class rate would be the same. With this problem in mind, the Central States and Rocky Mountain rate bureaus have devised five-digit zip-code class rate systems that are more sensitive to distance than the three-digit approach.

Weight Groups

Rail carriers publish class rates only for LCL- and CL-size shipments. Railway operations are capital-intensive and, therefore, require relatively large shipments for efficiency. Unit costs per pound or per mile decline significantly after the shipment size reaches a relatively large threshold weight (perhaps 20,000–30,000 lb). Thus, the railway industry publishes no finer breakdown for small (LCL) shipments.

Exhibit 4.4 **Relationship between motor carrier class rates and weight groups**

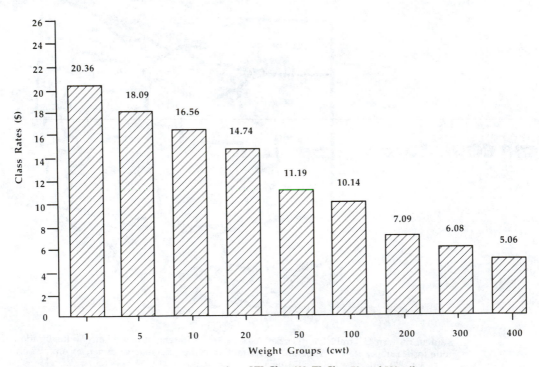

Assumptions: LTL Class 100, TL Class 70, and 500 miles

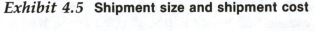

Exhibit 4.5 **Shipment size and shipment cost**

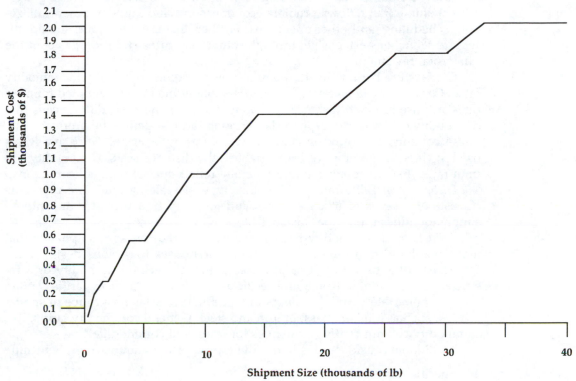

Shipment Size (thousands of lb)

Assumptions: LTL Class 100, TL Class 70, and 500 miles

By contrast, motor carrier costs are very sensitive to the weight of LTL shipments. Thus motor carriers establish class rates for five weight groups for LTL shipments weighing less than 10,000 lb and three groups for heavier TL shipments. As shown in **Exhibit 4.4**, class rates for these weight groups decline as the weight of a shipment increases. Larger shipments permit carriers to spread fixed terminal and related expenses over more units (pounds). This action creates a tapering relationship between shipment cost and shipment size like the one illustrated in **Exhibit 4.5**.

Commodity Rate Framework

Unlike class rates, which are comparable to retail list prices, commodity rates may be thought of as selective wholesale rate reductions that carriers have traditionally used to meet the needs of regional or local markets. Competition with other carriers or modes of transportation is the major reason for commodity rates. The transportation industry generally uses these rates to encourage traffic to move in volume that would not be shipped at the higher prevailing class rates. Volume traffic flows promote economies in handling, equipment utilization, scheduling, and other operations. Further, increased volumes should offset rate reductions established through commodity rates and provide about the same total revenue.

Carriers use both scale and point-to-point systems to construct commodity rates. *Commodity column* rates, for example, follow the pattern adopted for class rates and use product, distance, and weight scales. The basic difference is that commodity columns, rather than classification ratings, define the product scale. These columns are found in commodity column rate tariffs. The same RBNs used in class tariffs, however, still establish the distance scale. Commodity column rates, like exception ratings, generally reflect special regional conditions.

Distance commodity rates use another type of scale system. The rate table consists of a series of mileage blocks and rates. To find the rate, the shipper simply identifies the proper distance block.

Point-to-point commodity rates generally are used for large shipments that move regularly from mines, factories, or warehouses to various destinations. The rates apply only to the named points and a specific commodity or a restricted list of commodities. Both specific and general commodity tariffs contain point-to-point rates. Specific commodity tariffs have rates for groups of similar articles, such as lumber, glass, or iron and steel. General commodity tariffs, like commodity column tariffs, show rates for unrelated commodities.

Traditional considerations in deciding to publish a commodity rate include

- Specific route costs
- Proximity of carrier facilities to the route
- Transport characteristics of the commodity (density, stowability, handling, and liability)
- Balanced (bidirectional) traffic flows
- Competition from intramodal, intermodal, and private carriers

Today, however, many shippers simply negotiate contract rates or percentage discounts from class rates rather than bother with exception and commodity rates. Although these methods are attractive to shippers looking for an easy way to compute freight charges, they do not necessarily translate into the best deal. Commodity rates in bureau tariffs may offer lower rates that justify a more tedious rate search.

Information Technology in Pricing

The compilation, publication, maintenance, administration, and use of published tariffs imposes costs on the transportation industry and the shipping community that run into the billions of dollars.[9] Not surprisingly, the transportation community has looked to computers to rescue publishing agents and tariff users from the time and trouble of manual rate retrieval and maintenance.

The first computer-compiled tariff appeared in the late 1960s.[10] Shortly thereafter, shippers began to develop computer-driven rate retrieval systems, while third-party ventures initiated similar services. The impetus for these services came from efforts to reduce the overcharges and duplicate billings that were products of complex published tariff rate systems. Now shippers and carriers also see computers and telecommunications technology as vehicles for reducing the time and effort involved in rate retrieval, billing, accounting, and related activities.

At least a dozen firms currently offer automated, on-line rate retrieval services. Several offer expanded services that include features such as freight payment and auditing, report generation, and even simulation studies. Many shippers, especially the larger manufacturing concerns, either subscribe to on-line services or build their own in-house computerized rate systems. Commercial software is also available for in-house installation.[11]

Problem Areas

Although shippers and carriers have recognized the potential of computers since the 1960s, computerized pricing has been more elusive and difficult to attain than initially forecasted. Two reasons explain this difficulty. First, the complex system of tariffs presents a formidable challenge to computer systems analysts. Different tariffs, for example, may show class, exception, and commodity rates for the same shipment. Tariff rules normally will establish the priority of rates. Classification, rules, and rate tariffs, however, frequently contain different versions of a rule that addresses the same topic. Thus tariffs must also establish priorities for rules. Standard practice is to have the version of a rule found in a rate tariff apply instead of the version published in a rules tariff. Likewise, the provisions of rules tariffs apply instead of comparable provisions in classification tariffs.

Second, the diversity of published tariff rates makes both manual and automated rate retrieval systems expensive to install, update, and maintain. Interesting to note is that rate deregulation has significantly expanded the diversity and the number of rates. As one manager put it: "Rates now come in more flavors than an ice cream store offers, and carriers have a new 'rate of the week' comparable to a new 'flavor of the month.' "[12] For some carriers, the difficulty of updating computer files to reflect innumerable rate changes was unexpected.[13]

Promising Trends

Tariff computerization combined with electronic data interchange, where rates and related information are transmitted directly to users through telecommunication links, promises to make access to rates far easier, cheaper, and faster. In addition, the current competitive market environment appears to be forcing the transportation industry toward tariff simplification.

Electronic data interchange. Electronic data interchange (EDI) technology has tremendous potential for making transportation pricing methods more efficient and convenient. EDI allows firms with different computer systems to exchange business documents and information electronically. It reduces paperwork, increases accuracy, and transmits information rapidly.

The Transportation Data Coordinating Committee (TDCC), a nonprofit organization of shippers, carriers, and interested parties, has been working on the problem of electronic interchange for nearly two decades. The TDCC has focused its attention on all methods of doing business electronically, including billing, invoicing, warehousing, and purchasing, as well as transportation and distribution.[14] The principal aim of this organization is to develop and coordinate uniform standards for (1) electronic data interchange, (2) translation software that supports communication between different types of computer hardware, and (3) transmission software that will send, sort, and hold messages.

This effort is essential for the computerization and electronic data transfer of freight tariff and related billing information. Ideally, uniform codes and communication protocols should apply to all modes of transportation. From the shipper's perspective, such standardization would allow the traffic organization to eliminate the extra software now necessary to support separate modal rate systems.

New enterprises have begun to enter the transportation pricing arena to meet the needs of the transportation community. In the railway industry, DSI RAIL, a unit of Distribution Sciences, Inc., has established a National Railroad Ratemaking and Information System (see **Exhibit 4.6**). Participating railroads, for the first time, can electronically quote and maintain rates at a single data source. In the motor carrier industry, Capacity Exchange, Inc. (CAPEX) has initiated a *trading center* concept for truckload service. Shippers order service electronically. Carriers with equipment available bid for available loads. The bidding establishes a market rate, which shippers may review before they accept the carrier's bid. Thus rates are not stored but are created in real time.[15] Further, about thirty-five carriers have established EDI links with shippers, and another twenty are heading in that direction.[16] In addition, third-party vendors are providing translation software to facilitate EDI between previously incompatible computer hardware.

The ICC first addressed the issue of electronic data transfer in 1983, when it instituted a rule-making hearing on the electronic filing of tariffs.[17] The hearing raised the following questions: (1) Who should be responsible for setting technical standards and formats? (2) Should the commission maintain electronically filed tariff information? (3) If tariff information is transmitted directly to

In 1984, DSI Rail, a unit of Distribution Sciences, Inc., initiated a computer system that maintains and transmits rate information electronically. Participating railroads, for example, can now transmit a proposed rate together with a route. The system notifies other railroads with direct connections to the proposed route and requests a "concurrence." If the connecting railroads approve this request, all railroads receive an electronic copy of the new rate.

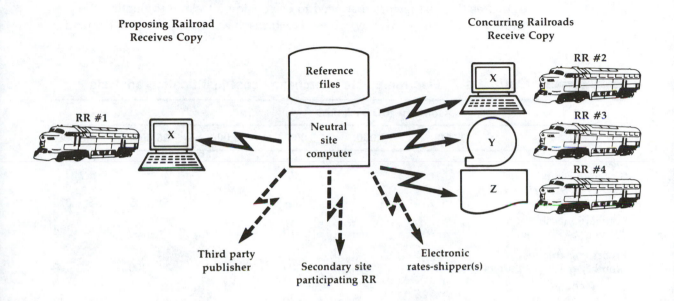

DSI Rail also offers shippers the opportunity to tap into this system, which is actually made up of the following seven subsystems:

1. Security: Maintains passwords that restrict access to authorized parties, such as directly connecting railway companies when evaluating proposed joint rates
2. Source data reference files: Contain information about geography (city, county, state, zip code), routes (carrier, junction, route code), mileage, base points, commodities, and equipment codes
3. Translation software: Provides communications support for computer-to-computer or computer-to-terminal exchanges
4. Menu options: Offer users menu options about basic information, commodities, geographic information, shipment conditions, and route-rate details
5. Rate retrieval: Permits on-line rate retrieval
6. Status: Monitors all rate items, including the carriers responsible for rate changes and rate status (effective and expiration dates)
7. Output: Establishes transmission protocols in standard Transportation Data Coordinating Committee formats

Source: Adapted from "Computerized Through-Rates," *Progressive Railroading* 26 no. 6 (June 1984). Reprinted by permission.

shippers, who should bear the transmission costs? (4) How might users without computer and telecommunications equipment gain access to the information? The ICC, however, is holding this hearing in abeyance until the issue of complete motor carrier deregulation is settled.

Although EDI offers tremendous benefits, companies must have enough shipping volume and computer expertise to make it work efficiently. **Exhibit 4.7** addresses the cost factors that need to be considered when evaluating the justification for EDI. Currently, about 125 firms have established EDI links for

■■■■ *Exhibit 4.7* **Electronic data interchange cost justification analysis**

	*Monthly Bill Volume**					
	100	**1,000**	**5,000**	**10,000**	**25,000**	**50,000**
Labor						
System design 500 hrs @ $45/hr over 5 yrs	$375	$ 375	$ 375	$ 375	$ 735	$ 375
Software modification 80 hrs @ $45/hr over 5 yrs	60	60	60	60	60	60
Processing charges Monthly participation fee	100	100	100	100	100	100
Log-on charge 8/month @ $2.00	16	16	16	16	16	16
Transaction charges # bills @ $0.14	14	140	700	1,400	3,500	7,000
Audit charges # bills @ $0.75	75	750	3,750	7,500	18,750	37,500
Phone expense # bills @ $0.10	10	100	500	1,000	2,500	5,000
In-house processing # bills @ $0.25	25	250	1,250	2,500	6,250	12,500
Software Charge Purchased software $15,000 over 5 yrs	250	250	250	250	250	250
Grand total	$925	$2,041	$7,001	$13,201	$31,801	$62,801
Cost per bill	$9.25	$2.04	$1.40	$1.32	$1.27	$1.26

*The figures used may vary by the company and are dependent on the software vendor and equipment being used.

Source: J. B. Vick, Vice-President Corporate Services and Traffic, Carolina Freight Carriers Corp., "Why EDI," *Motor Carrier Computer News* (Washington, D.C.: Regular Common Carrier Conference, December 1985).

transportation-related matters such as billing and rates. Nearly all these firms are in the *Fortune* 500 and have the volume to justify EDI setup and maintenance costs. Personal microcomputers (PCs) however, are beginning to make EDI feasible for many low-volume users. These desktop computers can accept EDI formats at about one tenth of the cost of transmissions through mainframes.[18]

Tariff simplification. The transportation community has sought tariff simplification for decades. In the 1980s, a freer pricing environment and new computer technology have enabled shippers and carriers to make significant progress. As previously discussed, many shippers insist on using simple-to-understand discounts from class rates. In addition, the use of contract rates has increased dramatically. These two approaches permit shippers to avoid the expensive task of building or purchasing computer systems to handle the complicated patchwork of class, exception, and commodity rates.

Electronic tariffs. The introduction of zip-code class tariffs on diskettes for microcomputers represents another significant source of progress in tariff simplification. This software provides a variety of rate maintenance features, such as the capability to make adjustments to rate scales to reflect negotiated discounts. **Exhibit 4.8** illustrates how easy it is to use one of these ''electronic tariffs.''

Nevertheless, zip-code class rate systems do not offer shippers uniformity. In other words, the same class, distance, and weight will produce different class rates in different zip-code tariffs. Thus traffic managers have difficulty comparing net rates. A 20 percent discount in one tariff might produce lower charges than a 30 percent discount in another.

Addressing this issue, the Rocky Mountain Motor Tariff Bureau has recently developed a nationwide uniform scale of class rates that also approximates the scales of other bureaus. Its tariff uses a five-digit zip-code rate base system and has been encoded onto a PC diskette. Consolidated Freightways (see Exhibit 4.8) and other motor freight carriers with nationwide authority have adopted this system. In addition, several major motor carriers, notably Yellow Freight and Roadway Express, have developed competing pricing systems that use a straight-line distance methodology to establish a uniform national scale of class rates.

Formula rates. The concept of formula rates is not new; indeed, several transportation leaders have long advocated the use of mathematical formulas in place of voluminous rate tables.[19] John T. McGraham, a railroad rate clerk, created one of the earliest applications. In 1871, the eastern railroads adopted ''McGraham's Formula'' (see **Exhibit 4.9**). This formula computes the rate to any point when the user simply inserts the distance of the shipment into the equation.

Users can build tapering relationships into such formulas by including a term that deducts a mileage discount multiplied by the distance squared (see

Exhibit 4.8 **Zip code class tariff on microcomputer**

```
            TM
OmniRate II - CFWY ICC TARIFF 555        222222222222  000000000000
Version 2.00                             2222222222222  000000000000
                                         2222          0000
                                         2222  000000000000000000
Tariff 555 Copyright 1985, 1986          2222  000000000000000000
Rocky Mountain Motor Tariff Bureau, Inc.(RMB)  2222          0000
                                         222222222222   0000
                                         222222222222   0000
                              APPLICABLE ORIGIN ZIP(S): ALL ZIPS


                     CFWY Tariff 555 - Issued by
                    E. V. Taylor, Vice President - Pricing
                 Consolidated Freightways Corp. of Delaware
                   P. O. Box 3062  Portland, OR 97208
                            (503) 226-4692


Issued: 03-19-86                          Effective: 04-01-86
                     ENTER INQUIRY DATE(MM/DD/YY): 07/21/86


1 - ENTER 5-DIGIT ORIGIN ZIP CODE.....: 16802   BASE: STATE COLLEGE    , PA

2 - ENTER 5-DIGIT DESTINATION ZIP CODE: 60606   BASE: CHICAGO          , IL

3 - SINGLE SHIPMENT CHARGE (Y/N)......: N

4 - ENTER % OF DISCOUNT(RATE/MIN CHRG): 20.00 / 50.00



LINE NO.          ITEM       CLASS        WEIGHT
6 -                1         100          2500
7 -                2         70           5000









                IS DATA ENTERED CORRECTLY (Y/N): Y

ORIGIN------------ 16802 BASE 16801-02 (STATE COLLEGE    , PA)  DATE
DESTINATION------- 60606 BASE 60601-99 (CHICAGO          , IL)  7-21-86
TARIFF AUTHORITY-- CFWY  555, RATE BASIS NUMBER 71410
=================================================================
   ITEM    CLASS         WEIGHT          RATE          CHARGE
    1      100.0          2500           1515      $   378.75
    2       70.0          5000           1076      $   538.00
                         -------                   ----------
              ACTUAL WT--  7500   CHARGE AT FULL RATE---- $   916.75
                                  DISCOUNT AT 20.00 PCT-- $   183.35
                                                          ----------
                                  DISCOUNTED CHARGE------ $   733.40
```

■■■■ *Exhibit 4.9* **Rate formulas**

McGraham's Formula
$$R = T + L(M)$$
Distance discount modification
$$R = T + T(M) - D(M^2)$$

where R = rate (cents/cwt)
  T = terminal costs (cents/cwt)
  L = linehaul costs per mile (cents/cwt)
  M = distance (miles)
  D = discount for distance (% per mileage interval)

■■■■ *Exhibit 4.10* **General Electric Company's guidelines for participating carrier LTL tariffs**

General Electric Traffic Councils cannot undertake evaluation of tariffs of every conceivable structure that might be invented as bureau LTL tariffs are supplanted by carriers' individual tariffs. Though not an immutable position, the following specification describes the tariff structures GE recommends to its approved LTL carriers.

1. Both origin and destination are expressed in terms of 3-debit ZIP Codes only, so all other factors being equal, the distance factor for all shipments between a pair of 3-digit ZIPs is the same.
2. Pairing of any two 3-ZIPs served produces either a rate basis or a standard distance in miles. Rate bases are discouraged and distances encouraged, because the latter facilitates production of prices from formulas rather than from tables that are inefficient for computer processing.
3. Published mileages between ZIP Code Sectional centers are used for distance-dependent calculations.
4. There is a table or formula for producing a base rate for each rate basis or distance.
5. There may be a table of percentage or dollar adjustments for origin 3-ZIPs, and/or one for destination 3-ZIPs, to permit adjustment of mileage-dependent base rates for directional, regional, or locational bias.
6. There are formulas for modifying base rates for
   - Magnitude of shipment weight
   - NMFC rating
   - Piece count (optional)
7. The order of application of the aforementioned modification formulas and tables is made clear.
8. There is a minimum charge per shipment; it is independent of all factors.
9. There is a maximum (i.e., truckload) charge per shipment determined from a formula and a widely accepted, published mileage data base that is available in computer file form; the charge is not affected by the kind or quantity of goods shipped but may be subject to a shipment value formula.
10. Corporate discounts and adjustments for specific customer facilities are documented in a separate publication or contract.

Source: Courtesy of General Electric Co.

■■■■■ *Exhibit 4.11* **Freight charge calculation in General Electric Company's tariff model**

Origin: Schenectady, New York Weight: 2500 lb		Destination: Louisville, Kentucky Rating: Class 85
ZIP 123 to 402	805 miles	
Base rate formula	$ 8.84 + $0.02/mile	
Base rate		$24.95
Direction multiplier	1.10	
Interface multiplier	1.00	
Weight group multiplier	0.63	
Class multiplier	0.85	
Modified base rate/cwt		14.69
Gross LTL charge		367.27
Minimum charge test	$41.00	
Discount multiplier	0.65	
LTL net charge		238.73
Maximum (TL) charge	$1610.00	
Pay		$238.73

Source: Courtesy of General Electric Co.

Exhibit 4.9). Squaring the distance produces increasingly larger discounts for longer hauls. Another alternative is to create separate formulas for different mileage groups. In 1979, the Southern Railway System followed this approach when it established a family of formulas to compute rates for aluminum articles shipped in lots of 50,000 lb or more.[20]

In 1983, General Electric developed a novel approach to the application of formula rates. Its transportation department created an LTL tariff model specifically tailored to the company's needs. As illustrated in **Exhibits 4.10** and **4.11**, this tariff uses a formula rate format that facilitates computerization. Further, it permits General Electric's traffic staff to (1) compare rates of the participating carriers and (2) isolate the discounts quoted by carriers. General Electric required motor carriers hauling its freight to adopt this tariff model.

■■■■■ ## Summary

The task of publishing rates for the astronomical number of price-service combinations derived from different commodities, city pairs, and weight groups presents a formidable challenge to the transportation industry. Carriers accomplish this task with published tariffs, negotiated contracts, and open-market arrangements.

Today's traffic managers spend little, if any, time searching through rate tariffs. Instead, they concentrate on documenting the nature of company traffic and on how to make it attractive to carriers. An understanding of the conceptual framework of tariff pricing facilitates these efforts and enhances the effectiveness of rate negotiations and cost analyses. Further, this framework helps us understand the problems confronting the computerization of freight rates and the promise of computer and telecommunications technology.

Class rate tariffs define rates in terms of uniform scales that reflect commodity, distance, and weight dimensions of shipments. Classification tariffs assign articles with similar transportation characteristics to relatively few classes of freight. The railway industry publishes the *Uniform Freight Classification* (UFC), whereas the trucking industry has two major classifications: the *National Motor Freight Classification* (NMFC) and the *Coordinated Freight Classification* (CFC). These tariffs contain standard rules for freight transportation, as well as article descriptions and ratings.

Carriers publish exceptions to classification ratings in separate class or exception rate tariffs. Exception ratings adjust class rate structures generally to reflect unusual regional operating conditions.

As a result of regulatory reform in the early 1980s, classification committees no longer use the fifteen traditional criteria to assign ratings to commodities. The ICC has eliminated ratemaking elements from the classification process and has condensed the twelve cost-of-service criteria into four distinct factors: density, stowability, handling, and liability.

Traffic managers have four important responsibilities in the classification process. First, they must understand the underlying principles of classification factors and use this knowledge in costing studies and rate negotiations. Second, as business conditions dictate, the traffic manager must negotiate changes in ratings, descriptions, and packaging requirements. Third, the traffic manager must assure that shipping documents show correct classification descriptions. Finally, to the extent that changes in classification may affect shipping operations, the traffic manager should be prepared to present the company's position on issues before classification committees.

Distance and weight scales also simplify the pricing task. Class tariffs organize rates by mileage block and by weight groups. Historically, the transportation industry has used the rate basis number system to determine the distance between shipping and receiving points. However, new zip-code systems that greatly simplify class rate tariffs have become popular.

Carriers publish commodity rates to meet the needs of shippers in selected markets. These wholesale rates encourage traffic to move in volumes that would not be shipped at the higher prevailing class rates (retail prices). Carriers use both point-to-point and scale systems to publish commodity rates, which generally take precedence over class rates. During the 1980s, the practice of discounting class rates and the use of negotiated contract rates have diminished the role of commodity rates.

Tariff publication is an expensive process for the transportation community. Since the early 1960s, shippers and carriers have sought ways to harness information processing technology to automate rate searches and related tasks. Yet the diversity of rates and the complexity of tariff systems have made automation difficult and expensive. Electronic data interchange (EDI), new pricing practices, electronic tariffs, and formula rates offer promising new directions.

Selected Topical Questions from Past AST&L Examinations

1. As traffic manager for a high-technology conglomerate, you must negotiate new classifications in the National Motor Freight Classification for your newly introduced products. The ICC recently identified four factors that it considers important to proper classification. Identify those four factors, and comment on them. (Fall 1982)
2. You are the logistics manager of a firm that ships a large volume of goods in cartons on pallets. Smedley, staff assistant to the vice-president of marketing, tells you he has a great accomplishment to report: He has been able to design a product package so that the cartons holding this product are perfect cubes, specifically one cubic foot in size. What would you tell Smedley? (Spring 1985)
3. The ICC has substantially revised the traditional way that motor carriers classify freight. The new rules eliminate the ratemaking dimension of classification and modify cost-of-service criteria. (Fall 1985)
   (a) What steps has the ICC taken to eliminate the ratemaking in the classification process?
   (b) What is the ICC's rationale for this new policy?

Notes

1. See Interstate Commerce Commission, *No. 36135, Rules Governing the Publication of Exception Ratings Higher than Classification Ratings*, April 6, 1984.
2. Interstate Commerce Commission, Ex Parte No. MC-98 (Sub-No.-1), *Investigation into Motor Carrier Classification*, decided February 25, 1983 and amended August 8, 1983.
3. Interstate Commerce Commission, *Investigation into Motor Carrier Classification*, p. 6.
4. For further reading by the author who coined the term "stowability," see Frank M. Cushman, *Transportation for Management*, (New York: Prentice-Hall, 1953), pp. 192–193.
5. See Cushman, *Transportation for Management*, pp. 194–197.
6. Robert C. Dart and Robert J. Kursar, "Truck industry plans major effort to save its classification system," *Traffic World* (June 15, 1981), p. 16.
7. Eastern Central Motor Carrier Association, "Statement of evidence in justification of zip code tariff publication," (Akron, Ohio: November 19, 1982, submitted to the Interstate Commerce

Commission in Ex Parte MC-82, effective January 3, 1983), p. 21; see also Interstate Commerce Commission, *Docket No. SR-45138 Conversion to Zip Code System*, issued August 23, 1982.

8. Eastern Central Motor Carrier Association, "Statement of evidence," pp. 32–33.

9. See Herbert O. Whitten, "Incorporating the transport pricing problem," paper presented before the Transportation Research Forum, Washington, D.C. chapter (Herbert O. Whitten and Associates, April 6, 1982), p. 12.

10. Charles A. Taff, *Management of Physical Distribution and Transportation*, 6th ed. (Homewood, Illinois: Richard D. Irwin, 1978), p. 431.

11. For a comprehensive listing of software vendors, see Bruce C. Arntzen, *Distribution Software from Top to Bottom* (Cambridge, Massachusetts: Arthur D. Little, Inc., 1985) presented to the Council of Logistics Management, annual meeting, 1985.

12. Walter A. Weart, manager, Marketing Services Support, Comtrac Rating Systems, as reported in *Traffic World* (June 10, 1985), p. 37.

13. L. L. Waters, "Deregulation — For better or for worse," *Business Horizons* 24 no. 1 (January–February 1981), p. 90.

14. Today the TDCC has expanded beyond its transportation origins and is referred to as the "Electronic Data Interchange Association." See Mark B. Solomon, "TDCC moves ahead to push EDI for all business transactions," *Traffic World* (December 30, 1985), p. 17.

15. "CAPEX: Transportation's answer to the stock exchange?" *The Private Carrier* (June, 1986), pp. 29–31.

16. For further reading, see Clifford Buys, *Motor Carrier/Shipper Electronic Data Exchange* (Washington, D.C.: American Trucking Association, Management Systems Department, 1985); see also Solomon, "TDCC moves ahead," p. 19.

17. Ex Parte No. 444, *Electronic Filing of Tariff, Advanced Notice of Proposed Rulemaking*, 48 Fed. Reg. 9762, March 8, 1983.

18. Solomon, "TDCC moves ahead," p. 19.

19. See Herb Whitten and Greg Whitten, "The E³ transport pricing system," *Traffic World* (January 15, 1979), p. 30; see also "The E³ transport pricing system: Corrections and amplifications," *Traffic World* (April 9, 1979), pp. 104–105; Herbert O. Whitten, *The Railroad and Motor Carrier Freight Rate Complex and Its Development* (Washington, D.C.: U.S. Department of Transportation, Office of Policy Review, 1972); Edward J. Marien, "Formula rates — Their time has come again," *Distribution* 81 no. 1 (January 1982), pp. 54–58.

20. "Southern devises rail-mileage rate formulas for simplifying tariffs," *Traffic World* (December 3, 1979), pp. 21–22.

Appendix 4.1 Procedures for Looking Up a Freight Rate

Finding the correct rate (class or commodity) actually involves two tasks. The first task is to identify and obtain the applicable rate tariff. Several tools will help. Rate bureaus, as well as individual carriers, publish tariff indexes arranged by type of tariff and by territorial application. In addition, commercial publications such as the *Official Guide to Railroads*, the *Motor Carrier Directory*, the *National Highway and Airway Carriers Routes* guide, and the *Air Cargo Guide* are especially useful.

The second task, of course, is to find the proper rate to use in the tariff. The procedures that follow assume the proper tariff has been identified and describe the steps for finding a class or a commodity rate.

Class Rate Procedures

In the class rate tariff, the procedure is to locate entries for the commodity, distance, and weight dimensions of the shipment in the rate tables to find the rate. For example, the procedure to find the class rate for a shipment of 700 lb of blank video disks transported by motor carrier from Willow Springs, Illinois to Minneapolis, Minnesota, is described next in relation to the tariff information shown in **Exhibit A4.1**. When looking at this illustration, note that scanning it from top to bottom is equivalent to thumbing through a tariff from front to back.

1. Find the article's rating and technical description in the appropriate classification tariff. In the NMFC (tariff NMF 100), the alphabetical "Index to Articles" shows Video Disks as Item 57320. When Item 57320 is found in the articles section, it shows the rating as 100.

The Middlewest Motor Freight Bureau also publishes exception ratings for various articles in several class rate tariffs. The tariff user has to consult the "Index to Articles" in the rate tariff to see if the article shipped (video disks) has an exception rating. In this illustration, class rate tariff MWB 501 provides an exception rating (92.5) for video disks that removes the application of the classification rating (100).

2. Find the rate basis number. The Rate Basis Section of a class rate tariff contains the rate basis numbers (RBNs). Since this section shows only major shipping and receiving base points, other group guide tariffs may have to be consulted. Such group guides identify the base points to use for shipping and

■ *Exhibit A4.1* **Class rate system**

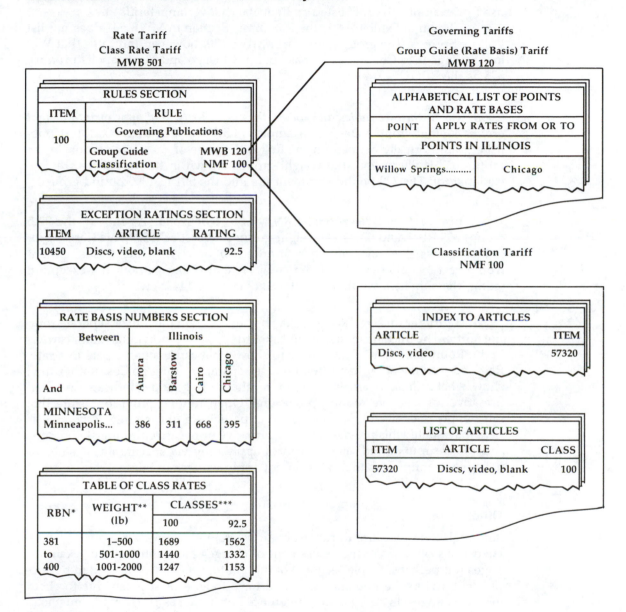

*Distance Scale
**Weight Scale
***Product Scale

receiving points. Rules in the class rate tariff will indicate what group guides to use (for example, the rule listed as Item 100 "Governing Tariffs").

As shown in Exhibit A4.1, the Rate Basis Section in MWB 501 does not list Willow Springs. The group guide tariff MWB 120, however, indicates that Willow Springs uses Chicago as its base point. Chicago and Minneapolis produce 395 as the RBN.

3. Go to the rate tables and locate the class, RBN, and appropriate weight group. Weight group codes (for example, "L5C" for less than 500 lb), if published, are normally defined on the first (title) page of each rate section in the tariff. In this illustration, the weight groups shown in the rate table are self-explanatory. The 700 pound shipment fits into the 501 to 1000 pound group.

4. Cross reference the rating, RBN, and weight group. As shown, the 92.5 rating, 395 RBN, and 501–1000 weight group intersect at 1332 cents (tariffs typically express rates as cents per hundred lb or cwt). Thus this shipment would cost $93.24 (7 cwts × 1332¢/cwt). Note also that without the exception rating, the shipment would cost $108.80 (7 cwts × 1440¢/cwt).

5. Verify the routing. Routing may be published in the rate or separate routing tariffs. The trucking industry publishes the codes for individual motor carriers in the Routing Section of rate tariffs to show what carriers participate in a rate. Railroads, on the other hand, tend to use separate routing guides. Rather than show what railroads participate in a rate, the routing guides indicate open, affirmative, or negative routing provisions. Open routing provisions state that rates apply over all routes of carriers that participate in the tariff. Railway carriers may qualify this provision with affirmative or negative routing statements. Affirmative routing indicates that rates apply only via specific lines, whereas negative routing notes that rates do not apply via certain lines.

Other Details

Although the procedure to construct a class rate is relatively simple, other related tasks often make the assessment of charges more difficult. For example, pages (or, perhaps, whole sections) of different tariffs may be amended by supplements. Thus the rate analyst would have to consult supplements to check for any changes. In addition, any reference notes such as (LU) (shipper must load and receiver must unload) shown in applicable items must be checked in the "Explanation of Abbreviations and Reference Marks" section found at the end of each tariff. Finally, the analyst might have to consult various rules that govern packaging, billing, collecting, and other similar matters.

■■■■■ *Exhibit A4.2* **Electronic zip code tariff procedure to rate a shipment**

```
MWB DISKRATER-II FOR TARIFF MWB 550
EFFECTIVE JANUARY 21, 1985
Copyright (C) 1985 by Middlewest Motor Freight Bureau
All rights reserved

                    CHOICES ARE AS FOLLOWS:

                    1. RATE SHIPMENTS
                    2. DISPLAY RATE SCALES

                    ENTER CHOICE ? _

ENTER A 3 DIGIT ZIP

ORIGIN ZIP? 530
DESTINATION ZIP? 750
NUMBER OF ARTICLES? 1
WEIGHT? 5000
CLASS? 100
ENTER DISCOUNT % ,ie 12 (IF NONE STRIKE ENTER KEY)? 30


                         TARIFF       MWB 550
                         RATE BASIS     30400
                         ORIGIN ZIP       530
                         DESTINATION ZIP 750

          WEIGHT      CLASS       RATE        CHARGE
          --------    --------    --------    --------

           5000 LBS   100         2187      $  1093.50
          --------                            ----------
           5000                             $  1093.50
                      LESS DISCOUNT OF 30%      328.05
                                              ----------
                                            $   765.45
                                              ----------

CONTINUE (Y/N) ?
```

Source: © Middlewest Motor Freight Bureau, Program author Jeff Michalson. Reprinted by permission.

Zip-Code Systems

A zip-code system follows the same basic procedure. The user finds the rating in the same manner. The RBN is found by using either three- or five-digit zip-code numbers of shipping and receiving locations. Separate "Points and Zip Code" tariffs will identify the zip codes to use for these points. Weight groups remain unchanged.

Zip codes make class rate tariffs easy to use. In addition, these tariffs come on diskettes for microcomputers. With the diskette version, the user has only to enter the origin and destination zip codes, the number of articles, the weight and class of each article, and the discount to retrieve the rate (see **Exhibit A4.2**). In addition, these electronic tariffs permit users to retrieve applicable portions of rate tables (see **Exhibit A4.3**), as well as accessorial charges.

■■■■■■■■■ *Exhibit A4.3* **Electronic zip code tariff procedure to display rate scales**

```
1. Display Class 100 to Class 50
2. Display Class 500 to Class 110

Enter choice ?  1


ENTER DISCOUNT   (ie 25, if none strike enter) 25



ENTER YOUR HEADLINE AND SIDELINE ZIPS
STRIKE ENTER KEY TO CONTINUE

1. HEADLINE ZIP ?  530
2. SIDELINE ZIP ?  750
3. SIDELINE ZIP ?
4. SIDELINE ZIP ?
5. SIDELINE ZIP ?
6. SIDELINE ZIP ?
7. SIDELINE ZIP ?
8. SIDELINE ZIP ?


        TARIFF MWB 550
      RATES BETWEEN ZIP 530

     REFLECTS A 25 % DISCOUNT
```

AND ZIP	WT GRP	100	92.5	85	77.5	70	65	60	55	50
	L5C	2831	2647	2463	2269	2079	1958	1839	1708	1583
750	5C	2470	2297	2131	1951	1782	1659	1545	1433	1310
	1M	2213	2057	1900	1736	1580	1472	1373	1260	1154
	2M	2148	1988	1834	1664	1508	1408	1298	1194	1088
	5M	1640	1525	1403	1274	1157	1070	983	908	821
	10M	1595	1482	1360	1235	1123	1041	956	879	801
	20M	1116	1037	951	865	785	728	670	615	561
MC	30M	957	889	816	741	674	625	573	527	481
4493	40M	798	741	680	617	562	521	478	440	401

```
        PRESS ANY KEY TO CONTINUE
CONTINUE (Y/N) ?
```

Source: © Middlewest Motor Freight Bureau, Program author Jeff Michalson. Reprinted by permission.

Commodity Rate Procedures

The procedure in the point-to-point system is to search the alphabetical indexes for the commodity, origin, and destination involved in the shipment. The indexes will list a series of item numbers next to each entry. If the user cannot find an entry in the commodity index, no rate for that particular commodity is published in the tariff. On the other hand, if the user can find the same item number in the three indexes, it means that a point-to-point rate is published. The user simply turns to that item to find the rate.

The scale system of publishing commodity rates may come into play in the same tariff when there is an entry in the Index to Commodities but not in one of the other two (origins and destinations) indexes. This result indicates that a commodity column rate or a distance commodity rate exists in a separate section of the tariff. The procedure is to look up the item number(s) next to the article found in the Index to Commodities. That item will provide information about how to find the required commodity, distance, or weight entries in the commodity column rate tables or in distance rate tables.

For example, using *Iron and Steel Tariff SMC 290* (see **Exhibit A4.4**), the procedure to find out how much it would cost to ship 38,000 lb of rough steel castings by truck from Aliquippa, Pennsylvania, to Greensboro, North Carolina, is as follows:

1. Check the indexes. Scan the tariff indexes to identify item numbers applicable to the commodity, the origin, and the destination. Although the Index to Commodities lists three numbers (Items 22610, 32925, and 33019), only Item 22610 is also found in both the Index to Origins and the Index to Destinations. Item numbers like 22610 that are found in the three indexes will provide point-to-point rates. If the tariff user can locate an entry in the Index to Commodities (such as Items 32925 and 33019), but not in one of the other indexes, it means that the tariff contains a commodity column rate or a distance rate for these articles.

2. Locate the item number(s). Tariffs list items in ascending order. Since point-to-point rates should take precedence over scale rates, start with Item 22610 in the point-to-point rates section of the tariff. The title page of each rate section will explain the application and precedence of rates contained in that section. Item 22610 shows 153 (cents per cwt) and 38M (38,000 lb) as the rate and volume minimum weight, respectively. Note that only Thurston Motor Lines (code THUR1— deciphered in the Routing Section of the Tariff) participates in this rate. Thus this shipment would cost $581.40, and the search is completed.

Suppose, however, that Item 22610 was not in the tariff. The tariff user would then have to look up the two item numbers found only in the Index to Commodities. As shown in Exhibit A4.4, item 32925 resides in a second, separate

Exhibit A4.4 **Commodity rate system**

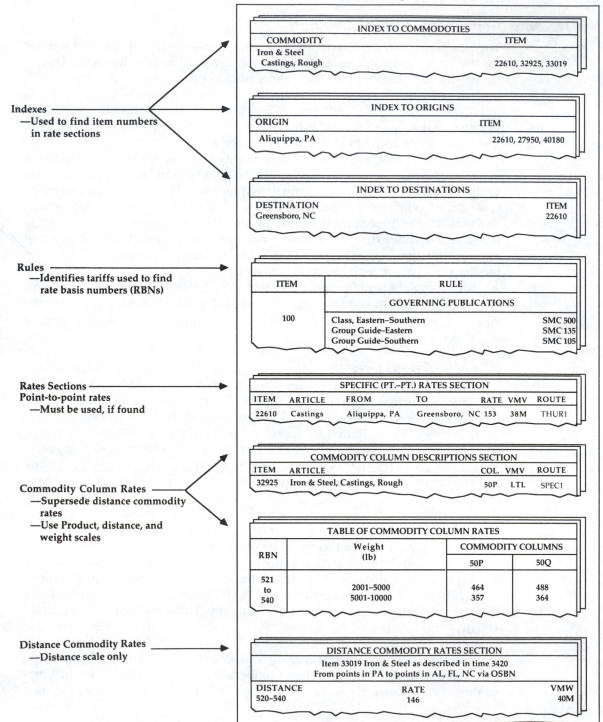

rate section that has commodity column rates. Item 32925 shows a "COL." of 50P as the entry to use in the rate table (just below the "Commodity Column Descriptions Section"). This entry serves the same purpose as the class or rating in the class rate system. In addition, note that 50P applies to less-than-truckload (LTL) shipment sizes hauled by Bowman Transportation (BOWM).

To locate the correct rate in the rate table, the user also has to identify the proper distance and weight group entries. The same rate basis number (RBN) used to find a class rate would also be used here. The "Governing Publications" rule (Item 100), as well as instructions at the beginning of the commodity column rate tables, will indicate the class rate tariff that has the RBN tables and the group guides that might be needed to identify base points.

As it turns out (but is not shown in Exhibit A4.4), Aliquippa bases on Pittsburgh, and Pittsburgh and Greensboro produce an RBN of 526. Thus, Bowman Transportation would charge 464 cents/cwt for a 2000 lb shipment.

Item 33019 is found in a third rate section that uses a distance scale only. The title page of this section states that the distance rates apply only when point-to-point or commodity column rates are not found. In this system, the mileage between Aliquippa and Greensboro, as shown in the *Household Goods Mileage Guide,* is the distance entry in the rate table. Given that the distance is 535 miles and Item 3420 contains the *rough steel castings* description (as required by Item 33019), Osborne Truck Lines (OSBN1) would charge 146 cents per cwt with a volume minimum weight of 40,000 lb (40M) for a total of $584 (38,000 lb as 40,000 × 146 cents/cwt).

chapter 5

Freight Rates

Introduction

Route-Related Rates
Interstate versus intrastate rates
Through rates
Arbitraries and differentials
Export-import
Continuous movement
Intermediate point
Paired and reload

Distance-Related Rates
Mileage
Group

Volume-Related Rates
LTL/TL and LCL/CL
Incentive

Any quantity
Aggregate tenders
Density

Miscellaneous Rates
Released value
Freight, all kinds
Container rates
Government rates
Minimum charges

Summary

Questions

Introduction

A wide variety of shipment sizes, distances, seasonal characteristics, and competitive circumstances cause the transportation industry to express freight rates in different ways. Alternative rate forms generally identify certain service characteristics or indicate the scope of application. Knowledge of the various forms of freight rates promotes better communication between shippers and carriers. This knowledge also enhances the traffic manager's ability to assess the impact of freight rates on total logistical costs.

Although rates may be stated in any measurable unit (for example, per vehicle, per mile, per ton), the transportation industry publishes the vast majority of domestic rates in cents per hundredweight (cwt) units. In addition, rates may apply to specific points, such as towns or cities, or to large areas, such as states or even a larger territory.

"Door-to-door" indicates that rates include pickup service at origin and delivery at destination. In contrast, "ramp-to-ramp" or "terminal-to-terminal" rates require the shipper to tender, and the receiver to pick up, the consignment at the origin and destination ramps or terminals, respectively. These terms find application mostly in express, piggyback, and air freight transportation. In the growing intermodal piggyback service area, railway carriers that offer complete door-to-door rates are in the retail business, whereas those that provide ramp-to-ramp rates are wholesalers.

The transportation industry expresses rates in terms of route, distance, volume, and miscellaneous characteristics. This chapter examines the major rate forms found in these four categories.[1]

Route-Related Rates

Interstate versus Intrastate Rates

What constitutes the legally applicable rate generally depends on whether the transportation is under federal or state jurisdiction. The Constitution specifically vests Congress with the power to regulate interstate commerce and trade with foreign nations. Freight traffic acquires an interstate character when carriers haul it across state or national boundaries.[2]

The simplicity of this working definition of interstate commerce belies the potential legal complexity involved in its practical application. For example, suppose (see **Exhibit 5.1**) a shipment moves from a plant to a warehouse in the same state and, subsequently, is moved out of the warehouse to a customer located in another state. The shipment from the warehouse to the customer, which crosses a state line, is part of interstate commerce. Yet without further information, it is not clear whether federal or state authority governs the move from plant to warehouse. To help resolve such jurisdictional issues, three guidelines have emerged from court and ICC decisions:

1. *Intent of the shipper.* If the shipment depicted in Exhibit 5.1 was continuous from plant to customer, its character would be interstate since it crosses a state line. The issue of jurisdiction arises from the interruption at the warehouse. The issue is whether that interruption is only temporary in an otherwise continuous shipment or if it is permanent and creates two separate movements.

The Supreme Court has ruled that it is the "original and persisting intention of the shippers" that determines the character of the commerce.[3] If the shipper intended all along to send the goods to a customer in the adjacent state, the stop at the warehouse is only a temporary interruption in a basically continuous movement across state lines.

For some commercial activity, however, the generalized intent of the trade so clearly embraces regional, national, or international markets that the shipper's specific intentions may be viewed as irrelevant; all transportation activity is seen as interstate commerce in the overall trade picture.[4] The grain trade, for example, generally falls into this category.

Exhibit 5.1 **Intrastate versus interstate commerce**

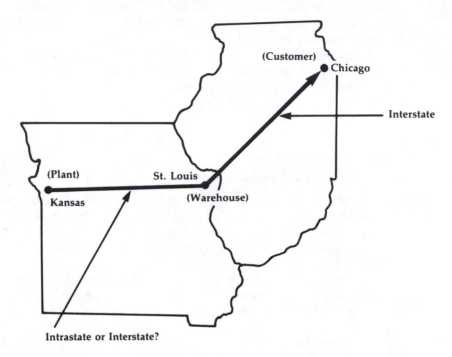

Changes of intent or changes in the character of freight, which take place while the goods are in transit, also affect the application of the intent of shipper principle. Suppose a shipper, who originally plans to serve an out-of-state customer, diverts that shipment to an intrastate location. In other words, the original intent changes from interstate to intrastate commerce while the goods are in transit. The guideline is that jurisdiction generally rests on whether the shipment actually crossed state lines. If the shipment entered another state en route, it becomes part of interstate commerce. If it does not cross state lines before it was diverted, the fixed and persistent intent of the shipper would be to make an intrastate shipment.

Another complication arises when a change in the character of the goods takes place en route. The ICC and courts have generally ruled that, unless transit privileges (see Chapter 11) apply, the act of subjecting articles to further manufacturing processes at an intermediate location breaks the continuity of the shipment.[5] Thus, even though the shipper clearly intends to deliver processed goods to interstate locations, an intrastate shipment of unprocessed material to the processing point is considered a separate shipment.

2. Intent of Congress. The Revised Interstate Commerce Act (RICA) does not extend ICC (federal) jurisdiction to railway transportation of property "entirely in a state," provided that the complete movement does not originate or terminate in a foreign country.[6] The jurisdictional issue here involves private carriage that originates in another state and precedes common carriage that is entirely within a state. Looking at Exhibit 5.1 again, if a shipper sends freight from Chicago to St. Louis by private truck and then by common carrier from St. Louis to Kansas City, is the common carriage entirely within Missouri subject to state or federal authority?

Addressing this issue in 1936, the Supreme Court held that Congress did not intend to have the ICC regulate a particular form of interstate commerce—that is, regulated common carriage entirely within a state interlined with preliminary interstate private carriage.[7] Clearly, the entire shipment encompasses a continuous movement between two states. For purposes of regulation, however, the private (unregulated) carriage is ignored; only transportation entirely in a state remains. Thus the intrastate rate applies to the common carriage between St. Louis and Kansas City.

Following the 1936 Supreme Court decision, other ICC and court cases have expanded the "particular form of interstate commerce" to include common carriage entirely within a state connected to preliminary or subsequent interstate private, exempt, or contract carriage, unless the noncommon carriage segment originates or terminates in a foreign country.[8] Currently, however, the ICC, the U.S. Department of Transportation, and the Private Carrier Conference, as well as several interested shippers and carriers, are attempting to overturn that 1936 Court doctrine and make this form of interstate commerce subject to federal jurisdiction.[9]

3. Intent of the carrier. The intent of the carrier sometimes appears as a key consideration when both intrastate and interstate routes connect two points in the same state. The ICC has indicated that when the carrier intends to use interstate routes to evade state regulation (for example, to apply higher interstate rates), the character of the commerce does not change.[10] For the lawful application of interstate rates, the carrier must demonstrate some valid operating reason for using interstate routing, such as the use of a more efficient terminal.

Regardless of whether the interstate routing is practical and reasonable or merely a subterfuge, the route actually used determines the legally applicable rate. The ICC has ruled:

> *The carriers must charge the applicable interstate rate . . . even though they could be liable for misrouting when they convert an intrastate shipment into an interstate shipment by moving it over an interstate route.*[11]

Thus the shipper must pay the legally applicable interstate rate; if the application of that rate constitutes an unreasonable practice, that is, an evasion of state regulation, the shipper must subsequently pursue a misrouting claim action (see Chapter 13).

Through Rates

A through rate refers to the total rate from origin to destination. It may be expressed as a single factor or as the combination of separately established rates.

Single factor. When one rate applies to the through route, that rate is a single factor. Unless the tariff indicates otherwise, this rate applies via named routes, even if lower combination rates also apply over the same routes. Single-factor rates take two forms:

1. Local. A local through rate applies over the lines of only one carrier. Local rates are also known as ''single line'' rates and may be published in class, exceptions, or commodity tariffs.

2. Joint. A joint rate is a single-factor rate that applies over the lines of two or more carriers. It is established by formal written agreement between the carriers. Joint rates have traditionally facilitated interline movements, especially by railway. Historically, railroads have interlined about 50 to 75 percent of their traffic.

As illustrated in **Exhibit 5.2(a)** and **(b)**, carriers establish joint rates for interline traffic for several reasons, one of which is competition. Two connecting carriers may publish a joint rate to compete with another carrier that offers single-line service. Another reason is efficiency. Often geographical factors and traffic flow patterns create opportunities to establish less circuitous interline routes that afford better service and lower unit costs.

In addition, shippers might instigate joint rates. Whenever the traffic volume is large enough to warrant interline economies, shippers may want to negotiate joint rates for several reasons. First, joint rates are usually less than the prevailing combination rates. Second, joint rates offer lower transaction costs to the shipper than the payment of separately established rates; only one bill of lading is necessary for the through movement, and only one freight bill for the services of all participating carriers is sent to the shipper. Third, the shipper that currently uses a local rate might look at interline alternatives for negotiating leverage.

Since the 1970s, however, regulatory reforms and deregulation have greatly diminished transportation industry incentives to establish joint rates.[12] Essentially, three fundamental changes in regulation discourage the use of joint rates, especially in the railway industry:[13]

1. Restricted antitrust immunity. As stated previously (see Chapter 4), in its implementation of reform legislation, the ICC has stripped away much of the

▬▬▬ *Exhibit 5.2* **Rationale for establishing joint-line rates**

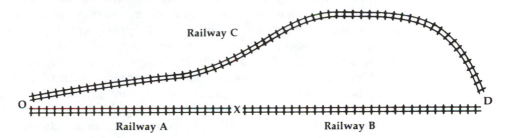

Railway C

O — X — D

Railway A   Railway B

Route O-X-D offers:
  Competition with Railway C's single-line route
  Shorter route

(a) Competition and shorter route

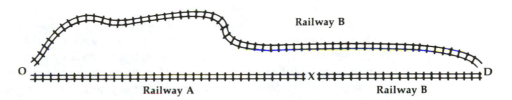

Railway B

O — X — D

Railway A   Railway B

Route O-X-D offers:
  Shorter route

(b) Shorter route

transportation industry's immunity from antitrust prosecution for joint pricing activity. For example, in Exhibit 5.2(b), the threat of antitrust surely will cool Railway B's enthusiasm to negotiate with Railway A to make a joint rate that will compete simultaneously with B's own local rate.

2. More freedom to cancel joint rates or add surcharges. The Staggers Rail Act of 1980 authorizes railway companies, under certain conditions that the ICC has laxly enforced, to cancel joint rates and reciprocal switching agreements or to add surcharges without the concurrence of other participating carriers.[14] Since 1980, many of the large railroads have taken such actions to eliminate inefficient routes and interchanges with other railroads, to improve transit-time performance, and to force more traffic to their own lines. Historically, joint rates have been "equalized" over all authorized through routes. In other words, shippers paid the same joint rate for alternative routes, regardless of differences in length or terrain. The rates, of course, reflected the costs of the least efficient routes. By closing inefficient joint routes and interchanges, Conrail claims it has eliminated more than 5 million car-miles and 128,000 interchanges annually and has improved transit-time per carload by an average of 1.2 days—results that have meant about $32 million annually in savings with virtually no rate increases.[15]

Nevertheless, shippers generally have been critical of such cancellations. They believe fewer pricing and routing options ultimately will increase transportation costs and impair the ability of shippers to negotiate rates with certain railway carriers. Two major shipper groups, the National Industrial Transportation League and the Chemical Manufacturers Association, initiated negotiations with the Association of American Railroads in 1983 to obtain a better balance between shipper and carrier interests. In 1985, the ICC unanimously approved the compromise agreement reached by the two sides earlier that year.[16] The new ICC rules will help shippers keep existing joint rates and through routes. In addition, it should help them win ICC orders forcing railway companies to agree to new joint rates, through routes, and reciprocal switching arrangements with their competitors.

3. More permissive policy toward mergers. The regulatory environment is much more permissive toward ''end-to-end'' mergers and abandonment of unprofitable lines. End-to-end mergers connect the lines of railway companies that operate in adjacent areas. As a result of such mergers, the railway industry has witnessed the development of regional megasystems and, in general, a rationalization of the national network in which the reach of single-line service is far greater. Likewise, since 1980, motor carriers have greatly expanded their operating authorities. The large carriers, of course, have a natural economic incentive to maximize the length of hauls and revenues.

Combination. A combination rate is a combination of separately established rates used to obtain the through rate. In the absence of single-factor rates, the lowest combination rate via the route of movement is the legal rate.[17]

The shipper normally pays considerably more for combination rates because the interchange arrangements are generally less efficient, and the overall expenses of handling, rating, billing, tracing, and so forth are greater. Traffic managers should give special attention to the minimum weights attached to the different rates that make up the through combination rate. For example, a shipping weight of 50,000 lb might meet a 50,000 lb minimum weight for one rate factor, but not, say, a 60,000 lb minimum weight for the other factor making up the combination rate. Thus the firm ends up paying for an extra 10,000 lb for part of the through movement.

Before reforms of the 1980s, the ICC generally considered single-factor rates that exceeded combination (or ''aggregate-of-intermediate'') rates as prima facie unreasonable. Likewise, the ICC generally prohibited railroads and water carriers from charging more for shorter than for longer hauls moving in the same direction and along the same route. After Congress passed reform legislation in 1980, however, the ICC has energetically pursued administrative deregulation. In 1983, the ICC effectively eliminated prohibitions against aggregate-of-intermediate or long-and-short-haul rate discrimination. In making its decision, the commission indicated that "the existing regulations are no longer necessary to carry out the national transportation policy" and that "changes in the transportation industry, including the rapid growth of intermodal competition, have all but eliminated the problems the long-and-short-haul clause was intended to remedy."[18]

Proportional rates. A proportional rate applies only to through traffic that moves over a certain segment of the through route. Shippers must add the proportional rate to one or more rates applicable to other segments of the complete route to form the through rate. Since there are at least two rate factors, the proportional rate also takes the form of a combination rate. Nevertheless, there is an important difference between proportional and combination rates, which is best explained by an example.

As shown in **Exhibit 5.3**, the combination of local rates for route O-X-D is $2.25, which is not competitive with the $2.00 local rate for the direct route O-D. By publishing a proportional rate of $1.15, Railway B can form a through rate ($1.15 + $0.85) that is the same as Railway A's local rate ($2.00). Note that

━━━━━━ *Exhibit 5.3* **Using proportional rates to meet competition**

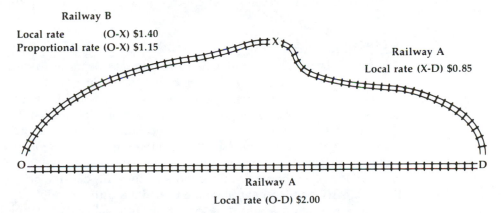

Railway B

Local rate   (O-X) $1.40
Proportional rate (O-X) $1.15

Railway A
Local rate (X-D) $0.85

Railway A
Local rate (O-D) $2.00

Through rate from O to D:
   via Railway A: $2.00
   via Railway B and A:
     without proportional rate: $1.40 + $0.85 = $2.25
     with proportional rate: $1.15 + $0.85 = $2.00

the $1.15 rate for route segment O-X applies only to through traffic. Railway A is not willing to establish a joint rate with Railway B because Railway A would rather capture all the traffic than share it with a competitor. Proportional rates allow a carrier to form competitive through rates for interline service without publishing joint rates.

In another major application, the transportation industry publishes proportionals for the through traffic that originates or terminates at points beyond the geographical scope of a particular tariff. Historically, the Mississippi River has defined a border between rate territories. To form a through rate for the shipments that originate in the West (for example, Los Angeles) but terminate at points throughout the East, a carrier might publish a proportional rate from Los Angeles to St. Louis to apply to all the eastern routes that converge at the St. Louis gateway. This publication action accomplishes two important things. First, it allows the carrier to comply with the legal requirement (for common carriage) to publish rates over all routes in which the carrier participates. Second, it permits an easier and cheaper method of publishing rates; that is, instead of publishing rates from Los Angeles to each of the myriad points east of the Mississippi, the carrier publishes only one proportional rate to apply to all through traffic.

The loss or dilution of antitrust immunity from collective pricing, the railway industry's merger activity, and the new statutory freedoms to cancel joint rates suggest that traffic managers will witness the increasing use of proportional rates for interline traffic.[19] Note, however, that although proportional rates may be competitive with single-factor through rates, they still impose additional transaction costs on the shipper in terms of shipping and billing documents, tracing, and information processing and control of shipper-owned equipment.

Arbitraries and Differentials

Sometimes tariffs contain rules that provide rates from or to designated points by the addition of arbitrary charges to, or the deduction of differentials from, the linehaul rate. For tariff simplification, rates are normally published for major competitive centers or base points (see Chapter 4). To establish rates to outlying communities and unnamed or unusual locations (for example, new construction sites or remote government facilities like the Lunar Exploratory Module in New Mexico), the carrier adds an arbitrary amount to the linehaul rate. Thus an *arbitrary* is a fixed charge usually expressed in terms of cents per hundredweight that the carrier adds to another rate to make a through rate.

Carriers often use arbitraries to offset unusual costs that might arise from inner-city congestion, route topography, or traffic imbalances. Traffic congestion in downtown and port areas delays pickup and delivery and makes transportation more expensive. Likewise, mountainous regions and steep grades increase operating expenses, and traffic imbalances may produce empty backhauls. The carrier passes on these costs to the shipper by publishing arbitraries applied to the base points that produce such unusual situations. The class or commodity rate still defines the through rate; however, an arbitrary charge is added to offset the higher costs of serving such points.

Further, in the railway industry, "relief line arbitraries" are sometimes published. These help certain short-line railroads earn more revenues than they would otherwise obtain through the normal division of joint-rate revenues.

By contrast, *differentials* usually permit the shipper to deduct (but sometimes add) an amount from the through rate to attract traffic to an inferior route or to stabilize rate relationships among ports, producers, or markets. Route-related differentials are frequently found in all-rail and rail-water routings to compensate for a route that is more circuitous, has a higher risk of loss or damage to lading, or provides only seasonal service. Differentials reduce the cost of transportation to counterbalance the higher logistical costs produced by inferior routes and less reliable transit performance. These logistical costs generally pertain to inventory, claims, or customer goodwill.

In the past, the railway industry often used differentials to stabilize rate structures among competing areas. Port differentials, for example, have had a long tradition in the transportation of export-import traffic. Rate bureaus would establish a base rate to one port, say Baltimore, and then use arbitraries or differentials to equalize rates to other ports along the same seacoast. This tactic permitted the less favorably situated ports to participate in traffic they would otherwise not receive.

Differentials have been a standard feature of railway pricing for many decades. Nonetheless, the rise of railroad megasystems, the use of computer-based costing technology to support ratemaking, the post-reform relaxation of published tariff rate regulation, and the freedom to engage in contract and exempt ratemaking are making differentials archaic.

Export-Import

An export rate applies only to traffic hauled from inland points to ports of export as part of through transportation by water carriage bound for foreign destinations. An import rate is just the reverse. The rate analyst often finds an export-import rate significantly lower than its domestic counterpart, although both rates apply to the same inland route of movement. The different characteristics of domestic and international markets, as well as the desire to promote foreign trade, generally explain such differences. (Export-import rates are discussed further in Chapter 15.)

Continuous Movement

Continuous movement rates apply when a single shipper exclusively and continuously uses the carrier's equipment for a specified period. Ordinarily, highway carriers do not guarantee exclusive use of a vehicle for single shipments unless the shipper pays an additional charge. When shipments do not reach vehicle capacity, carriers generally hold the vehicle until other freight is received to fill it. The charge for exclusive use compensates the carrier for the expected loss of revenue that would come from more fully utilized equipment.

The continuous service rate essentially extends the exclusive-use concept. In other words, shippers use continuous movement rates to assure exclusive use of carrier's equipment, say for faster service or more security, over an extended period.

Intermediate Point

The transportation industry establishes rates to virtually all locations in the United States, as well as to many points in Canada. Tariffs, however, usually do not show rates for some of the smaller shipping points. Instead, tariffs may contain rules to allow a rate listed for a specifically named origin and destination to apply to unnamed intermediate points along the route. The rate from or to the next more distant point shown in the tariff will apply to the unnamed point.

Paired and Reload

Some carriers now grant discounts to shippers that can coordinate balanced, two-way or "paired," moves between intrafirm locations or, with the cooperation of customers, interfirm points. Others grant discounts simply for reloading empty vehicles at destination and sending shipments the backhaul direction. The Burlington Northern Railroad, for example, has offered cash refunds ranging from $100 to $225 for reloads in westbound boxcars, flatcars, or gondola cars interchanged at the St. Louis or Chicago gateways.

Distance-Related Rates

Mileage

Rates that employ mileage scales are, by far, the most prevalent for the transportation of freight. As already illustrated in Chapter 3, tariffs usually define mileage scales in rate basis numbers. As the mileage scale increases, the amount of each additional rate increase normally tapers off because the fixed cost elements of terminal and linehaul operations remain unaffected by the length of the haul. In addition, longer hauls generally experience greater operating efficiencies. When no tapering effect appears, the rates have a proportional structure. In other words, successive increases in rates remain constant. Be careful not to confuse a "proportional rate structure" with the proportional rates already discussed.

In contrast to mileage rates, point-to-point rates hold distance constant. Further, "uniform" rates, which some courier or small package express companies charge for shipments delivered anywhere in the nation, ignore distance altogether. The group rate structure discussed next lies between these two extremes.

Group

The group rate structure expands base points to cover much larger areas, including entire regions or zones (see **Exhibit 5.4a**). Although more distant groups take higher rates, the size of the grouping area dilutes the importance of dis-

■■■■■■■ *Exhibit 5.4* **Blanket/group rate structure**

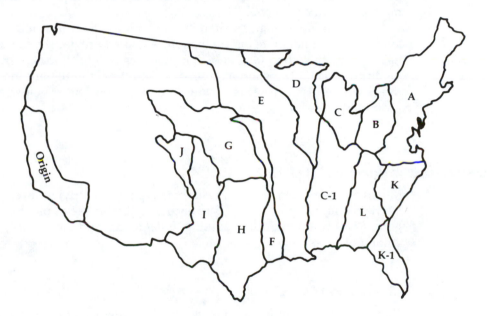

(a) **Blanket/Group Rate System from California**

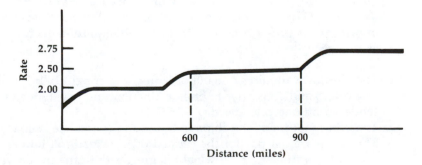

(b) **Blanket/Group Rate Taper**

tance. **Exhibit 5.4(b)** illustrates this effect. Note, for example, how the rate stays constant at $2.50 for all points between 600 and 900 miles.

Historically, several reasons explain the use of group rate structures. First, relatively rising rate scales restrict longhaul traffic. Group rates effectively extend and flatten the rate scale, thereby facilitating the flow of longhaul traffic. Not surprisingly, transcontinental traffic traditionally has found widespread application of this rate structure.

Second, in some regions, group rates support competitive market conditions. In the Appalachian Region, for example, coal fields define clusters of producers. Rail carriers developed group structures for the clusters to permit producers within each cluster to reach markets at the same price, thus encouraging more coal traffic to move.

Third, group structures simplify tariff publication and help carriers avoid long-and-short-haul rate discrimination. As noted previously, however, the ICC has effectively voided long-and-short-haul regulation. Furthermore, the growing list of exempt transportation services and the trend toward more cost-based rates, where distance plays a more important role, have helped to dismantle group structures.

Volume-Related Rates

Most volume-related rates express some type of underlying discount structure. Carriers offer volume discounts to encourage better use of plant and equipment generally through reduced handling, fuller loads, and more stable traffic flows, among other reasons.

LTL/TL and LCL/CL

Although the operating economies of the railway industry have made carriers virtually abandon the less-than-carload business, the distinction between less-than-carload (LCL) and carload (CL), as well as less-than-truckload (LTL) and truckload (TL), remains one of the most fundamental aspects of transportation pricing. As prescribed in tariff rules, the requirements for TL or CL rates are as follows:

1. The shipment requires one bill of lading and freight bill; it is made from one origin and consignor to one destination and consignee; and it is tendered on one calendar day.
2. The consignor loads the shipment, and the consignee unloads it.
3. The shipper must pay for the carload or truckload minimum weight, unless the actual shipping weight is greater than the minimum.

Carload and truckload rates offer shippers substantial discounts. Carriers also realize benefits. When making TL or CL shipments, carriers save loading and unloading expenses, avoid intermediate terminal handling, and generally experience less loss and damage. Furthermore, the trucking industry, which is in the business of hauling small shipments, offers a declining scale of rates for successively larger LTL weight groups (see Chapter 4).

This and similar discount structures make it essential to compare the value of holding orders to consolidate shipments with the value of meeting customer service requirements for rapid delivery. In addition, the traffic manager must determine whether to make a relatively small shipment, say 17,000 lb, a truckload by paying for the truckload minimum weight, say 20,000 lb. The traffic manager, of course, can make the same kind of assessment for adjacent LTL weight groups. Such tradeoff calculations actually involve the concept of a *weight-break,* which identifies the breakeven quantity that makes the cost of shipping a smaller quantity and paying a higher rate the same as the cost of shipping the minimum quantity required for a discount and paying a discount rate. (See Chapter 10 for further discussion of this concept.)

Incentive

Incentive rates take several forms. Some forms may be used by only one particular mode of transportation. Others find more widespread application. The major incentive rate forms are as follows:

1. Alternating rates and minimum weights. The concept of alternating rates and minimum weights is best shown by example. In **Exhibit 5.5**, the shipper has a choice between 105 cents/cwt and a 20,000 lb minimum, and 94 cents/cwt and a 36,000 lb minimum. If the cargo actually weighs 20,000 lb, the shipper would use the 105 cent rate. Likewise, the shipper would use the 94 cent rate for a shipment weighing 36,000 lb. What rate applies, though, for a 33,000 lb shipment? The shipper should select the rate and minimum weight that will produce the lowest total charges. Here the shipper finds it less expensive to pay for 36,000 lb at the 94 cent rate than for 33,000 lb at the 105 cent rate.

Exhibit 5.5 **Alternating rates and minimum weights**

Article	Origin	Destination	Rate*	MW†
Bakery goods group	Memphis, Tenn.	Little Rock, Ark.	105	20000
			94	36000

*Cents per cwt
†Minimum weight in lb

The railway industry frequently uses alternating rates and minimum weights to compete with motor carriers for hauls generally longer than one hundred miles and for tonnages greater than 50,000 lb. When shippers have such tonnage, they may find rail rates substantially lower than competing truck rates.

2. Multitier (base-excess). Two- or three-tier incentive rate structures encourage shippers to load vehicles to full visible capacity. To illustrate, the multitier incentive rate structure might provide a railway base rate of 300 cents/cwt for the first 60,000 lb, 275 cents/cwt for the next 40,000 lb, and 260 cents/cwt for any additional weight over 100,000 lb loaded in the same car. Given this information, the shipper would compute the charges for a 110,000 lb shipment as follows:

First 60,000 lb: 600 cwt $\times$ 300 cents/cwt = $1800
Next 40,000 lb: 400 cwt $\times$ 275 cents/cwt = 1100
Next 10,000 lb: 100 cwt $\times$ 260 cents/cwt =   260
                                          $3160

Thus the shipper pays an effective rate (actual cost of cents/cwt) of 287 cents per cwt for the entire 110,000 lb shipment.

This kind of incentive rate structure finds application in the transportation of commodity groups such as lumber, fertilizer, canned goods, and iron and steel. Carriers generally offer such multitier rates when the average-size shipment is less than the capacity (cubic or weight) of the vehicles used for a particular type of service (fifty-foot boxcars, forty-foot vans, and so on).

3. Multiple car. Railway tariffs sometimes contain a scale of declining rates for single shipments when the volume of freight requires multiple freight cars. These rates, which often apply to the transportation of grain, lumber, and some chemical commodity groups, yield substantial discounts over regular single car rates.

4. Trainload. Single shipment incentive structures also extend to trainload volumes. Railway companies offer reduced rates connected with minimum weights for entire trainload combined with minimum weights for each freight car in the train. These rates are designed for the transportation of high-volume bulk commodities such as coal, grain, stone, cement, and similar articles. Frequently, unit trains, which consist of roughly one hundred cars or more, operate as a single unit from origin to destination for the purpose of dispatching, loading, unloading, and billing. These trains operate very efficiently hauling coal for utility companies. They bypass classification yards and load and unload quickly.

5. Time-volume. Carriers may file rates that apply only when shippers tender, or consignees receive, certain quantities of freight during specific periods. For example, in an early application of this type of rate structure, several railway companies offered to reduce the single-car rate from about $4.00 to $3.65 per net ton when, during a twelve-month period, they hauled and delivered at least 1.5 million net tons of coal to one consignee.[20]

In the 1960s, the Illinois Central introduced a new twist with its "rent-a-train" concept. The railway company gave shippers exclusive use of entire trains and guaranteed a minimum number of round trips for a fixed yearly fee. Thus, during the lease period, shippers can significantly reduce the cost per ton for transportation by efficiently scheduling and utilizing equipment to maximize the number of round trips and minimize any empty backhauls.

"Agreed rates" represent another variation. They offer discounts to shippers that agree to use a specific railway carrier to haul a certain percentage of their total annual tonnage. Historically, the ICC has ruled out the use of agreed rates in the United States mainly on the grounds that such ratemaking discriminated against the small firms that could not produce enough traffic volume to obtain the agreed rates. The administrative reforms to rate regulation in the late 1970s, followed by the more extensive 1980 reforms, have paved the way for the widespread use of agreed rates, especially in contract carriage.

Carriers have expanded the time-volume concept to include discounts that apply only after the firm tenders a minimum tonnage or dollar value of traffic to a carrier during a month, quarter, or year. For example, each quarter the shipper might earn an 8 percent discount on all LTL movements made after the first $10,000 worth of business is tendered to the carrier. The discounts encourage firms to remain loyal to the carriers that offer them.

6. *Conditional sliding scales.* The General Electric Company initiated this incentive plan in the wake of the transport pricing reforms that took hold in the early 1980s. This incentive structure uses a sliding scale of discounts that increases as traffic volume increases from one period to the next. After the shipper meets a minimum volume of LTL business during the initial period, participating carriers offer a discount applied to all similar traffic shipped during the next period. The size of the discount earned by the shipper in the later period depends on the size of the increase in traffic volume realized in the earlier period. For example, the scales might appear as follows:

Volume increase from one period to next	Amount of discount
25%	5%
50%	10%
100%	20%
200%	30%

The plan may apply to a firm's entire logistical network or to specific traffic lanes. For example, the discount scale in General Electric's plan, which many carriers have accepted and incorporated in tariffs filed with the ICC, applies to the entire network. If one plant can increase the amount of traffic tendered to a certain carrier and invoke a higher discount for the next period, all the other company plants would participate in that discount as well.

As a large company, General Electric has the leverage to make it worthwhile for carriers to participate in its plan. Other firms, of course, may not have that bargaining strength. In such situations, carriers are likely to restrict this form of discounting to particular traffic lanes. Only plants tendering the increased volume over a certain route would obtain the discounts that apply only over that route.

The lane approach generally is more economic for the carrier. Although some economies should accrue from handling incremental traffic over the prime traffic lane, the carrier does not have to haul traffic over other lower density routes at the same discount level earned by the prime lane.

Any Quantity

Any quantity (AQ) rates do not allow volume discounts. Instead of specifying a minimum weight for a reduced truckload or carload rate, carriers publish "AQ" in place of the minimum weight and charge the same rate regardless of the quantity tendered to the carrier in a single shipment.

Carriers generally assign AQ designations to bulky articles that are difficult to handle or stow. These items quickly fill a vehicle to capacity but weigh comparatively little, perhaps 10,000 lb or less. They create few operating economies for the carriers when moved in large quantities. Thus carriers are likely to hold AQ rates at higher LTL or LCL levels to generate revenues on par with amounts earned from normal loads.

Aggregate Tenders

Shippers earn aggregate tender rates when they can aggregate two or more LTL shipments and tender them to a carrier at one time and place. Usually, a 5000 lb minimum weight is necessary to invoke the discount. Except for the pickup, the shipments remain independent of each other and take separate bills of lading.

LTL rates ordinarily build in a cost-recovery factor for the pickup of relatively small shipments. The aggregate tender, however, removes the need to apply that factor to each shipment when only one pickup is required.

Density

Rate tariffs may show rates for different density groups. As the density (pounds per cubic foot) of the article increases, rates for the different density groups

Exhibit 5.6 Bumping application for density-based rates

NMFC Article Description	Density (lb/cf)	Rating	Actual Measures per Package			Bumping Weight Cube × minimum level of next density group
			Density (lb/cf)	Weight (lb)	Cube (cf)	
Electrical equipment group Item 63270 telephone housing						
	<1	400				
	1	300				
	2–3	250				
Density group-------->	4–5	150	5	20	4	4 cf × 6 lb/cf = 24 lb
Bump to next group------>	6–11	100				
	12–14	85				
	>14	70				

Computation of charges

Given rates as follows: Class 150 — $34.67/cwt   Class 100 — $21.27/cwt

Without bumping option: 20 lb @ $34.67/cwt = $6.93 per package

With bumping option: 24 lb @ $21.27/cwt = $5.10 per package decrease. Likewise, classification tariffs sometimes list article descriptions with subgroups that relate to a density scale.

In addition, tariffs may have rules that allow shippers to bump the article into the next higher density group to obtain a lower total charge. The bumping concept, illustrated in **Exhibit 5.6** is similar in application to making an LTL shipment a TL shipment by paying for the minimum truckload weight. For density groups, though, the process works as follows:

1. Use the density of the shipment to determine the group. As shown in Exhibit 5.6, the density of each package is 20 lb/4 cf or 5 lb/cf, which places it in the ''4–5'' group.
2. Multiply the cubic measurement of the article by the initial level of the next higher density group. For example, 4 cf times 6 lb/cf (from the ''6–11'' group) is 24 lb. The shipper uses this weight to compute charges when bumping the shipment into the next higher group.
3. Compute the tradeoff. Assuming rates of $34.67 per cwt and $21.27 per cwt for the ''4–5'' and ''6–11'' density groups, respectively, the tradeoff is as follows:

Without bumping: 20 lb @ $34.67 = $6.93 per package
With bumping: 24 lb @ $21.27 = $5.10 per package

Usually, to take advantage of bumping rules, the shipper must declare its application on the bill of lading, as specified in the rules before the shipment.[21]

As the pricing environment gets more competitive, rates become more closely related to costs. Since density has a significant impact on equipment utilization and other elements of operations, even greater reliance on density in ratemaking appears likely. Such a trend has important ramifications for marketing, packaging, and traffic management coordination. Specifically, it emphasizes the need to assess tradeoffs between the value of consumer-oriented packaging that adds bulk and the extra transportation cost for hauling bulkier articles. For example, larger, but not heavier, breakfast cereal boxes use more shelf space and provide greater brand name visibility. On the other hand, the company must pay higher transportation costs essentially to move extra air in the cereal boxes. Similarly, density-based rates place greater significance on the tradeoff between protective packaging and transportation costs. For example, styrofoam padding reduces damage as well as package density.

Miscellaneous Rates

Released Value

A released value rate applies when the shipper agrees in writing to place a limit on the amount that can be claimed if loss or damage occurs. The concept is to charge less for transportation in exchange for a reduction in the amount that the carrier would have to pay the shipper for loss or damage. Released rates are typically established for commodities that have (1) a wide range of values, which makes claims difficult to estimate, (2) a comparatively high susceptibility to loss or damage, (3) a high claim-to-freight revenue ratio, and (4) a high frequency of claims.[22]

Both the Staggers Rail Act and the Motor Carrier Act of 1980 make it easier for common carriers to limit their liability through the use of released value rates and deductibles. The ICC has the authority to approve applications from railway companies and truckers for the publication of these rates. In contrast, for contract carriage, shippers and carriers negotiate liability issues, including released value conditions. (Released value rates are discussed further in Chapter 12.)

Freight, All Kinds

The description "freight, all kinds" (FAK) allows the rate to apply to shipments of one or more commodities without naming the individual article(s). Freight forwarder, piggyback, and air freight services frequently use FAK rates for merchandise traffic.

FAK rates are not practical for all commodities. As previously discussed in Chapter 4, commodities with diverse characteristics create different costs for the carrier. Thus a single FAK rate may not be suitable.

Nevertheless, some shippers apparently have negotiated "single-factor" pricing programs that adopt the general FAK concept for diverse LTL traffic.[23] One freight rate (factor) is negotiated for all shipments from an origin point to a predefined geographical territory, regardless of weight or freight classification. To establish this kind of program, a shipper needs to have a large volume of LTL freight that moves regularly over well-defined traffic lanes. The assumption is that over time transportation costs will consistently reach an identifiable average cost per hundredweight, despite variations in shipment size and freight classifications.

Container Rates

Container rates encompass two fundamental types of pricing activity. On one hand, carriers may charge a fixed rate for transporting a container between origin and destination terminals. Shippers may stuff the container with all kinds of freight, but they must pay for pickup and delivery and for return of the container to the destination terminal. The air freight industry primarily sponsors this method of pricing.

On the other hand, container rates also apply to traffic moving in the large trailer-sized containers that follow an international intermodal itinerary. (See Chapter 15 for further discussion of this pricing activity.)

Government Rates

Section 10721(b)(1) of the RICA permits common carriers subject to ICC jurisdiction to transport property for all levels of government without charge or at reduced rates. This section carries on a tradition first established in Section 22 of the original Interstate Commerce Act to compensate the federal government for its railroad land grants. Such grants were made during the latter part of the nineteenth century to promote the development and settlement of the West.

The vast majority of Department of Defense (DOD) traffic benefits from reduced rates filed in tariffs called military rate tenders. Often motor carriers simply file one or two page tenders that indicate all traffic originating at a government contractor's plant will go at, say, class 50 instead of at the higher actual ratings found in the classification tariffs.

Minimum Charges

Occasionally, inexperienced shippers encounter an unpleasant surprise when they receive a freight bill requesting payment for an amount greater than the rate times the weight of the shipment. As a customary practice, carriers establish minimum charges to recoup certain costs, such as pickup, delivery, and billing, which occur regardless of the shipping weight. These charges apply whenever they exceed the amount constructed from the rate times the weight.

Minimum charge rules are a standard element of class and commodity rate tariffs. Although minimum charges come into play mostly with small shipments, their application may extend to carload and truckload shipments.

Summary

The transportation industry expresses rates in many different ways. The various rate forms identify certain service elements or indicate the scope of application. Four categories of rates were investigated: route, distance, volume, and miscellaneous. The investigation focused on the ways in which domestic common carriers express freight rates and the managerial implications.

Freight acquires an interstate character when carriers haul it across state or national boundaries. Three guidelines for determining whether *intrastate* or *interstate rates* apply to certain routes included (1) intent of the shipper, (2) intent of Congress, and (3) intent of the carrier.

A *through rate* may be expressed as a single factor or as a *combination* of separately established rates. Single-factor through rates may apply to *local* (one carrier or single-line) or to *joint* (two or more carrier) routes to form local and joint rates. Changes in regulation, however, have diminished transportation industry incentives to establish joint rates. A *proportional rate* applies to a segment of a through route and must be combined with another rate to create a through rate for the traffic moving over the entire route. Deregulation has increased the use of this rate form.

Carriers adjust the linehaul charge with *arbitraries* to make adjustments for service to unnamed or unusual locations and to offset abnormal costs. *Differentials* usually permit the shipper to deduct an amount from the through rate to compensate for an inferior route.

An *export rate* applies only to traffic hauled from inland points to ports of export as part of a through movement by water carriage. *Import rates* apply to the reverse situation. Other route-related rates discussed briefly included *continuous movement, intermediate point,* and *paired-and-reload* rates.

Distance-related rates describe *mileage* and *group* rate structures. Mileage rate structures were defined as tapering, proportional, constant, and uniform. A group rate structure expands base points to cover larger areas or zones.

Volume-related rates usually involve a discount structure. The major rate forms in this category included carload (CL), truckload (TL), incentive, any-quantity (AQ), aggregate tender, and density. Incentive rates also take several forms as follows: (1) alternating rates and minimum weights, (2) multitier or base–excess, (3) multiple car, (4) trainload, (5) time–volume, and (6) conditional sliding scales.

Other *miscellaneous* rates forms include released-value, freight all kinds (FAK), container, and government rates. In addition, shippers have to be alert for minimum charges.

Selected Topical Questions from Past AST&L Examinations

1. What are "export" and "import" rates charged by U.S. railroads and trucking firms? How and why do they differ from domestic rates? (Spring 1980)
2. The BVG Company shipped a truckload of amplifiers from their State College, Pennsylvania, manufacturing plant via their private truck fleet to the terminal facility of Rocket Express, Inc. (a motor common carrier) in Ft. Wayne, Indiana. At Ft. Wayne, the trailer was dropped in Rocket's yard and immediately hooked up to one of the Rocket's linehaul tractors for continuous transportation and delivery to a customer in Chicago, Illinois, the intended destination of the shipment at the time it left State College.
   (a) What portion(s), if any, of the service, rates, and charges of the shipment are subject to the rules and regulations of the ICC and the Interstate Commerce Act? Explain.
   (b) Assume that BVG had interchanged the shipment with Rocket at Hammond, Indiana. Would your answer to (a) be the same or different? Explain. (Fall 1981)

3. Please define any three of the following terms: (Fall 1981)
   (a) blanket rate
   (b) FAK rate
   (c) Density based rate
   (d) Tapered rate
4. Name at least five recent TL and LTL common carrier pricing innovations that have the effect of lowering transportation expense to the shipper. (Spring 1981)
5. Carloads of rare hardwood lumber are ordered from a lumber mill in Portland, Oregon, to a Lincoln, Nebraska, plant for the specific purpose of manufacturing and marketing fine wooden tables. These tables are all customer made and re-shipped in TL quantities to Nebraska destinations. The customer locations in Nebraska were known at the time the lumber moved from Portland. Are the TL movements within Nebraska interstate or intrastate in nature? Why? (Fall 1982)
6. Many observers have stated that since the Staggers Rail Act of 1980 there has been a decided decrease in the use of blanket rates. Define this rate form. How would elimination of this form of pricing affect the locations of some industries? (Fall 1984)
7. Accra and Phobia are two adjoining states. Inside Phobia are the cities of Nelly and Bly. A firm has been making shipments from the capital of Accra via Nelly for ultimate delivery in Bly through an interline arrangement by two carriers. The interstate movement of this traffic just became exempt by the ICC. A zealous state utility commission employee in Phobia now claims that the Nell-to-Bly move is subject to that state's regulation. Comment. (Fall 1984)
8. The Revised Interstate Commerce Act still provides authority for carriers to establish through routes and joint rates. Nonetheless, several fundamental changes, which have evolved during the past ten years, discourage the use of joint rates, especially in the railroad industry. Discuss at least two of these major changes. (Spring 1985)

Notes

1. For the official wording of many of the rate definitions used in this chapter, see 49 CFR 1312.
2. See U.S.C. 10521(a)(1).
3. *B&O. S.W.R. Co.* v. *Settle*, 260 U.S. 166, 173–4. For cases that address the issue of continuity, see *A.C.L.R. Co.* v. *Standard Oil Co.*, 275 U.S. 257; *Petroleum Products Transported Within a Single State*, 71 MCC 19; *Bausch Contract Carr.* App., 2 MCC 4; *Eldon Miller, Inc. Ext. — Illinois*, 63 MCC 133. For informative discussion of specific applications, see *Traffic World's Questions and Answers* 27–28 (Washington, D.C.: Traffic Service Corp., 1979).
4. See *Farmers Union* v. *Kansas Commission*, 302 F. Supp. 778, 783 (1969); *Indiana Transit Service, Inc.* v. *Feature Film Service, Inc.*, 111 MCC 544 (1967); see also Kenneth U. Flood et al., *Transportation Management* 4th ed. (Dubuque, Iowa: Wm. C. Brown), 1984, p. 48.
5. See *Southern Pacific Transportation Co.* v. *ICC*, 565 F. 2d 615 (1977); see also Flood, *Transportation Management*, p. 55.
6. 49 U.S.C. 10501(b)(1).
7. *Pennsylvania R. Co.* v. *Public Utilities Commission of Ohio*, 298 U.S. 170, 177 (1936).

8. See *Motor Transportation of Property Within Single State*, 94 MCC 541 (1964); *Commercial Carrier Corporation Extension*, 103 MCC 787, 795 (1967); *Behnken Truck Service Extension*, 103 MCC 787, 795 (1967); *Long Beach Banana Distributors* v. *Atchison, T. & S. F. Railway Co.*, 407 F. 2d 1173 (1969); see also Flood, *Transportation Management*, pp. 49–52.

9. Robert M. Butler, ''Private carriers support DOT call to expand 'Armstrong' definition,'' *Traffic World* (October 13, 1986), pp. 33–34; see also, ICC MC-C-10963, served April 23, 1986.

10. *Penn. Pub. Util Comm.* v. *Hudson Transp. Co.*, 83 MCC 729; *Hudson Transp. Co.* v. *U.S.* and *Arrow Corp.* v. *U.S.*, 219 F. Supp. 43, aff'd 375 U.S. 452; *Wooleyhan Transport Co.* v. *George Rutledge Co.*, 162 F. 2d 1016.

11. *Applicable Rates on Intrastate Shipments Moving via an Interstate Route*, 335 ICC 472 (1969).

12. For a comprehensive and provocative discussion of this point as it applies to the railroad industry, see John C. Spychalski, ''Paradoxes of the revolution in railway intramodal relationships,'' paper presented before Transportation and Public Utilities Group, American Economic Association (San Francisco: December 29, 1983); see also ''Progress, inconsistencies, and neglect in the social control of railway freight transport,'' *Journal of Economic Issues* XVII no. 2 (June 1983).

13. See 49 CFR 10703, 10705.

14. See 49 CFR 10705a; Spychalski, ''Paradoxes,'' pp. 11–12; *Chesapeake and Ohio Railway Co., et al.* v. *ICC*, 704 F. 2d 379 (1983); Interstate Commerce Commission, *Joint Rates Study*, pp. 14–15; Porter K. Wheeler and Ronald L. Freeland, ''Joint rate cancellations since the Staggers Rail Act of 1980,'' *Proceedings—Twenty-seventh Annual Meeting Transportation Research Forum* 27 (1986), pp. 122–130.

15. ''Deregulation: A new era for rail marketing,'' *Progressive Railroading* (October 1983), p. 36.

16. Ex Parte 445 (Sub-No. 1) *Intramodal Rail Competition*, September 11, 1985; see also Ex Parte 456, *Staggers Rail Act of 1980*.

17. See ''Statement of rates and fares'' 49 CFR 1312.14.

18. 367 ICC 235 (1983).

19. For further discussion, see Richard S. M. Emrich and Vernon J. Haan, ''Implementing Staggers: A free rail transportation market through proportional rates,'' *Traffic World* (December 3, 1984), pp. 111–120.

20. *Coal from Kentucky, Virginia and West Virginia, to Virginia* 308 ICC 99 (1959).

21. For further details about this and other tariff rules applications, see Ray Bohman, Jr., *Guide to Cutting Your Freight Transportation Costs Under Trucking Deregulation* 2d ed. (Gardner, Massachusetts: Bohman Industrial Traffic Consultants, Inc., 1982).

22. William J. Augello, *Freight Claims in Plain English*, rev. ed. (New York: Shippers National Freight Claim Conference, 1982), pp. 183–184.

23. See James H. Stone and Linda M. Yunashko, Stone Management Consultants, ''Single factor rate structures can reduce administrative costs for both shipper and carrier,'' *Traffic World* (December 2, 1985), pp. 102–105.

part *two*

Planning and Operations

*O*ur attention shifts in Part Two from the external business environment in which traffic must operate to the internal decision environment of the traffic organization. The eleven chapters that constitute Part Two focus on planning and operations. Chapter 6 introduces transportation planning and discusses how this process can be helpful to traffic management. It anchors the remaining chapters, which present elements of traffic planning and operations supportive of the overall strategic direction of the firm.

Generally speaking, traffic management has three major areas of strategic opportunity today: contracting, consolidation, and information processing. The first two of these areas are addressed in Part Two. Although information-processing issues are embedded throughout the book, Part Three gives special attention to them.

Chapter 7 investigates the shipping process and related topics, including documentation, terms of sale, rate retrieval, and freight payment. Chapters 8 and 9 concentrate on contracting, costing, and negotiation concepts and issues.

Chapter 10 looks at freight consolidation programs. The material presents consolidation techniques but focuses mainly on the concepts, principles, and issues related to program design. Chapters 11 through 13 explore special carrier services, carrier liability for loss and damage, and claims administration. Chapter 14 focuses on the important decisions and management responsibilities that confront traffic managers when they consider private carriage. Finally, Chapters 15 and 16 introduce two special topics of interest: international traffic management and hazardous material goods movement.

chapter 6

Transportation Planning

chapter objectives

After reading this chapter, you will understand:
The transportation planning framework.
The difference between strategic, managerial, and operational planning.
Methods for developing an effective strategic business plan for traffic management.

chapter outline

Introduction

General Planning Framework
Strategic planning
Managerial planning
Operational planning
Managerial and operational control

Developing a Strategic Business Plan for the Traffic Area
Step 1: Management overview
Step 2: Assumptions and issues

Step 3: Strategies and major programs
Step 4: Resource requirements
Step 5: Key indicators
Step 6: Contingency plans
Step 7: Implementation
Step 8: Control

Summary

Questions

Introduction.

The word "strategy" has been a business buzzword for more than a decade, and rightly so. In simple terms, a strategy is a means to an end. In a business sense, it represents any action or set of actions designed to help achieve an organization's goals. "Strategic planning" not only helps the firm to adapt to changing environments but represents the firm's effort to influence the future course of events. It establishes the guidelines for managerial action.

Today's shippers cannot afford to miss the significant opportunities that appropriate planning can identify. Traffic managers, moreover, cannot confine their attention to planning issues related only to the traffic and logistics functions. Rather, they must look beyond these boundaries to address the strategic elements of transportation that affect the firm as a whole. In other words, today's traffic managers must be competent in strategic, managerial, and operational planning.

This chapter amplifies the traffic management decision-making framework presented in Chapter 1. The first part of the chapter investigates the general nature of planning as it relates to the development of transportation strategies.

144

The second part discusses how to develop a strategic business plan for the traffic area.

General Planning Framework

Exhibit 6.1 identifies the relationship between the framework of traffic management and the basic kinds of planning. The left side of the exhibit contains each of the elements that constitute the firm's internal environment; it shows traffic planning, operations, and control as integral parts of a firm's logistical and corporate strategies. The right side relates the planning continuum to traffic management.[1]

The concept of a planning continuum embodies three important guidelines. First, managers must coordinate planning at all levels (strategic, managerial, and operational). Second, planning should involve managers at all levels; it ought not to be reserved exclusively for some high-level, elite group of strategic planners. Although some degree of expert guidance may be useful, ultimately, any planning process will depend heavily on effective communication with, and the cooperation of, many people throughout the firm. Third, managerial and operational control needs to be an integral part of this coordinated effort.

Strategic Planning

Strategic planning sets broad guidelines for managerial planning. Functional managers in logistics, marketing, manufacturing, and finance devise coordinated strategies, plans, and policies that are consistent with overall corporate goals and objectives.

Historically, the traffic function has not been among the front-runners in participating in the corporate strategic planning process. As discussed in Chapter 1, traffic managers have had little opportunity to exercise creative decision making before the 1980s. They generally reacted to objectives set by marketing, manufacturing, and finance. The business environment of the 1980s, however, has made many firms recognize that the effective use of transportation can result in a significant competitive advantage in either lowering the cost of doing business or providing a level of customer service better than that offered by the competition. As a result, both logistics and traffic managers are becoming involved more frequently in the overall corporate strategic planning process.

Although traffic managers may contribute to corporate strategic planning, they are finding that strategic planning is a meaningful tool for establishing suitable directions for the traffic area itself. As such, this task represents a relatively new and important responsibility for traffic managers. As one of the coauthors of a major study of transportation strategies for the 1980s has observed:

> *New attitudes, skills, and resources are required. Through systematic transportation management, companies can turn a traditional regulatory staff function into an opportunity to achieve competitive advantages in cost and service.*[2]

Exhibit 6.1 Transportation planning framework

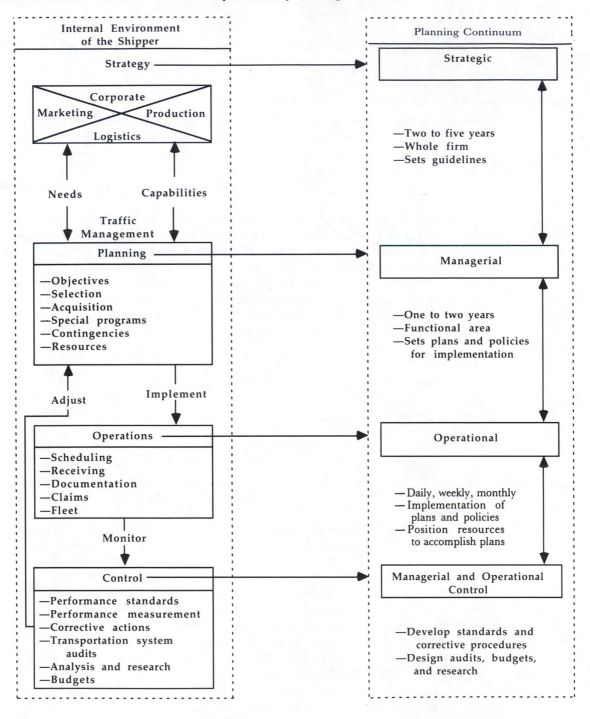

Planning horizon. Strategic planning extends over a relatively long period, two to five years, for example. The long-range nature of strategic planning means managers generally face considerable uncertainty about forecasts of future events. This uncertainty requires an element of planning flexibility that permits continual updating and revision.

Planning focus and issues. In general, the planning process relates to the firm as a whole and to broad, general issues such as (1) achieving superior financial performance and relevant market share; (2) creating barriers to entry; and (3) developing a balanced, effective portfolio of lines of business. Specific areas of concern to the traffic manager will be discussed in the section on the development of a strategic business plan for the traffic area of the firm.

Managerial Planning

Included here is the development of plans, policies, and programs for implementation and control of transportation operations. The focus is specifically on the traffic area and includes plans and priorities for a number of important traffic functions.

Planning horizon. Managerial planning takes place during a shorter period, about one to two years. By definition, this aspect of the planning process will need to be consistent with the overall directions and priorities established in the strategic planning phase.

Planning focus and issues. At issue here is the need to direct attention to the objectives and resources of the traffic area itself. Included may be areas such as (1) planning and administration, (2) rates and services, (3) traffic operations, and (4) proprietary transportation.

Operational Planning

This level of planning is related directly to the performance of necessary daily activities within the traffic management function. The principal purpose of operational planning is to see that all the resources are positioned to accomplish the activities that must take place.

Planning horizon. This part of the planning process encompasses a daily, weekly, or monthly time horizon. In comparison to strategic and managerial planning, operational planning involves a far shorter time period for planning purposes.

Planning focus and issues. The operational planning function within the traffic area has three principal components: (1) implementing operational changes to meet the goals and objectives identified in the managerial and strategic planning processes, (2) managing resources, primarily labor and fixed assets; and (3) evaluating performance and, if necessary, replanning.[3]

Managerial and Operational Control

Effective implementation of the strategic, managerial, and operational planning processes requires a control system consisting of the following elements: (1) a set of standards to measure progress toward organizational goals, (2) a system of measurements to compare actual performance against standards, and (3) a mechanism for correcting deviations from standards. The control process may include annual reviews of all matters pertaining to transportation planning and operations, as well as analysis and research relating to new alternatives for transportation services.

The purpose of managerial and operational control is both to assist in providing feedback to update and revise portions of the planning process and to address problem areas. It also links the entire planning framework into a coordinated whole. In addition, planning horizons are not to be regarded as fixed but as flexible to the needs and situations of individual companies. Depending on the circumstances at hand, shorter or longer time frames may be advisable.

The next section of this chapter focuses on the task of developing a workable business plan for the traffic area of the firm. The approach will recognize the strategic importance of the traffic function to the firm and will provide a framework for implementing strategic planning within the traffic function.

Developing a Strategic Business Plan for the Traffic Area

Before discussing the business planning process itself, a few important points regarding terminology are in order. First, the term "strategic management" may be thought of as the overall process by which an organization's strategies are formulated, evaluated, implemented, and controlled in a manner consistent with its mission, goals, and objectives, as well as its present and projected environment. Strategic planning is viewed as a forerunner of strategic management, and developing a strategic business plan is today regarded as one of the most important elements of the strategic management process.[4] The term "strategic planning" is commonly used by people in business today to refer to the recurring set of activities that are performed as part of the planning continuum discussed in reference to Exhibit 6.1.

The tangible output of the strategic planning process is sometimes referred to as a "strategic business plan," a term often shortened to simply "business plan." This section focuses on the need for the traffic function to document its planning effort in a formal business plan. The goal is not to be so consumed by planning that attention is directed away from the tasks of operations and control, but to see that overall directions and priorities in traffic and transportation are thought out well in advance of implementation.

A business plan, properly conceived, will serve well as a daily guide for those involved in operations and control. Effective business plans, unlike those designed simply to appear bulky and gather dust on the shelf, will be of continual assistance to managers throughout the traffic and logistics areas.

In addition, the business plan should be used as an educational vehicle to prove to logistics and corporate management that traffic management decisions are being made in a manner consistent with a predetermined set of priorities, and that alternatives and consequences have all been given advance consideration.

Another valuable use of an effective business plan will be apparent during the conduct of periodic budget hearings or at the time that requests for resources are being made to corporate and logistics management. An understandable, logical business plan can be an exceptionally effective tool in attempting to secure funding and support for needed projects in the traffic area.

Outlined in this section is an eight-step approach to the development of a business plan for the traffic function. Included among the steps are the essential elements for developing a business plan suitable for any area of activity within the firm. The title of each step is also an appropriate title for each major section of the business plan. **Exhibit 6.2** summarizes the various steps.

Step 1: Management Overview

Any business plan should begin with a number of introductory or background comments to provide an initial basis for understanding the scope and importance of the functional area being considered. When developing a business plan for the traffic area, it is usually helpful to highlight the importance of activities in traffic and distribution, and to stress their relevance to the attainment of goals and objectives that have been set for marketing, logistics, and the corporation as a whole. This provides the plan's users with a sound understanding of how decision making in the traffic area can have a broad impact on the pursuit of other goals throughout the firm.

The management overview should also include a precise statement of the principal purpose served by the traffic function, in effect, its reason for existence. This statement of purpose, or "mission statement," establishes an overall direction for the traffic function and makes it easier to determine the contributions that must be made in the traffic area to achieve corporate and logistical goals. **Exhibit 6.3** shows how a mission statement might evolve for traffic management. Although traffic executives might like to think their firms are at the stage III or IV level, many firms remain relatively unsophisticated at the stage I or II level.

■■■■ *Exhibit 6.2* **Steps in developing an effective strategic business plan**

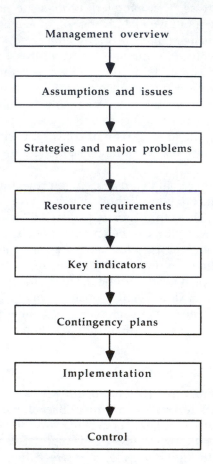

■■■■ *Exhibit 6.3* **Evolutionary stages of a mission statement**

Stage I	To see that the purchase and use of transportation services are at minimum cost and in support of the firm's logistical mission and overall corporate goals and objectives
Stage II	To be a significant profit center within the logistics area of the firm
Stage III	To provide responsive and profitable transportation services of sufficiently high quality so that activities in the traffic and logistics areas may be viewed by the firm's customers as a key competitive strength
Stage IV	To be one of the areas of significant strategic activity within the firm—one around which other strategic priorities of the firm revolve

This section should identify the principal goals established for the traffic function. Generally speaking, the goals should be divided into two categories: (1) those relating to service and (2) those relating to financial resources. Service-oriented goals should include the following:

- Transit time performance
- On-time pickup and delivery
- Over, short, and damage (OS&D), and claim activity
- Tracing and expediting capabilities
- Customer satisfaction

Financially oriented goals should be structured so as to include the effect of traffic decisions on

- Cost
- Profit
- Investment
- Asset utilization

As discussed in a later section of this chapter, the selection and use of appropriate "key indicators" will facilitate keeping track of the extent to which the goals are being met.

Sometimes a business plan distinguishes between "goals" and "objectives." When this occurs, a goal generally represents a broad statement of what is to be achieved, whereas an objective tends to be more specific. Sometimes, objectives are regarded as subsets of goals. Regardless of the terminology used, sound planning requires that goals and objectives for the traffic area be consistent with the overall mission for the traffic area, as well as with the mission, goals, and objectives for the logistics area and the firm as a whole.

Finally, efforts should be made to logically subdivide traffic activities into a number of "strategic business units" (SBUs), each of which can have its own goals and objectives. This subdivision might involve the segmentation of traffic responsibility by product type, freight lane, carrier type, or some other means. The benefit is that a number of goals and objectives may be identified and each customized to a set of specific types of activities.

Exhibit 6.4 shows the transportation mission and the transportation objectives of the Chevrolet-Pontiac-Canada Group of General Motors Corporation.

Step 2: Assumptions and Issues

It is necessary to identify the environmental factors, both internal and external, that will affect the ability of the traffic function to carry out its activities and achieve its goals. Although there are a number of ways of expressing these factors (for example, assumptions, issues, risks, opportunities), they should repre-

TRANSPORTATION MISSION

Plan, manage, and co-ordinate a quality transportation network that will assure the delivery of products to our customers in the most effective manner; thereby providing the CPC group a significant competitive advantage

signature

DIRECTOR OF TRANSPORTATION

TRANSPORTATION OBJECTIVES

- **Improve the level of service and preserve the quality of products in the transportation network**

- **Effect the most economical transportation system consistent with our customer requirements.**

- **Encourage the participation of individuals to contribute to the achievement of CPC objectives.**

- **Expand transportation involvement in forward planning activities.**

- **Encourage innovation and the implementation of new technologies to maximize profit contribution.**

- **Improve productivity through the expanded use of information systems in transportation.**

- **Utilize strategic planning to ensure transportation support of the CPC mission.**

Source: Courtesy of General Motors Corporation, Chevrolet-Pontiac-Canada Group. Reprinted by permission.

sent all conditions beyond the immediate control of the traffic area. Among the most noteworthy of these environmental factors are the following:

- Market for transportation services
- Price-service offerings by suppliers of transportation services
- Customer demand patterns for outbound finished products
- Production schedules for inbound raw materials, assemblies, component parts, and so on
- Labor productivity
- Overall logistical priorities within the firm
- Competitors' strategies in traffic and logistics
- General economic conditions

The real value of this step, once the factors have been identified, is to assess the impact that each factor would have on the ability of the traffic area to achieve its goals. If severely adverse effects are possible, it may be useful to develop programs to minimize these. For example, if a strike or work slowdown among certain motor common carriers is a possibility, developing alternative transportation arrangements is a good idea. Similarly, if a company depends heavily on a private fleet of leased (rather than owned) vehicles, monitoring the likelihood of tax law changes as they apply to the leasing of transportation equipment is necessary.

In addition to an examination of factors external to the firm's traffic area, a comprehensive environmental assessment should include a formal transportation audit, which has been identified as "the underpinning of the transportation strategic planning process."[5] **Exhibit 6.5** shows a transportation audit checklist, which directs attention to the areas of operations, policy and planning, systems and procedures, and management and organization.

Developing sound and responsive contingency plans, as discussed in step 6, will also help greatly in reacting to anticipated, and unanticipated, occurrences.

Step 3: Strategies and Major Programs

Once the mission, goals, and environmental factors have been identified for the traffic function, attention should be directed toward the development of formal strategies and major programs. Many of the decisions made in the traffic area truly are of strategic significance and have far-reaching effects on the ability of a firm to achieve its logistical and corporate goals and objectives. As shown in **Exhibit 6.6**, strategic decision areas involve the selection and acquisition of transportation services and the availability of resources. Specific issues not only important to the traffic area, but also linked closely to a firm's logistical and marketing strategies, are: (1) mode choice; (2) carrier selection; (3) use of a private fleet versus for-hire transportation services; (4) method of purchase; (5)

Exhibit 6.5 **Transportation audit checklist**

Operations
— Freight flows
— Modal/carrier selection and use
— Costs
— Service performance

Policy and Planning
— Customer service requirements
— Financial objectives
— Terms of trade

Systems and Procedures
— Management information needs
— Administrative support systems
— Analytic tools

Management and Organization
— Staffing and training
— Coordination mechanisms
— Management controls
— Organizational design

Source: Temple, Barker and Sloane, Inc., *Transportation Strategies for the Eighties* (Oak Brook, Illinois: National Council of Physical Distribution Management, 1982), p. 176. Reprinted by permission.

terms of trade; and (6) nature and availability of information, personnel, organizational, facility, and financial resources. Principal concerns closely related to these issues include the following:

- Maintenance of appropriate service levels
- Unit cost control
- An intensive versus a broad-based approach to carrier selection
- Contracting methods and length of commitment
- Lease or purchase of equipment
- Use of intermediaries
- Productivity of personnel, equipment, and financial resources
- Control over loss and damage, and over claims
- Freight bill payment, audit, and analysis
- Carrier evaluation

Although the importance of each of these strategic decision areas varies from firm to firm, this list is helpful because it suggests the kinds of transportation strategies that formal programs need to evaluate. In addition, **Exhibit 6.7** provides a framework for assessing the impact of environmental influences on strategic decision areas within the traffic area. Most important, however, is to re-

■■■■■■ *Exhibit 6.6* **Strategic transportation issues**

Strategic Decision Area	Issues
Selection	Mode choice (including multimodal and intermodal options)
	Carrier selection — broad-based v. intensive approach
	Private fleet operations
Acquisition	Methods of purchasing transportation
	— Negotiate tariff rates
	— Contracts — competitive bid or negotiations; length of commitment
	— Private fleets — lease or buy
	— Use of intermediaries — agents, brokers, freight forwarders, shipper associations
	Terms of trade
Resources	Information systems
	Organization — staffing, centralized v. decentralized; dedicated transportation group
	Investment in facilities and equipment
	Financial

Source: Based on Temple, Barker, and Sloane, Inc., *Strategies for the Eighties* (Oak Brook, Illinois: National Council of Physical Distribution Management, 1982), pp. 180–189.

member that strategies and major programs are simply the means by which the mission and goals will be accomplished.

Strategy evaluation. Once the business strategies for the traffic area have been formulated, each must undergo a careful examination. The first issue to raise is whether adopting the strategy is likely to lead to achieving the goals and objectives of the traffic function or of specific activities within traffic.

Second, it is useful to evaluate each strategy in light of possible adverse environmental factors and to test each for sensitivity to likely states of the environment. Several examples of appropriate criteria to use when evaluating strategic decision areas are shown in **Exhibit 6.8.**

It is also necessary to compare the strategies with one another to search for possible areas of conflict or contradiction. For example, assume a firm wants to increase market share by maintaining high service levels but, at the same time, wishes to keep costs at a minimum. As a result, the firm chooses to move volume shipments by rail to a large number of public warehouses in areas where customers are concentrated. Service levels are high because of the availability of

■■■■■■ *Exhibit 6.7* **Key issue identification matrix**

	Influences on Transportation Strategy		
Strategic Decision Area	**Internal Strategies**	**External Pressures**	**Carrier Strategies**
Selection Mode choice Carrier selection Private fleet operations			
Acquisition Tariff or contracts Lease or buy Use of intermediaries Terms of trade	Rate each impact: 1 = Critical impact 2 = Some impact 3 = Little or no impact		
Resources Informations systems Organization Facilities and equipment Financial			

Source: Adapted from Temple, Barker and Sloane, Inc., *Transportation Strategies for the Eighties*, Oak Brook, Illinois: National Council of Physical Distribution Management, 1982, p. 181. Reprinted by permission.

inventory, but the shipment damage by rail may not be acceptable. In other words, the benefits of maintaining the high levels of inventory may be more than offset by the dollar value of the damage. A preferable alternative is to use higher cost truck transport to company-owned or operated facilities or perhaps directly to customers. A well-executed evaluation would not only identify such a contradiction in advance but also help to resolve the issue. If the strategies themselves

■■■■■■ *Exhibit 6.8* **Strategic transportation elements and evaluative criteria**

Strategic Decision Area	*Typical Evaluative Criteria*
Selection and acquisition	Transportation cost and service attributes Stability of supply Leverage with carriers Flexibility and risk of commitments Ease of administration
Resources	Ability of staff Ability to monitor performance Ability to plan for changing environment

cannot be reconciled without great difficulty, it is necessary to consider restating the goals and objectives.

Step 4: Resource Requirements

Each of the major programs should be subjected to a careful analysis of its (1) initial investment needs in terms of funding, people, and equipment; (2) continuing and periodic resource needs; and (3) overall financial viability. In the case of this last one, it is necessary to justify the use of resources for each project and to show conclusively that the value to the firm of going forward with the project exceeds that of other investment opportunities for the traffic area and for the firm as a whole.

While each project undergoes scrutiny, it is appropriate to develop pro forma financial statements for the traffic function. Prospective balance sheet and income statement information are helpful for managing overall activity as well as for communicating financial details of the traffic area to upper-level management.

Step 5: Key Indicators

To measure, monitor, and control the extent to which major goals are being achieved, it is particularly useful to identify a concise set of "key indicators" that can be observed regularly over time. These performance gauges, which should be as simple and straightforward as possible, serve three main purposes. First, they represent an excellent source of feedback information for line managers having various responsibilities within the traffic area. Second, they are capable of determining over time whether strategies and major programs are on target and helping to achieve traffic goals. Third, they are useful for keeping top management regularly informed about whether progress is being made toward achieving these goals. In effect, the key indicators represent a mechanism for identifying and spotting changes that can have significant effects on the traffic function, as well as the firm as a whole. (This topic will be treated more fully in Chapter 17.)

Step 6: Contingency Plans

Another component of a well-developed business plan is a section devoted to the topic of contingency planning. Contingency plans reflect advanced preparations for handling emergencies not considered in the regular planning process."[6] Since the regular planning process is based principally on the most probable events (as discussed above), it makes good business sense to make some attempt to be prepared for the unexpected.

To be effective, contingency planning should first identify the range of emergency situations that could occur and that would justify the development in advance of feasible, defensive strategies. Then, in the event the contingency does occur, it is important to develop a calculated plan for action.

Candidate areas for contingency planning are often identified by postulating a number of "what if" questions. For example: "What if organized labor were to go on strike? What if we were to suffer a major accident involving one of our private fleet vehicles? What if our communications system were to become totally inoperative? What if one of our distribution centers were to be hit by a tornado? These occurrences may not be very likely, but they might spell disaster for a business that had made no advance preparation.

Two important attributes of contingency plans are (1) that they help to eliminate fumbling, uncertainty, and delays in making the needed response to an emergency, and (2) that they make such responses more rational.

Step 7: Implementation

Perhaps the most critical juncture in the business planning process is the translation of plans into action, the making of real-world business decisions in a manner consistent with the mission, goals, strategies, and programs that have been formulated for the traffic area. Many studies have been done on why plans sometimes fail (see **Exhibit 6.9**). Perhaps the most common reason is that attention is focused on the planning process itself, to the exclusion of effective implementation.[7]

Actions that should help to improve the chances for successful implementation include[8]

- The use of budgets and flexible budget systems
- Project plans
- Management by objectives (MBO)
- Other tactical planning tools such as Gantt milestone charts for scheduling tasks and network charts for more complex interrelationships

Prerequisites to the successful implementation of traffic strategies include the consent and approval of corporate top management, as well as the interest, involvement, and support of various key people throughout the logistics area. Implementation should be conducted using a time-sequenced plan, and the overall timing of the effort must be conducive to success. Given that effective management of the traffic function is an essential ingredient of a responsive logistical system, the process of implementation will certainly be a challenging one.

Step 8: Control

The process of control requires comparing achieved results with expected results and, to the extent that discrepancies exist, taking corrective action. Once the

■■■■■■ *Exhibit 6.9* **Reasons that planning fails***

1. Corporate planning has not been integrated into the firm's total management system.[†]
2. There is a lack of understanding of the different dimensions of planning.
3. Management at different levels in the organization has not properly engaged in or contributed to the planning activities.
4. The responsibility for planning is often wrongly vested solely in the planning department.
5. In many companies, management expects that the plans as developed will be realized.
6. In starting formal planning, too much is attempted at once.
7. Management fails to operate by the plan.
8. Extrapolation and financial projections are confused with planning.
9. Inadequate inputs are used in the planning.
10. Many companies fail to see the overall picture of planning and get hung up on details.

*From a study of 350 companies in Europe and the United States.

[†]Although the reasons cited are phrased in terms suitable for corporate planning purposes, they are equally applicable to the areas of traffic and logistics.

Source: Adapted by permission from Kjell A. Ringbakk, ''Why Planning Failed,'' *European Business*, Spring, 1971.

elements of the plan are implemented, it is important to see that expected results are actually achieved and, if they are not, to understand why. Over short periods, the key indicators should be the measures observed most carefully. Over longer periods, attention should be focused directly on the extent to which traffic function goals are achieved.

By using effective tools, such as summary and exception reports and budget variance analysis, the traffic manager can receive important information on a day-to-day basis, as well as be kept informed about the progress toward meeting long-term goals. This monitoring may provide information for (1) identifying weaknesses and suggesting changes in the methods used for strategy implementation, (2) reformulating or reorientating certain strategies, or (3) revising the goals or objectives that were to be achieved. Thus, it is clear that the feedback process relies heavily on a timely comparison of actual versus expected results and on the ability of management to determine when changes to goals, strategies, or implementation mechanisms are in order.

Although this definition of control is simple and straightforward and is widely used in industry today, one of its liabilities is that it offers little guidance in determining when the difference between actual results and expected results is large enough to justify corrective action. A growing constituency therefore believes this concept of control can be improved significantly by incorporating the additional steps included in statistical process control (SPC). (Chapter 17 includes additional detail on this topic.)

Although adherence to these eight steps should provide the basis for a well-developed, comprehensive business plan, each traffic manager should adapt the steps to the unique aspects of his or her company. For example, if there were well-defined statements of mission, goals, and objectives for the firm as a whole and for other functional areas (including logistics), adding a section that deals specifically with this subject matter would be useful. This effort would allow the users of the business plan to have a better understanding of the context in which the priorities and directions for the traffic area have been established.

The business plan should be used daily as a blueprint for action and will need to be updated regularly. For this reason, a quarterly review of activity and progress should provide sufficient opportunity to compare actual performance with the goals and objectives set. At the time of this review, minor modifications and adjustments to the plan may be accommodated easily. On a broader scale, the overall plan needs to be reviewed at least once yearly so that any major changes that may be necessary can be incorporated. In this way, the plan continues to be current and reflective of what the traffic function is attempting to accomplish.

Summary

This chapter relates the key components of strategic planning to traffic management. Strategic planning has been used for some time on a firmwide basis and in the major functional areas, but there has been limited application of this valuable tool in relation to the traffic function.

Following a brief discussion of the role and importance of the planning process as it relates to the traffic function, this chapter identifies three specific types of planning: strategic, managerial, and operational. Each of these is discussed in terms of purpose, planning horizon, and planning focus and issues.

A planning continuum is defined as consisting of these three types of planning, along with the issue of managerial and operational control. A transportation planning framework is developed by understanding the relationship between the planning continuum and the internal environment of the shipper.

To understand how to implement the strategic planning process, this chapter includes an approach for developing a strategic business plan for the traffic area of the firm. The strategic business planning process is defined as consisting of the following steps: management overview, assumptions and issues, strategies and major programs, resource requirements, key indicators, contingency plans, implementation, and control.

Although the traffic area of the firm has not been among the front-runners of implementing formal approaches to planning, there are signs that strategic planning is becoming more accepted as a valuable, effective tool. To the extent that this occurs, the traffic functions will be in a better position to achieve desired goals and objectives amid the pressures of changing internal and external environments.

Selected Topical Questions from Past AST&L Examinations

1. Assume you are the traffic manager of a large manufacturer of household appliances. What information would you want to know before making your carrier selection decision? (Spring 1980)
2. What, if any, is the relationship between deregulation of transportation and logistic management, or what future changes are possible between carriers and logistics (materials) management? (Fall 1981)
3. You are the vice-president of logistics for a large auto firm. Recently, you received a memo from the chief executive officer asking you to prepare a report:
   (a) Briefly outlining the major provisions of the recent deregulation laws in the trucking and railroad industries.
   (b) Discussing the impact these regulatory changes are having on your job. (Fall 1982)
4. The chief executive officer recently said that his firm had to react to two key distribution influences since 1973. "High fuel costs," he said, "made us react against fast premium transportation, while recent high prime interest rates made us react against our warehouse system." Comment and elaborate. (Fall 1984)

Notes

1. The framework of this section generally follows the scheme established by James A. Constantin in "Planning perceptions and concepts," chap. 5, and "A framework for logistics planning," chap. 6, both of which appeared in James F. Robeson and Robert G. House, eds., *The Distribution Handbook* (New York: The Free Press, 1985).
2. Lewis M. Schneider, "New era for transportation strategy," *Harvard Business Review* 63 no. 2 (March–April, 1985), p. 126.
3. Paul H. Zinszer, "Operational planning," chap. 7 in James F. Robeson and Robert G. House, eds., *The Distribution Handbook* (New York: The Free Press, 1985).
4. For further discussion about the relationship between strategic management and strategic planning as they apply to transportation and physical distribution, see C. John Langley Jr., "Strategic management in transportation and physical distribution," *Transportation Journal* 22 no. 3 (Spring, 1983), pp. 71–78; C. John Langley Jr., "A framework for applying the strategic management process to logistics management," *Logistics: Concepts and Applications, Proceedings of the Logistics Resource Forum, 1982* (Cleveland: Leaseway Transportation

Corporation, 1982); and George A. Steiner, *Strategic Planning* (New York: The Free Press, 1979).

5. Temple, Barker and Sloane, Inc., *Transportation Strategies for the Eighties* (Oak Brook, Illinois: National Council of Physical Distribution Management, 1982), p. 176.

6. Dr. Bernard J. Hale, ''Contingency planning,'' chap. 8 in James F. Robeson and Robert G. House, eds., *The Distribution Handbook* (New York: The Free Press, 1985), p. 118.

7. See Michael A. McGinnis, ''Strategic planning— Dodging the traps,'' *Canadian Transportation and Distribution Management* (May, 1985), pp. 41–42.

8. George A. Steiner, ''Translating strategic plans into current decisions,'' chap. 13, in *Strategic Planning* (New York: The Free Press, 1979), pp. 215–228.

chapter 7

Making the Shipment

chapter objectives

After reading this chapter, you will understand:

The purposes and applications of the four basic shipping documents.

The delivery terms of sale and the implications of these terms for traffic management.

How to use transportation services and special documents to secure payment of goods before delivery.

The nature of, and the options for, basic rate retrieval and freight payment systems.

The key considerations for carrier credit arrangements.

Introduction

Documentation
Bill of lading
Waybill
Freight bill
Delivery receipt
Documents used for contract,
exempt, and private carriage

Terms of Sale
Delivery terms
Planning considerations

*Using the Carrier to Secure Payment
for the Goods*

*Rate Retrieval and Freight
Payment Systems*

Carrier Credit Arrangements

Summary

Questions

*Appendix 7.1—Bill
of Lading Preparation*

Introduction

Several legal and institutional elements affect the task of preparing, initiating, and paying for a shipment. Standard shipping documents represent a key part of the entire shipping process and the terms of sale. Shippers often use carriers and special documents to secure payment from the consignee for the goods. Rate-retrieval and rate-audit systems play significant roles in the shipment process, and carrier credit is an important consideration. These elements comprise the fundamental considerations for making the shipment and are critical to sound traffic planning.

Documentation

Like other commercial enterprise, the transportation industry uses certain kinds of documents to conduct its business. A shipment requires four main transportation documents: (1) the bill of lading, (2) the waybill, (3) the freight bill,

164

and (4) the delivery receipt. These documents, which come in a variety of forms, constitute the basic paperwork found in all shipments. Other important traffic management documents are discussed in the chapters dealing with accessorial services (Chapter 11), claims (Chapter 13), export-import shipments (Chapter 15), and hazardous goods carriage (Chapter 16).

Exhibit 7.1(a–c) identify the specific tasks involved in making inbound and outbound shipments, and **Exhibit 7.1(d)** shows how major shipping documents relate to the flow of goods. Each of the four main documents plays an important role in one or more movement activities.

■■■■■■■■■ *Exhibit 7.1* **Shipment process**

Physical Activity	*Task*
1. Traffic informed of outbound move from order processing. Obtain goods from warehouse. Obtain shipping documents from order processing or warehouse.	
	2. Preplan move if needed.
3. Order or schedule equipment from carrier.	
	4. Schedule and assign dock for loading.
5. Equipment arrives.	
6. Assign crew for loading.	
7. Plan method of loading.	
8. Plan method of bracing.	
	9. Check shipment against bill of lading, packing list, and any other documents.
10. Affix packing list to shipment.	
11. Load.	
12. Obtain signed bill of lading from carrier.	
	13. Confirm back to order processing or inventory control.
14. Vehicle or car departs.	
	15. Trace if necessary.
16. Obtain freight bill.	
17. Audit freight bill.	
18. Send bill to accounting with voucher to authorize payment.	
19. File claim if necessary.	

(a) Routine Day-to-Day Traffic Operations — Outbound Movements

(continued)

Exhibit 7.1 (continued)

Physical Activity	Task
1. Traffic informed of goods to arrive. 2. Goods arrive. 3. Determine if goods are proper to receive against authorization in step 1.	
	4. Assign unloading crew.
5. Unload. 6. Sign carrier delivery receipt. Note any loss or damage. Obtain copy.	
	7. Assign inspector or tally person if used. 8. Determine where goods are to be moved to within facility. 9. Assess movement manpower.
10. Move goods. 11. Inform purchasing or other receiving department of goods arrival. 12. Receive freight bill if done. 13. Audit freight bill. 14. Send bill to accounting with voucher to authorize payment. 15. File claim if needed.	

(b) Routine Day-to-Day Traffic Operations — Inbound Movements

Physical Activity	Task
1. Receive order for service from shipper.	
	2. Assign equipment for pickup. 3. Dispatch equipment to shipper location.
4. Equipment moves to shipper.	
Shipper loads vehicle or car.	
5. Either shipper calls for vehicle to be taken away or driver waits with equipment and leaves when loaded. 6. Bill of lading is signed with copy left with shipper.	
	7. Assign routing.
8. Weigh shipment if necessary. 9. Prepare waybill and freight bill from bill of lading.	

(continued)

Exhibit 7.1 (continued)

Physical Activity	Task
	10. Dispatch linehaul movement. 11. Monitor movement.
12. Send freight bill copy to shipper, receiver, or both. Send copy to collection point with bill of lading. 13. Collect freight bill. 14. Shipment arrives at destination terminal.	
	15. Assign delivery movement.
16. Provide delivery personnel with delivery receipt. 17. Deliver goods. 18. Obtain signed copy of delivery receipt.	
Goods unloaded.	
19. Match up bill of lading, delivery receipt, and freight bill for document storage. 20. Handle claim if necessary.	

(c) Routine Carrier Operations

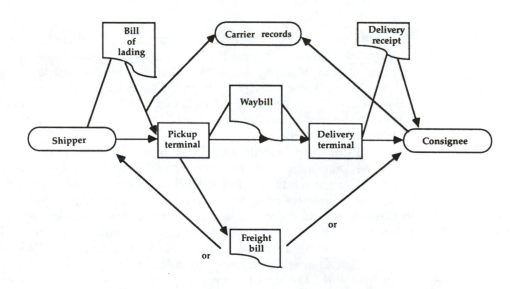

(d) Flow of Goods and Paperwork for a Shipment.

Bill of Lading

The bill of lading initiates the shipment and is the most important single document of the shipping process. It performs two key functions. First, it serves as the contract of carriage. By accepting the goods and signing this document, the carrier contractually agrees to haul the goods described from the named pickup point, via the routes indicated, to the consignee's delivery address. When a carrier does not complete the haul successfully, the offense is a breach of contract, which (as discussed in Chapter 12) constitutes the legal basis for most freight claims actions.

Second, the bill of lading serves as a receipt for the goods shipped. Without a copy signed by the carrier, the shipper would have little or no proof of carrier liability in the event the shipment was lost or destroyed.

In addition, in case of disputes, the bill of lading may provide documentary evidence of title to the goods. The presumption is that the holder of the original bill of lading is the owner of the goods. Historically, the issue of ownership for goods in transit has been significant for traffic management, because the legal custom was to tie the risk of loss and, thus, the claims administration task, directly to the passage of title. The adoption of the Uniform Commercial Code (UCC) by nearly every state, however, has diminished the significance of this issue. As discussed later in this chapter, the UCC allows the parties to establish provisions in the sales contract that decouple the risk of loss from the passage of title. Therefore, the sales contract may allow title to pass to the buyer after loading at origin, but the seller still assumes the risk of loss or damage to the shipment in transit.

Issuance. Although the Revised Interstate Commerce Act requires transportation companies performing common carriage to issue, that is, supply and prepare, the bill of lading, shippers will often preprint, prepare, and utilize their own forms (see **Exhibit 7.2**). Even with carrier supplied forms, shippers will generally fill out the bill of lading for the carrier so as to ensure accuracy of the required items and to not allow the pickup to be delayed by the driver's having to fill out the form.

Bills of lading supplied by the shipper offer several advantages. One, they can be preprinted with the shipper name, location, and specific commodity names. Incorrect description of articles shipped is, therefore, minimized. Having the proper description is especially important for hazardous materials. Two, the shipper enjoys the benefit of uniformly sized bills of lading for filing and retrieval. Three, many shippers use computers to prepare bills of lading on special computer forms. Four, all these factors lead to savings in clerical expenses. Although shippers incur these expenses and perform this function for carriers, carriers still have legal responsibility for the information on the bill of lading. Appendix 7.1 contains guidelines for preparing a bill of lading.

Exhibit 7.2 Customized bill of lading

STRAIGHT BILL OF LADING–SHORT FORM
ORIGINAL—NOT NEGOTIABLE

SHIPPER'S B L NUMBER	PART	OF

CARRIER _____ Carrier's No _____

AT _____ DATE _____ **ARMOUR-DIAL, INC.**

PROTECT FROM FREEZING

					CAR INITIAL		CAR NUMBER	
TRANSMISSION DATE	SEQUENCE NUMBER	ORDER NUMBER	SHIPMENT TO ARRIVE DATE	SHIP TO ARRIVE TIME	SHIP TO NUMBER	ACCOUNT NUMBER	CUSTOMER ORDER NUMBER	SEAL NO

CONSIGN TO   ROUTE

QUANTITY	UPC CASE CODE	WEIGHT	FRT CODE	DESCRIPTION

Subject to Section 7 of Conditions of applicable bill of lading, if this shipment is to be delivered to the consignee without recourse on the consignor, the consignor shall sign the following statement.

The carrier shall not make delivery of this shipment without payment of freight and all other lawful charges.

ARMOUR-DIAL INC

_____
(Signature of consignor)

If charges are to be prepaid write or stamp here "To be Prepaid"

TO BE PREPAID

**SHIPMENT TENDERED
SORTED AND SEGREGATED**

EXPLANATION OF FREIGHT CODES

01 Cleaning, Scouring or Washing Compounds (noibn), liquid or other than liquid STCC 28-419-15 & 20
02 Soap, noibn, liquid or other than liquid STCC 28-419-90 & 91
03 Compounds, ammonia or Ammonia, cleaning, liquid STCC 28-422-10
04 Sizing, fabric, in pressurized dispensing containers STCC 28-911-72
06 Compounds, buffing or polishing, NOI, including Boat, Floor, Furniture or Vehicle Polish or Wax STCC 28-423-10
14 Toilet Preparations—Shampoo. See Note STCC 28-441-45
15 Toilet Preparations, NOI. See Note STCC 28-441-90
20 Meats, Cooked, Cured or Preserved, with or without Milk, Eggs, Vegetable or Fruit Ingredients (noibn). STCC 20-134-14
21 Bouillon Cubes. See Note. STCC 20-999-13
23 Pizza Pie Mix, consisting of flour, yeast and sauce, with or without other ingredients. STCC 20-452-15
24 Pulp or Puree, Tomato, Canned STCC 20-336-15
28 Glycerine, other than crude, in bulk, in drums or in tank cars or tank trucks. See Note. STCC 28-185-20

NOTE: Where rate is dependent on value the agreed or declared value of the property is hereby specifically stated by the shipper to be not exceeding 50 cents per pound.

PALLET OR DUNNAGE WEIGHTS

NO OF PACKAGES	WEIGHT	FRT CODE	FRT RATE	FREIGHT EXTENSION	FREIGHT CLASSIFICATION DESCRIPTION (SEE LEGEND IF BLANK)
			TOTAL		

	NO	AVERAGE	TOTAL WTS
PALLETS			
CORR BOX TYPE			
CORR SHEETS			
PLYWOOD			

THIS SHIPMENT IS PROPERLY DESCRIBED AND IS SUBJECT TO VERIFICATION IN ACCORDANCE WITH WEIGHT AGREEMENTS ON FILE WITH THE WESTERN, EASTERN, SOUTHERN AND TRANSCONTINENTAL WEIGHING AND INSPECTION BUREAUS

THE CORRECT WEIGHT IS _____ POUNDS
ARMOUR-DIAL INC
SHIPPER'S IMPRINT IN LIEU OF STAMP, NOT A PART OF BILL OF LADING APPROVED BY INTERSTATE COMMERCE COMMISSION

ARMOUR-DIAL, INC. Shipper

Per _____
PERMANENT POST OFFICE ADDRESS OF SHIPPER
TRANSPORTATION DEPT STA 3214
GREYHOUND TOWER
PHOENIX, ARIZ 85077

CARRIER - IMPORTANT NOTICE
THIS NUMBER MUST APPEAR ON
YOUR FREIGHT BILL
Mail Charges To: ARMOUR-DIAL, INC.
Dept. F. — Phoenix, Az. 85077

_____ Agent

Per _____

INVOICE FILE   **33196**

Source: Armour-Dial, Inc. Reprinted by permission.

Types. The transportation industry uses several types of bills of lading. These documents serve the special needs of the different modes of transportation or address special situations.

The *straight* bill of lading is the most basic rail and motor common carrier form. It is used for domestic shipments and for international movements to Canada and Mexico. It can be either a "long form" or a "short form" bill of lading, the only difference being its appearance. The long form has all ten bill of lading provisions printed on its back, whereas the short form omits the ten provisions. The short form is printed this way only for convenience; it still carries all the legalities of the long form.

Shippers use the *order* bill of lading to retain title to the goods and to withhold delivery until the consignee makes good on the balance due in the purchase agreement. The carrier's and the consignee's bank participate in the purchase transaction to collect the payment for goods and, often, the freight charges as well. The order bill of lading can be conveniently used when goods are shipped, but a specific consignee is not determined until the goods are en route.

The *uniform export* or *through export* bill of lading is used for shipments requiring an overland move, an ocean haul, and an overland delivery in a through movement. Although the initial carrier signs the document for the others that participate in the haul, the liability for freight loss and damage applicable to each mode (as described in Chapter 14) does not change. Export rates often apply for the overland portion when this bill of lading is used. In addition, port-handling charges are often absorbed by both the land and ocean carriers. This bill of lading exists in straight form and order form.

The *livestock* bill of lading is specially used in rail transportation of live cattle. Railroads must adhere to fairly strict regulations dealing with cattle movements, and many of these are embodied in this bill of lading. Similarly, with this document the carriers are not held liable for certain forms of loading negligence by the shipper. The decentralization of the slaughtering industry, as well as the high cost and low profitability of livestock traffic for the railroad industry, has caused this traffic and this bill of lading to virtually disappear. Now, livestock mostly moves by exempt motor carriage.

The *ocean* bill of lading is used in overseas trade by water transportation companies (see Chapter 15). The *air* bill of lading is used in both terminal-to-terminal and door-to-door air freight moves. Each carrier supplies bills of lading to shippers who are generally required to have them fully prepared by the time of pickup. These bills of lading are used in a through move where air carriers often arrange with local motor carriers for the shipper-to-air terminal and the air terminal-to-consignee legs of the move. The motor carrier leg of these through moves using an air bill of lading is generally exempt from ICC regulation, so the surface bill of lading is not appropriate here. Since air freight is exempt from economic regulation, no single air bill of lading form is required, and no single liability is standard. The shipper must be familiar with each air carrier tariff and bill of lading.

The *government* bill of lading was created by the General Accounting Office (GAO) for use by government agencies and the U.S. military when commercial carriers are used. Prepared by the government-agency shipper, the executed bill of lading constitutes a type of draft on the U.S. treasury.

Waybill

Transportation companies use waybills to control shipments. At the originating terminal, carrier personnel use information gleaned from the carrier's copy of the bill of lading to prepare the waybill. At the time of rating and billing, carriers write the waybill number on the bill of lading. This number acts as an internal control item in the carrier system; it is often used for tracing or billing control and in loss and damage processes.

A railway company waybill is shown in **Exhibit 7.3.** The majority of railroads do not send this document in paper form; instead, they transmit it electronically to computers at intermediate or destination stations where printouts are made. Railway companies use these forms to record operations-related information such as vehicle and train numbers, routing, and special instructions about hazardous materials, stopoffs, diversion orders, inspections, and so on. The waybill also serves as a receipt between interline carriers.

In the motor carrier industry, a photocopy of the bill of lading normally fulfills the function of the waybill. For control purposes, motor carriers assign a *pro number* to each shipment. This term, derived from the word *progressive*, indicates that control numbers are in consecutive order.

In the Consolidated Freightways system, for example, drivers carry pads of ''PRO'' labels. A driver sticks these labels on copies of the bill of lading and on the freight. A photocopy of the bill of lading moves with the freight and is the waybill. In addition, Consolidated Freightways' centralized computer installation sends daily manifests to each terminal. Outbound manifests, listing all outbound shipments arranged alphabetically by shipper name, include information as follows: PRO number, consignee name and location, terminal, pieces, weight, commodity description, and charges. Inbound manifests show the same information, but are sorted in consignee sequence.

Freight Bill

The freight bill is the carrier's invoice for freight charges. It is either prepaid by the shipper or collected from the consignee. Although shippers generally have a choice, carriers may insist on prepayment for certain traffic or services. In addition, shippers may choose to pay only part of the total charges.

Exhibit 7.3 A freight waybill

PLACE SPECIAL SERVICE PASTERS HERE

724 — SOUTHERN RAILWAY SYSTEM — 724
FREIGHT WAYBILL

FORM 487-C-6 PT.
REV. 11/72

STOP THIS CAR | AT | FOR | LENGTH OF CAR | | MISC. PRO. NO.
ORDERED | FURNISHED

CAR INITIALS AND NUMBER | TRAILER INITIALS AND NUMBER | CHASSIS INITIALS & NUMBER | PLAN | WAYBILL DATE | WAYBILL NUMBER

CAR TRANSFERRED TO | TRAILER INITIALS AND NUMBER | CHASSIS INITIALS & NUMBER | PLAN

TIME: | BY:

TO STATION | STATE | FROM NO. | STATION | STATE

BILLED AT NO. | STATION | STATE

ROUTE | ROUTING SHOW "A" IF AGENT'S SHOW "S" IF SHIPPER'S | FULL NAME OF SHIPPER AND ADDRESS | CUSTOMER CODE NO.

RECONSIGNED TO | STATION | STATE | **RWC**

PER

ORIGIN AND DATE, ORIGINAL CAR, TRANSFER FREIGHT BILL, AND PREVIOUS WAYBILL REFERENCE AND ROUTING WHEN REBILLED.

AUTHORITY

CONSIGNEE AND ADDRESS | CUSTOMER CODE NO.

BILL OF LADING DATE | WEIGHED

AT

BILL OF LADING NO. | GROSS

TARE

FINAL DESTINATION AND ADDITIONAL ROUTING | ALLOWANCE

INVOICE NO. | NET

SPECIAL INSTRUCTIONS — C. L. TRAFFIC

If charges are to be prepaid, write or stamp here, "TO BE PREPAID"

Subject to Section 7 of Conditions:

☐ YES | SIGNATURE OF CONSIGNOR

NOTE—Where the rate is dependent on value, shippers are required to state specifically in writing the agreed or declared value of the property. This agreed or declared value of the property is hereby specifically stated by the shipper not to exceed

PER

DATE REV. W'B PREP. | COMMODITY CODE NUMBER | FOR THE CARRIER

PER

NO. PKGS.	DESCRIPTION OF ARTICLES SPECIAL MARKS AND EXCEPTIONS	WEIGHT	RATE	FREIGHT	ADVANCES	PREPAID

DESTN. AGENT'S F/B NO.

Outbound Junction Stamps Here, Yard Stamps on Back.

DESTINATION AGENT STAMP

FIRST JUNCTION | SECOND JUNCTION | THIRD JUNCTION | FOURTH JUNCTION

724 — SOUTHERN RAILWAY SYSTEM — 724

Source: Southern Railway System. Reprinted by permission.

Section 7 of the straight bill of lading contains the "without recourse clause," which may play an important role in determining whether shippers or receivers must pay the freight bill. By signing this block, the shipper instructs the carrier to withhold delivery of the goods until the consignee pays any outstanding balance. If the carrier fails to comply with these instructions, the shipper is not liable for the freight bill. Suppose, for example, the carrier delivers the goods and later learns the consignee will not pay for the shipment. In this situation, the carrier has no recourse against the shipper. On the other hand, if the carrier attempts to collect from the consignee before the tender of delivery, the carrier has met its Section 7 obligations; therefore, it has recourse against the shipper if the consignee refuses to pay. In other words, the "without recourse" clause does not give shippers absolute protection against liability for the charges.

It is permissible to sign Section 7 when making prepaid, as well as collect, shipments. Although the shipment is prepaid, the traffic manager may want to execute Section 7 to protect the firm against any excess charges, say, from re-consignment, switching, or weighing. Carrier rules generally exclude Section 7 applications from the shipments that involve perishables, nonagency stations, or stopoff services.

Motor carriers generally combine the freight bill and the delivery receipt into a single document. As shown in **Exhibit 7.4**, for example, Yellow Freight's freight bill is the #1 copy in a set of six documents produced at the same time at considerable savings. After the driver returns to the original terminal, "billing" takes place whereby computers or pricing specialists determine the class or the commodity rate and assess charges. The billing process also initiates invoices (freight bills) for prepaid shipments. Some carriers, however, may send copies of the bill of lading to a regional or central point for freight bill preparation and for mailing or electronic transmission.

A growing number of shippers and carriers have begun to realize the promise of electronic data interchange (EDI) in this area. In fact, K Mart, one of the first companies to adopt EDI of freight bills in 1981, estimated that it reduced its operating expenses by 0.8 percent of $18.6 billion in sales.[1] About fifty motor carriers currently transmit freight bills and bills of lading electronically.[2] In addition, third-party vendors now support this process.

Delivery Receipt

The signed delivery receipt (see Exhibit 7.4) triggers the freight bill for collect shipments and is used as prima facie evidence that the carrier has met its contractual obligation to deliver the goods. The consignee, when accepting delivery of the goods, should note any visible loss or damage on the delivery receipt. Taking notes makes subsequent claims proceed more smoothly than claims based on an inspection made several days later by the carrier's agent and after the unloading crews have moved the goods into the warehouse.

Exhibit 7.4 Freight bill and delivery receipt

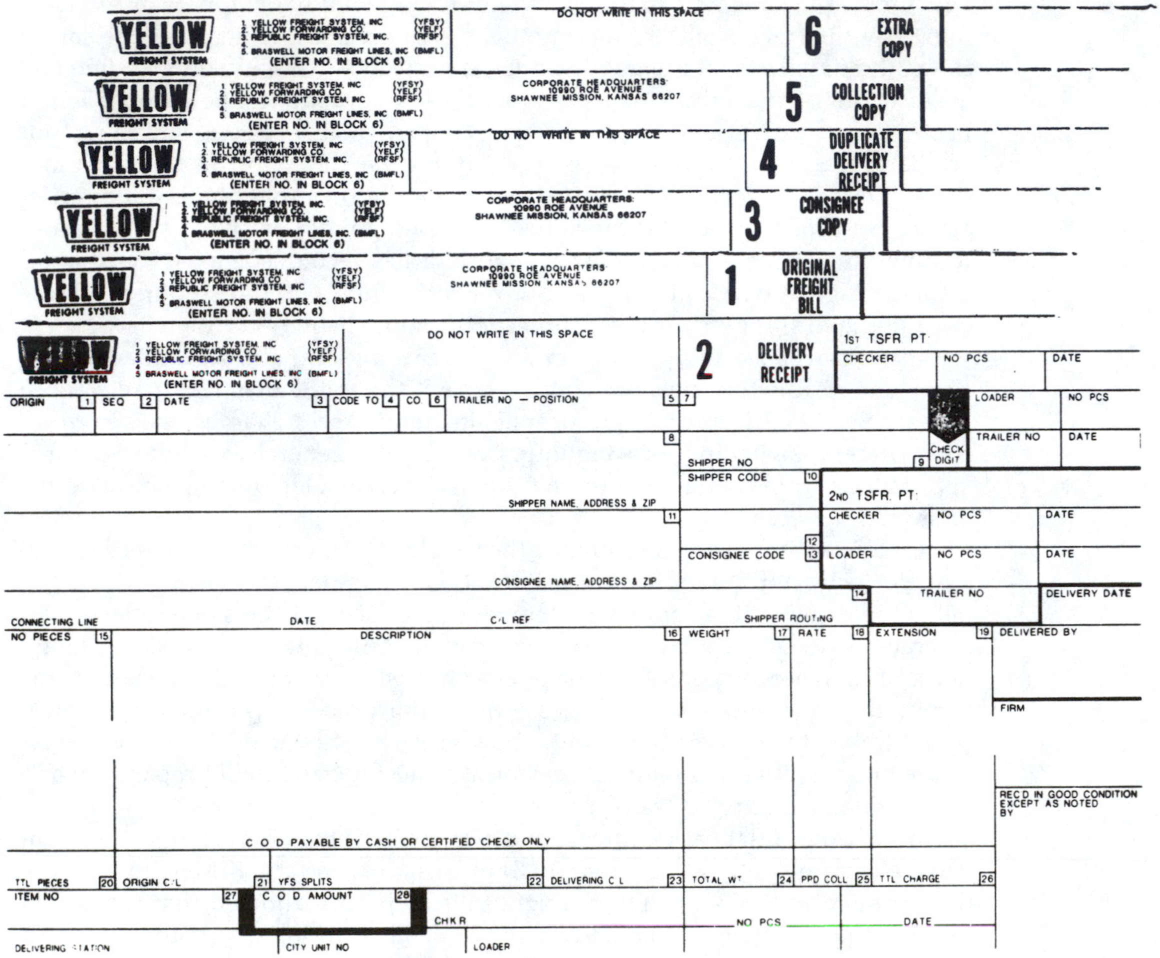
YELLOW FREIGHT SYSTEM
1 YELLOW FREIGHT SYSTEM, INC (YFSY)
2 YELLOW FORWARDING CO (YELF)
3 REPUBLIC FREIGHT SYSTEM, INC (RFSF)
4
5 BRASWELL MOTOR FREIGHT LINES, INC (BMFL)
(ENTER NO. IN BLOCK 6)

DO NOT WRITE IN THIS SPACE

CORPORATE HEADQUARTERS: 10990 ROE AVENUE SHAWNEE MISSION, KANSAS 66207

6 EXTRA COPY
5 COLLECTION COPY
4 DUPLICATE DELIVERY RECEIPT
3 CONSIGNEE COPY
1 ORIGINAL FREIGHT BILL
2 DELIVERY RECEIPT

1st TSFR PT: CHECKER | NO PCS | DATE
LOADER | NO PCS
TRAILER NO | DATE
CHECK DIGIT 9

ORIGIN | 1 SEQ | 2 DATE | 3 CODE TO | 4 CO | 6 TRAILER NO — POSITION | 5 | 7

8 SHIPPER NO
SHIPPER CODE 10
SHIPPER NAME, ADDRESS & ZIP 11

2ND TSFR. PT: CHECKER | NO PCS | DATE
12 LOADER | NO PCS | DATE
CONSIGNEE CODE 13
CONSIGNEE NAME, ADDRESS & ZIP
14 TRAILER NO | DELIVERY DATE

CONNECTING LINE | DATE | C/L REF | SHIPPER ROUTING

NO PIECES 15 | DESCRIPTION | 16 WEIGHT | 17 RATE | 18 EXTENSION | 19 DELIVERED BY

FIRM

REC'D IN GOOD CONDITION EXCEPT AS NOTED BY

C O D PAYABLE BY CASH OR CERTIFIED CHECK ONLY

TTL PIECES 20 | ORIGIN C/L 21 | YFS SPLITS 22 | DELIVERING C/L 23 | TOTAL WT 24 | PPD COLL 25 | TTL CHARGE 26

ITEM NO 27 | C O D AMOUNT 28 | CHKR | NO PCS | DATE

DELIVERING STATION | CITY UNIT NO | LOADER

Source: Yellow Freight System, Inc. Reprinted by permission.

A distinction between a refused and a returned shipment should be made. When the consignee refuses delivery of a shipment, the carrier temporarily stores the cargo and requests disposition instructions from the shipper. A returned shipment requires the consignee to accept delivery first. The consignee is responsible for new shipment documentation and for making arrangements to return the cargo to the shipper.

Documents Used for Contract, Exempt, and Private Carriage

All the documents discussed thus far constitute the main thread of information and control needs in the movement of goods by common carriage. Contract carriers generally use various documents that serve similar purposes. Even in private carrier settings, firms find it good practice to utilize some form of pickup and delivery receipt as well as a movement control document.

Terms of Sale

Besides price, warranty, credit, and other similar considerations, the sales contract should address four major issues related to traffic management: (1) Who pays the freight? (2) Who controls the shipment in transit? (3) Who assumes the risk for loss, damage or delay? (4) Who has the burden of filing claims? These issues represent contractual matters, which the courts must settle in the event of disputes. The law of sales, which has been codified in Article 2 of the UCC, is the primary source of law that guides the courts and the buyers and sellers on these matters. The basic principle of this article is that the provisions in the sales contract decide such issues.

Sometimes buyers and sellers fail to include these provisions. In the absence of an explicit agreement, the shipping terms (such as F.O.B., free on board) determine when the title to the goods, as well as the risk of transportation, passes from seller to buyer. The shipping terms, combined with two closely related qualifying provisions, comprise the basic delivery terms of sale. These terms are examined next.

Delivery Terms

The transportation community has developed a variety of shipping terms. The F.O.B. designation, for example, is normally used in domestic commerce. By contrast, many different designations are used in international commerce (as discussed in Chapter 15). Despite such a variety, the UCC views these shipping terms as forming either a "shipment" or a "destination" contract.

In the shipment contract, the seller retains title to the goods and bears all costs and risks until the goods are tendered to the carrier at the seller's place of business. After delivery to the *carrier,* which acts as the buyer's agent, the buyer takes title and assumes the risks and responsibilities for the goods in claims situations. In the destination contract, the seller must deliver the goods to the *buyer* at a particular destination. On delivery, title passes along with the risk of transportation.

As already indicated, convention calls for the F.O.B. named point (that is, origin or destination) to define shipping arrangements in domestic commerce.

More precisely, F.O.B. origin and F.O.B. destination form shipment and destination contracts, respectively. As shown in **Exhibit 7.5**, the title to the goods, as well as the risks and responsibilities for goods in claims situations, transfers from seller to buyer at the named point. Moreover, unless qualifying sales terms indicate otherwise, the expense of transportation passes from the seller to the buyer at the named point.

Many firms frequently use two qualifying delivery terms to determine each party's responsibilities for the freight charges. The first term (see Exhibit 7.5), either ''prepaid'' or ''collect,'' establishes which party must pay the carrier for transportation services. The seller prepays the freight charges, or the carrier collects them from the buyer.

The second qualifying delivery term, usually either ''allowed'' or ''charged back,'' indicates which party will ultimately pay for transportation. The phrase ''collect and allowed'' means the buyer pays the freight but is allowed to deduct the freight charges from the sales invoice. Conversely, the phrase ''prepaid and charged back'' requires the seller to pay the carrier for the transportation and to add that amount to the sales invoice.

Planning Considerations

The delivery terms of sale have significant implications for traffic management. They affect carrier selection and routing control, freight payment cash flows, claims administration, and shipment consolidation opportunities.

Control. When firms have the option, they must consider the merits of retaining control over carrier selection and routing. Generally speaking, such control is obtained by taking title to the goods while in transit. Control is desirable usually when the traffic function is well organized and staffed and has a good deal of clout with carriers. Traffic managers can do a better job of planning the selection, acquisition, and control of transportation services—in terms of costs or services—than less sophisticated buyers or sellers can. For example, the selection of carriers and the methods of acquiring their services might be part of a long-range program that is closely coordinated with marketing and production strategies.

The customs of the trade usually dictate the terms of sale. For example, the chemical and the clothing industries prefer collect terms, but the steel industry relies on prepaid arrangements. Regardless of method of payment, most buyers take control over inbound shipments. By one estimate, about 90 percent of all shipments are purchased with F.O.B. origin terms.[3]

Nonetheless, relinquishing control to the supplier (F.O.B. destination) may have merit. Suppliers are sometimes better situated than the traffic function for purchasing transportation. Some suppliers are better informed about local traffic conditions; others have more negotiating leverage with carriers.

Exhibit 7.5 **Delivery terms of sale**

	Shipping Terms		*Freight Payment Terms*			
	Who assumes title and bears the risk of transportation?		**Who pays the carrier?**		**Who ultimately bears the cost of transportation?**	
	Shipment	Destination	Prepaid	Collect	Charged back	Allowed
Type of contract Terms	FOB origin	FOB destination	Prepaid	Collect	Charged back	Allowed
Responsible party	FOB origin → Buyer	FOB destination → Seller	Prepaid → Seller	Collect → Buyer	Charged back → Buyer	Allowed → Seller
Passage of title and risk	Title and risk pass to the buyer → Seller — Goods in transit → Buyer	Title and risk pass to the buyer → Buyer				

Occasionally, traffic or purchasing managers automatically assume that F.O.B. destination, prepaid (delivered-pricing) terms offer lower freight costs. They reason that these delivery terms place the burden of carrier selection on the supplier. The supplier, moreover, pays the freight and assumes responsibility for claims actions. Relinquishing control to the supplier, though, does not necessarily guarantee superior routing. In addition, suppliers usually include the transportation cost as a factor in determining the selling price.

The traffic manager may make a contribution to the firm by taking control of inbound transportation of materials. For example, in coordination with the traffic manager, either the purchasing manager negotiates F.O.B. origin, collect terms in exchange for a reduction in the selling price. The supplier's experience with transportation cost determines the amount of the reduction. The traffic function then applies its clout with carriers to negotiate freight rates and services better than those previously provided by the supplier. The control of carrier selection and routing, moreover, might offer opportunities to enhance or to initiate shipment consolidation programs.

As part of these plans and programs, traffic organizations may develop detailed routing guides for their vendors. A misconception often arises in connection with this effort. Shippers may assume that the F.O.B. origin designation gives the buyer routing control over the entire shipment, and therefore the vendor must be legally obligated to follow the routing instructions. In other words, the assumption is made that the party paying the freight has title to the goods in transit, and that ownership of the goods in transit gives owners the unrestricted right to select carriers and route shipments.

This assumption is wrong for two reasons. First, the responsibility for freight charges is not necessarily tied to the transfer of title. With the F.O.B. origin, prepaid terms, for example, the buyer has title to the goods in transit, but the seller pays the freight bill. Second, with the F.O.B. origin designation, the control of transportation transfers to the buyer *after* the goods are loaded at origin. The buyer, as the owner of the goods, can then route the shipment from that point, provided the buyer also assumes liability for freight charges. Alone, these delivery terms offer no remedy to the problem vendor that disobeys routing instructions and tenders the goods to an unauthorized carrier. The solution here is to include a separate clause about compliance with routing instructions in the sales contract. In addition, if routing is specified on the purchase order, and the seller accepts the order, the routing becomes part of the sales contract.

The case for adopting delivery terms of sale that permit control of outbound traffic rests on reasons similar to those for inbound traffic. For many firms, consolidation opportunities are generally greater for outbound than for inbound traffic. Moreover, by assuming the responsibility for carrier selection and routing and the claims responsibilities, the traffic function may add an important dimension to delivery service from the customer's perspective.[4]

Cash flow. The delivery terms of sale affect the flow of capital tied up for goods payment. The shipment contract (F.O.B. origin) makes the sale invoice due and payable when the carrier has possession and control of the shipment at origin. By contrast, the balance due for the sale of the goods does not become payable until delivery of the goods to the buyer when destination contracts (F.O.B. destination) are used.

In addition, the delivery terms can favorably or adversely affect cash flows generated by the payment or the reimbursement of freight charges. For example, if the customs of the trade require delivered pricing terms, quoting collect and allowed establishes a delivered price without tying up the firm's funds in prepaid freight.

Claims administration. The selection of shipping terms can have an important effect on the level of claims administration. Generally the party that assumes the risk of transportation bears the burden of claims actions. As a goodwill gesture, however, some firms with experienced staff file claims on behalf of their customers. This service is performed even though the firm is not required to.

In a deregulated transportation environment, claims staff will have to watch carefully for changing practices. For example, with the advent of deductibles, some firms sell F.O.B. origin, prepaid, while paying reduced rates that carry deductibles. Thus, the seller gets a discount, but the buyer has to absorb the deductible amount in the event of loss or damage.

Traffic managers can also use the delivery terms of sale to make tactical adjustments to eliminate claims problems. Take the case of the vendor that persistently misdescribes articles on the bill of lading, an act that causes the carrier to erroneously charge rates higher than those based on correct descriptions. By using prepaid and charged back terms, the buyer could add only the correct amount to the sales invoice and pass the burden of filing overcharge claims to the vendor.

Freight rate discounts. The widespread practice of carrier discounting has confronted both shippers and receivers with a new problem—specifically, how to account for the discounts in sales transactions. Using prepaid and charged back, a shipper may add freight charges based on the full class rate and then save the difference between the charge computed from full class rate and the actual discounted outlay. Similarly, a receiver using freight collect and allowed, may claim the full class rate as the basis for its allowance but actually pay the carrier a discount rate. Whether sellers or buyers are legally or ethically required to pass discounts along to the party ultimately responsible for freight charges is an ambiguous issue. For example, not passing along the discount may be justified because the seller or the buyer earned it by performing certain services such as loading, unloading, or multiple shipment tenders. Both sides, of course, first need to be aware of this situation. The best course of action may be for buyers and sellers to resolve this matter through direct negotiations.[5]

Using the Carrier to Secure Payment for the Goods

Shippers often rely on transportation companies and documents to secure payment from consignees for the goods. These methods include shipping the goods by C.O.D. (collect on delivery) or by order bill of lading. In addition to extending credit to consignees, these options include selling on consignment or using a letter of credit.

The C.O.D. system requires that the bill of lading stipulate the sum of money that the carrier is to collect from the consignee before or at delivery. This sum, which normally includes the charge for the sale and the freight charges, can be paid in cash, certified check or banker's check, or money order. The tariffs to which each carrier is a party will indicate the exact form and process for handling C.O.D.s. The carrier will not deliver a shipment until payment is made and will retain a portion of the sum as a fee for providing this service.

Several factors should be considered when using this service. One is the sum to be collected. If it is large, the carrier or driver might not wish to be responsible for such funds. A carrier might be slow in remitting the collected funds to the shipper. A further problem is that most motor carriers will not accept C.O.D. shipments for interline movements, because of possible complications in collecting from other carriers.

An alternative method to assure the collection of the goods' invoice value is to use the order bill of lading. **Exhibit 7.6** illustrates an order bill of lading form, which is similar in content to a straight bill of lading form with the exception of the following statement:

> *The surrender of this original order bill of lading properly endorsed shall be required before the delivery of the property. Inspection of property covered by this bill of lading will not be permitted unless provided by law or unless permission is endorsed on this original bill of lading or given in writing by the shipper.*

The process for an order bill of lading is shown in **Exhibit 7.7.** The shipment is initiated when the carrier signs the order bill of lading and leaves copies with the shipper (step 1). The shipper, in turn, sends completed copies of the bill of lading to a bank designated by the consignee (step 2). On receipt (step 3), the bank notifies the consignee of the amount that the goods are worth and usually the freight charges as well. The consignee pays the bank the amount involved (step 4). The bank delivers a fully "verified" or "endorsed" (paid) order bill of lading to the consignee (step 5). The carrier arrives with the goods but does not relinquish them to the consignee until surrender of the fully executed order bill of lading (steps 6 and 7). The bank then deducts a processing fee and remits the bulk of the funds for the goods to the shipper (step 8).

■■■ *Exhibit 7.6* **A sample order bill of lading**

OD-138 6/73

THIS MEMORANDUM is an acknowledgment that a bill of lading has been issued and is not the Original Bill of Lading nor a copy or duplicate covering the property named herein, and is intended solely for filing or record.

Shipper's No._____

Agent's No._____

YELLOW FREIGHT SYSTEM, INC.

RECEIVED, subject to the classifications and tariffs in effect on the date of the issue of this Bill of Lading.

From_____, Date_____, 19_____

At_____Street,_____City,_____County,_____State,

the property described below, in apparent good order, except as noted (contents and condition of contents of packages unknown), marked, consigned, and destined as shown below, which said company (the word company being understood throughout this contract as meaning any person or corporation in possession of the property under the contract) agrees to carry to its usual place of delivery at said destination, if on its own railroad, water line, highway route or routes, or within the territory of its highway operations, otherwise to deliver to another carrier on the route to said destination. It is mutually agreed, as to each carrier of all or any of said property over all or any portion of said route to destination, and as to each party at any time interested in all or any of said property, that every service to be performed hereunder shall be subject to all the conditions not prohibited by law, whether printed or written, herein contained, including the conditions on back hereof, which are hereby agreed to by the shipper and accepted for himself and his assigns.

The surrender of this Original ORDER Bill of Lading properly indorsed shall be required before the delivery of the property. Inspection of property covered by this bill of lading will not be permitted unless provided by law or unless permission is indorsed on this original bill of lading or given in writing by the shipper.

Consigned to ORDER OF_____

Destination_____Street,_____City,_____County,_____State

Notify_____

At_____Street,_____City,_____County,_____State

Routing_____

Delivering Carrier_____ Vehicle or Car Initial_____ No._____

No. Packages	KIND OF PACKAGE, DESCRIPTION OF ARTICLES, SPECIAL MARKS AND EXCEPTIONS	*Weight (Sub. to Cor'tion)	Class or Rate	Check Column	Subject to Section 7 of conditions, if this shipment is to be delivered to the consignee without recourse on the consignor, the consignor shall sign the following statement:

Subject to Section 7 of conditions, if this shipment is to be delivered to the consignee without recourse on the consignor, the consignor shall sign the following statement:
The carrier shall not make delivery of this shipment without payment of freight and all other lawful charges.

_____
(Signature of Consignor.)

If charges are to be prepaid write or stamp here, "To be Prepaid."

Received $_____
to apply in prepayment of the charges on the property described hereon.

Agent or Cashier

Per_____
(The signature here acknowledges only the amount prepaid.)

Charges Advanced:

*If the shipment moves between two ports by a carrier by water, the law requires that the bill of lading shall state whether it is "carrier's or shipper's weight."
NOTE—Where the rate is dependent on value, shippers are required to state specifically in writing the agreed or declared value of the property.

The agreed or declared value of the property is hereby specifically stated by the shipper to be not exceeding_____per_____ $_____

_____Shipper _____Agent.

3 Per_____ Per_____

Permanent Address of Shipper:_____Street,_____City,_____State

Source: Yellow Freight System, Inc. Reprinted by permission.

Although the order bill of lading system avoids many of the problems of C.O.D. shipments, a potential problem is that it might slow actual goods delivery when the mail is used to forward the documents. Problems also arise if the documentation gets lost; the buyer may have to post bond to receive the shipment, for example. In addition, some carriers refuse to accept order-notify shipments.

■■■■■ *Exhibit 7.7* **Process of using an order bill of lading**

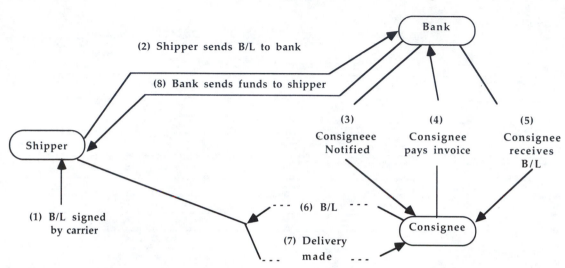

On the whole, the order bill of lading is frequently and successfully used when firms ship goods from their plants in the general direction of expected markets but without known consignees. The seller can "play" various markets and sell the goods to the highest bidding buyer through this system. The seller uses the carrier and the banking system to guarantee payment for the goods.

■■■■■ *Rate Retrieval and Freight Payment Systems*

Determining the proper charge for a purchase and verifying the bill received for it is a basic job in purchasing. Likewise in traffic, major efforts are made to determine the proper charges and to verify the received freight bill against this information. The task is not an easy one for traffic personnel for several reasons. First, many (more than one million) new and revised tariffs are filed with the ICC each year. Second, tariffs are complex, and many rates might appear applicable for a particular move but only one is legally correct (as discussed in Chapter 4). Third, there are several types of retrieval and payment systems available to the firm.

An efficient rate retrieval and freight payment system is crucial to a corporate need for optimum cash flow, minimum process expenditures, and reduction in customer service problems.[6] An incorrect charge causes the firm to have cash outstanding with a carrier that could otherwise be used or invested by the firm. Further, an error in billing, because the rate is too high, leads to the company's paying more for a move than the customer reimburses for the amount of the original freight bill. Such an error can also lead to customer ill will when the firm issues the bill a second time. A $100 overcharge discovered and paid two years later could represent more than $27 in lost cash opportunity in a 13 percent money market period.

Firms use five types of rate retrieval and payment systems. **Exhibit 7.8** presents each of these systems from the standpoint of features, advantages, disadvantages, optimum use range, and short-term back-up capabilities. The five systems are as follows:

1. Manual: In house. This system consists of company personnel using printed tariffs to analyze rates and verify freight bills. This requires obtaining tariffs from carriers and publishing bureaus, and training people in their use. The in-house manual system provides the firm with the capability of performing nearly any task needed in this realm. Only the degree of commitment and cost determine the extent to which preaudit, audit, report generation, and other objectives are served. Some firms have large staffs to manually preaudit bills, whereas others have smaller staffs to selectively audit bills of certain dollar amounts. Tariff simplification, in the wake of deregulation, has made the training and retrieval processes much simpler than in the past. The greater use of contract carriage also makes these processes simpler.

2. Manual: Consulting firms. This system also uses personnel to manually retrieve rates and verify freight bill charges. Shippers may request carriers to send bills directly to the consulting firm for payment. These firms charge for their services through a flat fee per bill, on the basis of a percentage of overcharge obtained, or both. The use of these firms is limited for preaudit capabilities. Often, it is not possible to receive a bill, transmit it to the firm, receive it back with verification or correction, and pay it within a short time. With deregulation of the required carrier credit period, these services will be easier to use for preauditing.

3. Computer: In house. This operation can be fast and efficient for a large volume of freight bills. These systems require extensive programming, inputting continuous changes in tariffs, and a commitment to the computer hardware. Before the 1980s, only large operations could efficiently use these systems, because they often took many years to bring on-line. Although the computer can be used to produce reports in almost any format for analytical use, one disadvantage of these systems is that the manual system is dropped, thereby leaving the firm without a sense of market intelligence gained from having people handle the tariffs. People can often detect subtle market changes where the computer cannot, unless it is specifically programmed.

Deregulation has helped spur tariff simplification. Where some carriers previously used a several-hundred page complex tariff, many of these are now re-

Exhibit 7.8 **Comparative analysis of alternative rate retrieval and freight payment systems**

System	Pros	Cons	When Appropriate	Short-term Back-up
Manual: In house	Full control Confidentiality Training feature	Cost Commitment to manpower	Good for small to medium volume of freight bills Sometimes difficult to preaudit, unless have large staff and tariff capabilities Volume of 5000 or more bills a month seen by many as maximum for this system	Manual consultant
Manual: Consulting firm	Low cost Can audit bills already audited in house Firms can also file for overcharges	Preaudit often not possible Fees	Can be used for any volume of freight bill activity Seek other systems if this takes too long to audit bills	Another manual consultant
Computer: In house	Confidentiality Ease of report generation Preaudits Release employees Can handle freight bill payments	Long-term commitment and large investment Long start-up time Only simple rates can be handled	For 5000 or more bills a month volume When need for preaudit is high If manual systems too slow	Manual in house Manual consulting firm
Computer: Consulting firm	All above, except confidentiality Reduce manpower and space needs in firm	Same as above, except this requires less start-up time	Same as above	Manual consulting firm Manual in house
Bank payment plans	Few, except if "no-balance" service used	Loss of cash flow opportunity	Can be used with any volume	In house

placed with ones having few commodity groupings, origin and destination in the form of three-digit zip codes, and simple rate tables. Today, twenty-page national tariffs are not uncommon. Tariff simplification also makes the use of desktop computers feasible. These two trends—deregulation and desktop computers—are shifting the minimum five thousand bills a month for computer use to lower levels.

4. Computer: Consulting firms. Consulting firms offer the same advantages and disadvantages as in-house computer operations, though consulting firms may have faster start-up and less need for manpower.

5. Bank payment plans. These plans deal with how the carrier is paid once a freight bill is authorized for payment. Most of the original plans consisted of the firm's placing a cash balance at a bank with direct computer links to the carrier and with the amount of the bill to be paid daily. With this plan, check writing expenses are reduced, but the lost cash opportunity from the cash balance is a major drawback. A few banks offer balanceless checking accounts for small fees, where the shipper can write and send a check to the carrier, have the check returned to the bank, and pay the bank when the check arrives. This eliminates the cash balance problem.

The selection of a freight rate retrieval and payment system is a key decision for traffic managers. It involves a systems analysis, an evaluation of corporate and traffic objectives, and a review of specific choices. (Chapter 16 discusses specific criteria to consider in the selection process.)

Carrier Credit Arrangements

For nearly forty years, regulations required a maximum credit period for common rail and motor freight bill payment. Title 49 Code of Federal Regulations, Part 1300–1325 basically required a five-day credit period for rail bill payment, and a seven-day collection period for common motor carrier bills. This regulation was a major exception to the practice of other nontransportation industries, which typically permit thirty-day and longer credit periods for bill payment. Revised ICC rules now establish a uniform fifteen-day credit period for all modes of transportation subject to its jurisdiction. Carriers may extend this period to thirty days if they charge for the privilege. Moreover, they may offer shippers discounts for early payments or assess penalty charges for late payments.

The credit term, or number of days in which a shipper or receiver may pay the freight bill, has great cash flow and opportunity cost implications for both the party responsible for payment and the carrier. One eastern chemical manufacturer has motor carrier bills averaging $1 million per day. Under the old seven-day rule, this company always had $7 million in cash with carrier accounts payable in the same amount. In a 10 percent money market period, this arrangement earned the firm $700,000 a year, less financial transaction costs. A thirty-day credit term affords an opportunity to invest $30 million and earn $3 million a year in interest. What is gained here by the shipper, however, is lost by the carrier that used to obtain payment in the shorter period.

The new credit rules mean (1) the credit period, (2) the discount for early payment, and (3) the late payment penalty may be used in the terms of payment. For example, a motor carrier may request payment by sending a shipper a freight bill for $500 with payment due in thirty days, 1/2 percent discount for fifteen-day payment, and 1 percent a month carrying charge for payments after thirty days. Thus, the question of how a traffic manager determines if early, regular, or late payment is best arises.

A method to evaluate the payment alternatives is as follows:

1. Calculate the cost of credit (C).

$$\frac{365}{\text{Gross days} - \text{Discount days}} \times \text{Discount percent} = C$$

where gross days = number of days until payment is due
discount days = number of days allowed for early payment discount

2. Compare C to the firm's cost of money.

   If C is greater than the cost of money, pay early with discount.
   If C is equal to the cost of money, decide either.
   If C is less than the cost of money, pay at the end of the credit period.

In the above example, the formula appears as $365/(30-15)$ times 0.005, which equals 12.2 percent. If the firm's cost of capital is 10 percent, the carrier should be paid early. To use the carrier's funds for the sixteenth through the thirtieth day would be more costly than to borrow required funds from a bank at 10 percent. If a carrier offers 1/2 percent, 7 days, net thirty day terms, and the shipper firm's cost of capital is 10 percent, the formula computes to 7.9 percent compared to 10 percent. In this instance, using the carrier's funds for a full thirty days (at full freight bill without the discount) is cheaper than borrowing the funds from a bank. The traffic manager also can use this tool to determine the discount or the length of the discount period that will induce early payment.

Sometimes carriers levy late payment penalties. The penalty might be a 1 percent fee for every thirty days the payment is late. This penalty translates into a 12 percent annual rate of interest, which must be compared to the firm's cost of capital to determine whether delaying payment to the carrier with penalties is less costly than paying early and seeking funds from a bank.

A problem can arise between traffic and the firm's accounting and treasury office. A rate negotiation might be settled with the agreement to pay the carrier its gross bill amount in ten days. The bill is received, verified, and sent to accounting with authorization to pay it. If, however, the treasury office delays payment because favorable interest can be earned in money markets, the carrier may not be paid for thirty or forty days. The carrier may return and raise its rate to cover its lost cash flow. In one instance, a traffic organization saw its contract rate raised by $15,000 a year for this reason, while the corporate treasury department earned $8000 in interest off the funds. The key here is for traffic to determine when the carrier is to be paid. The traffic manager has to coordinate with the other involved managers to make sure such tradeoffs are clearly understood.

Summary

The traffic manager must consider five fundamental elements of the shipment process. The first element, documentation, focuses on the bill of lading, the waybill, the freight bill, and the delivery receipt. The bill of lading initiates the shipment and functions primarily as a contract of carriage and a receipt for the goods shipped. Although ICC-regulated common carriers have the duty to issue the bill of lading, shippers will often supply and prepare customized bills to cut clerical expenses. The main types of bills of lading include: straight, order, uniform or through export, livestock, ocean, air, and government.

A waybill is a carrier document used to control shipments. It contains information such as vehicle number, route, shipment weight, customer code, and special instructions. For control purposes, carriers assign the same waybill number (rail) or PRO number (truck) to the waybill, the bill of lading, and (if appropriate) the individual pieces of freight.

The freight bill is the carrier's invoice for freight charges, which may be prepaid by the shipper or collected from the receiver. By signing section 7, the without-recourse clause, of the straight bill of lading, the shipper instructs the carrier to collect any balance due for freight charges from the consignee. This clause, however, does not provide absolute protection from the payment of freight charges; in certain instances, carriers may still collect from the shipper.

The delivery receipt triggers the freight bill for collect shipments. It also serves as prima facie evidence that the carrier has delivered the shipment in good order.

The second element involves the terms of sale. These terms should settle the issues of whether the buyer or seller will (1) pay for the freight, (2) control the routing of the shipment, (3) assume the risk of transportation, and (4) assume the claims administration burden. An explicit agreement in the sales contract should provide the answers. In the absence of such agreements, the shipping terms of sale govern.

The third element relates to the ways in which shippers can use carriers and shipping documents to secure payment from the consignee for the sale. The two fundamental methods are (1) collect on delivery (C.O.D.) and (2) order bill of lading or order-notify shipments.

The fourth area relates to rate retrieval and freight payment systems. There are five basic system choices: (1) manual: in house, (2) manual: consulting firms, (3) computer: in house, (4) computer: consulting firms, and (5) bank payment plans.

Finally, traffic managers must consider carrier credit arrangements, which have important cash flow and opportunity cost implications. The credit period, the discount for early payment, and the late payment penalty have to be evaluated when deciding whether early, regular, or late payment is best for the firm.

Selected Topical Questions from Past AST&L Examinations

1. Define what is meant by F.O.B. origin pricing, and discuss the respective advantages and disadvantages of such a policy. What viable alternatives are there to F.O.B. origin pricing? Discuss. (Spring 1980)
2. Identify and briefly discuss the three principal functions of the bill of lading. What is the most important distinction between a straight and an order bill of lading? (Spring 1981)
3. Why would a traffic department be necessary for a firm that has all of its inbound materials, supplies, and equipment purchased F.O.B. delivered and all of its finished products sold F.O.B. origin? (Spring 1981)
4. Many industrial firms are using so-called bank freight payment plans for the payment of some or all of their transportation bills. What advantages may a shipper experience from participating in such a plan? (Spring 1981)
5. The regulated common motor carrier industry has a credit regulation that basically requires a bill to be represented in seven days and a collection within seven days of the bill presentation. What will be the likely financial statement effects on shippers and motor carriers if this regulation is changed to a thirty-day credit period? (Spring 1981)
6. Deregulation has brought with it a significantly increased usage of private trucking, and many firms are now using their own trucks to pick up merchandise previously shipped to them by common carrier freight on a delivered price basis. Customers now using their own trucks naturally want a better price, and this is having a complicating effect on your delivered pricing system. Your management asks you to evaluate the pros and cons of eliminating the delivered pricing system altogether and substituting an F.O.B. shipping point system, freight charges collect. Please indicate how you would deal with such problems as charging customers for transportation from shipping points more distant than those customarily serving them. (Spring 1981)

7. Assume you are the logistics manager of a firm that manufactures peanut butter. Your product is sold nationally, and sales in 1981 were $4.5 million. You have a logistics staff of six people, not counting clerical staff. You are considering using a computerized rate retrieval system. Assuming you decide to use computerized rates, should you perform the function internally, or should you use the services of an outside rate retrieval firm? Discuss the advantages of both. (Fall 1982)

8. Describe the key elements of and distinguish between the following two shipping terms. Be sure to include any comments relevant to claims.
    (a) F.O.B. origin, freight prepaid and charged back
    (b) F.O.B. destination freight collect (Fall 1984)

9. You are negotiating with two motor carriers: Carrier A and Carrier B. Carrier A is a firm with which your western division has been doing business for several years. The western division has been paying its freight bills to Carrier A within ten days of receipt. The eastern division, on the other hand, has been taking more than fifty days to pay Carrier B.
    (a) What effect does fast or slow freight bill payment have on a motor carrier?
    (b) You are thinking of asking the carriers for a discount for early payment, in say ten days. What are the pros and cons of this for both of your divisions as well as for the carriers?
    (c) Which carrier would be most likely to be interested in the early payment discount? Why? (Spring 1985)

Notes

1. Joan Feldman, ed., "The bytes that bind: Electronic data interchange," *Handling and Shipping Management* (May 1984), pp. 62–64.
2. Mark B. Solomon, "TDCC moves ahead to push EDI for all business transactions," *Traffic World* (December 30, 1985), p. 19.
3. Ernest Gilbert, "Problems in purchasing transportation," *National Association of Purchasing Managers 67th International Purchasing Conference*, Los Angeles, California, May 10, 1982.
4. See Lynn Edward Gill, "Delivery terms—Important element of physical distribution," *Journal of Business Logistics* 1 no. 2 (1980), pp. 60–82.
5. A good discussion of this issue, which has been excerpted from the *Barrett Transportation Newsletter,* appears in "Questions and answers," *Traffic World* (June 3, 1985), pp. 4–5.
6. Lester A. Probst, *The Freight Payment Problem: An Analysis of Industry Applied Solutions in the Nineteen Eighties,* 1st rev. (Metuchen, New Jersey: Transportation Concepts and Services, Inc., 1980).

Appendix 7.1 **Bill of Lading Preparation**

The bill of lading is prepared by completing items 1–13, shown in **Exhibit A7.1** and described as follows:

1. *Shipper's number.* This entry supports file management and is used for identification purposes in correspondence with consignees and carriers. It also serves as a control element in the freight bill accounts payable system, and shippers often insert a control number here.
2. *Shipper's name/location/date.* These entries contain the shipper's name, the pickup address, and the shipment date. The date should indicate exactly when the goods were tendered to the carrier, because the rates in effect on this date are legally applicable.
3. *Consignee's name/location/mailing address.* The carrier must deliver the goods to the consignee at this location. Accurate information is essential to avoid unnecessary delays or expense.
4. *Routing and/or carrier.* This item stipulates the carrier(s) and the possible interline routing that the shipper desires.
5. *Vehicle number.* In carload and truckload shipments, the vehicle-owner code (or name) and number should be indicated.
6. *Number of packages, truckload, or carload.* This information is used for counting, control, inspection, loss and damage inspection, and rating purposes.
7. *Description of articles, packaging, marks, or exceptions.* The exact article descriptions shown in tariffs must be used. If the shipment requires dunnage (discussed in Chapter 11) or other allowances or special instructions, they should be noted here. This entry provides space to indicate whether the shipper, carrier, or receiver will load or unload the motor freight, as is required by the Motor Carrier Act of 1980.
8. *Weight.* This part shows the gross weight of the shipment, which includes the packing material, the container (such as boxes, barrels, bags, and similar items), pallets, and dunnage.
9. *Rate.* If the proper rate is known, it is good practice to enter it onto the bill of lading.
10. *Name of traffic manager, supervisor, or person preparing the bill of lading.* This name is in addition to the required signature (on the ''Per'' line) and the name of the company. This is the person that carriers will contact if any errors or discrepancies on the bill of lading are detected.
11. *Section 7.* This section contains the ''without recourse clause.''
12. *Prepayment.* This portion of the bill of lading identifies the party (shipper, consignee, or both) that will pay the carrier for services rendered. The

■■■■■ *Exhibit A7.1* **Sample bill of lading form**

STRAIGHT BILL OF LADING - SHORT FORM- Original-Not Negotiable
RECEIVED, subject to the classifications and tariffs in effect on the date of issue of this Original Bill of Lading

YELLOW FREIGHT SYSTEM

SHIPPER NO. ①_____

CARRIER NO. _____

YELLOW FREIGHT SYSTEM, INC. YFSY

(NAME OF CARRIER)   (SCAC)

DATE ②_____

TO: CONSIGNEE ③	FROM: SHIPPER ②
STREET ③	STREET ②
DESTINATION ③ ZIP ③	ORIGIN ② ZIP ③
ROUTE: ④	VEHICLE NUMBER ⑤

NO. SHIPPING UNITS	KIND OF PACKAGING, DESCRIPTION OF ARTICLES, SPECIAL MARKS AND EXCEPTIONS	Weight (Subject to Correction)	Rate	CHARGES (for Carrier use only)
⑥	⑦	⑧	⑨	

REMIT C.O.D. TO: ADDRESS

COD   AMT $

C.O.D. FEE: PREPAID ☐ $ COLLECT ☐

TOTAL CHARGES: $

Note—Where the rate is dependent on value, shippers are required to state specifically in writing the agreed or declared value of the property.
The agreed or declared value of the property is hereby specifically stated by the shipper to be not exceeding

Subject to Section 7 of the conditions, if this shipment is to be delivered to the consignee without recourse on the consignor, the consignor shall sign the following statement.
The carrier shall not make delivery of this shipment without payment of freight and all other lawful charges.

⑫ FREIGHT CHARGES
FREIGHT PREPAID except ☐ Check box if charges when box at right is checked ☐ are to be collect

$ ⑬ _____ per _____

⑪ _____ (Signature of Consignor)

RECEIVED, subject to the classifications and tariffs in effect on the date of the issue of this Bill of Lading, the property described above in apparent good order, except as noted (contents and condition of contents of packages unknown), marked, consigned, and destined as indicated above which said carrier (the word carrier being understood throughout this contract as meaning any person or corporation in possession of the property under the contract) agrees to carry to its usual place of delivery at said destination, if on its route, otherwise to deliver to another carrier on the route to said destination. It is mutually agreed as to each carrier of all or any of, said property over all or any portion of said route to destination and as to each party at any time interested in all or any said property, that every service to be performed hereunder shall be subject to all the bill of lading terms and conditions in the governing classification on the date of shipment.
Shipper hereby certifies that he is familiar with all the bill of lading terms and conditions in the governing classification and the said terms and conditions are hereby agreed to by the shipper and accepted for himself and his assigns.

This is to certify that the above-named articles are properly classified, described, packaged, marked, and labeled, and are in proper condition for transportation, according to the applicable regulations of the Department of Transportation.	SHIPPER
	PER ⑩
	CARRIER
	PER
	DATE

OD 136

Source: Yellow Freight System, Inc. Reprinted by permission.

freight bill is either prepaid by the shipper or collected from the consignee. Although shippers generally have a choice, carriers may insist on prepayment for transporting certain services or commodities. In addition, shippers may choose to pay only part of the total charges. In these cases, this entry includes the amount prepaid.

13. *Released valuation.* If the goods are moving under released valuation, the shipper must insert the specific valuation per pound (or per applicable unit), as shown in the tariff.

Both the shipper and the carrier sign the original bill of lading, which the shipper may send to the consignee. The shipper also signs the first copy, known as the shipping order, and the agent takes it to the originating terminal point. The carrier will use its copy of the bill of lading to prepare a waybill and a freight bill. The shipper retains any additional copies and files them for record keeping and report generation.

chapter 8

Contracting for Transportation Services

chapter objectives

After reading this chapter, you will understand:
The rationale for contracting.
The key tasks in each phase of the contracting cycle.
The processes, advantages, and disadvantages of competitive bidding and the negotiation methods of contracting for transportation services.
Some of the pitfalls in contracting.

chapter outline

Introduction

The Contracting Cycle
Background research
Solicitation
Evaluation and award
Administration

Potential Pitfalls in Contracting
Traditional carrier duties
Standard tariff provisions
Contingencies
Reciprocity and conflict of interest

Summary

Questions

Appendix 8.1—Checklist for Contracts Between Shippers and Railroads

Appendix 8.2—Background Research for Contracting

Introduction

Since the 1970s, when Congress and the ICC began to relax regulation of contract carriage, contracting has become an increasingly popular and important method of acquiring transportation services. In the field of traffic management, contracting refers to the tasks, skills, and activities necessary to conduct contractual relationships in the acquisition of transportation services. Developing and managing contracting programs now constitutes a greatly expanded and important responsibility for the traffic function.

In general, carrier pricing managers and senior traffic executives forge agreements and submit them for review by their respective legal staffs. To create effective agreements, traffic managers must know contracting methods, elements of contract law, cost concepts, and cost estimating techniques, and they must have good negotiating skills.

This chapter investigates the different phases of the contracting cycle, the rationale and the fundamental methods of contracting, and some of the pitfalls of contracting. Appendix 8.1 illustrates annotated contract clauses for railway carrier agreements. Chapter 9 focuses on the topics of costing and negotiation.

No attempt has been made to cover the elements of contract law, so separate study of that subject is recommended.

Although the discussion concentrates primarily on long-term contracts lasting a year or more, many of the concepts, principles, and issues included in this chapter also apply to short-term contracts.

The Contracting Cycle

Activities that constitute the entire contracting process generally go through well-defined phases. As shown in **Exhibit 8.1**, these phases include background research, solicitation, evaluation and award, and administration. All together, they define the *contracting cycle*. Actual activities, of course, may vary considerably among firms and in different situations.

Background Research

Thorough preparation is especially important for successful contracting. The contracting action, like the other methods of purchasing transportation services, originates in the organization's planning activity after needs have been identified and contracting has been chosen as the best purchasing option.

What transportation needs? Before selecting and acquiring transportation services, the traffic manager must clearly identify the firm's present and projected needs. IBM provides a good illustration of the scope of this task in **Exhibit 8.2**. In most instances, the shipper should be prepared to specify transportation needs by origin and destination. The types of questions a traffic manager needs to answer include:

Exhibit 8.1 **Contracting cycle in traffic management**

Background research	Solicitation		Evaluation and award		Administration
	Sourcing → Notification		Opening → Closing		
What needs?	What partners?	Invitation for bid?	Responsive?	Block in details	Performance
Why contract?	How many?	Request for proposal?	Price/cost criteria	Legal review	Changes
What contracting method?		Request for negotiation?		Preaward survey?	Termination
					Renewal
					Contingencies

■ Should the shipper choose door-to-door, terminal-to-terminal, stopoff, refrigeration, or some other kind of service to meet its needs?

■ What levels of service—same-day, next-day, or one-week delivery—are acceptable?

■ How much is the shipper willing to pay for such service?

■ What amount of revenue will the traffic generate for the carrier during the contract period?

■ Is the traffic suitable for contracting?

Appendix 8.2 presents a comprehensive list of questions like these. Although this list is extensive, the traffic manager should view it as suggestive rather than definitive. The variety and form of contracting elements are limited only by the knowledge and creativity of the parties to the contract.

More sophisticated organizations conduct transportation system audits to support traffic planning (see Chapter 6). Reviewing past policies and practices that determine transportation requirements is an important part of these audits. Such reviews, for example, helped IBM discover that some customers did not

---

■■■■■■ *Exhibit 8.2* **Data IBM gathers for carriers before requesting bids**

- What its volumes are
- What frequencies are required
- What timing is needed
- What types of vehicles are necessary
- What the product description is
- What level of service is expected
- What the measurements are
- What sizes are involved (dimensions and weight)
- What the pickup address is
- What the delivery address is
- What extra services are required (e.g., security)
- What the billing instructions will be
- What cargo and warehousing liabilities are wanted
- What the warehousing requirements are
- What the claims inspections and waiver requirements are
- What reports are needed
- What days service is not required (e.g., floating and statutory holidays)
- What the guidelines are for application of rates
- What the award rationale will be on bids submitted
- What sensitivity and environmental controls are needed
- What packaging will be used
- What third-party carrier involvement there will be
- What is expected regarding shipments to remote locations

Source: Douglas W. Seip, ''How IBM's Bidding Program Reduced Rates by 20%,'' reprinted with permission from *Canadian Transportation and Distribution Management* (Canada's national physical distribution magazine, 1450 Don Mills Road, Ontario), February 1983.

need three-day delivery service for certain high-technology products and that less expensive six-day service would suffice.[1]

Why contract? Background research actually leads up to a decision about which method is best for purchasing transportation services: (1) paying the published tariff rate for common carriage or the going rate for exempt carriage, (2) leasing or buying transportation equipment and labor, (3) using intermediaries, or (4) contracting. For contracting to work, each party should gain something (see **Exhibit 8.3**).

The *carrier* expects a commitment of traffic that assures freight revenues, contributes to its profitability, supports planning, and helps to secure more favorable financing. During the 1981 recession, many carriers used short-term contracts to halt traffic erosion by trading rate concessions for volume commitments. Carriers also use short-term contracts as flexible instruments that permit rapid adjustments to changing market conditions. On the other hand, during periods of general economic growth, the long-term contracts that build stable relationships and place greater emphasis on service elements win more favor.

Although volume commitments represent the centerpiece of transportation contracts, shippers should not overlook the other important elements of freight traffic that affect carrier operations and contribute to carrier profitability. For example, more fully loaded vehicles or higher load factors contribute to better equipment utilization. Likewise, more reload opportunities at destination or more balanced traffic flows, perhaps from carefully coordinated inbound and outbound traffic patterns, reduce empty backhauls or "deadheading." Further, the frequency of shipments or the use of less desirable equipment can help the carrier operate more efficiently and profitably, especially in the capital-intensive railroad industry.

The carrier can also link performance to revenues. For example, performance clauses in contracts may link slower or less consistent transit times to lower rates and less revenue. This approach is relatively new to railway management. Traditional railway operating department responsibilities have emphasized operating efficiency over service considerations, such as that generated by long, slow trains. The carrier, though, is not likely to alter its operations to meet the shipper's needs, unless a profit can be made.

For the *shipper*, contracting not only opens the way for securing lower rates, it also permits maximum flexibility to mold transportation services to fit the firm's overall operations and strategic plans. Contracting hands the traffic manager a powerful tool for both solving distribution problems and creating new ways to improve even well-managed logistics systems. Specifically, contracting may benefit the shipper in some or all of the following ways:

1. Reduced rates. Getting reduced rates is probably the most important single reason for contracting. Negotiating leverage and the potential size of rate

■■■■■■ _Exhibit 8.3_ How contract features affect shippers and carriers

Feature	Carrier Impact	Shipper Impact
Lower rate for guaranteed tonnage	Usually means high total revenue. Guarantee of business enhances efficient operations and assists in financing.	Ties shipper to carrier. Reduces rate. Reduces shipping options
Lower rate for guaranteed percentage of shipper's traffic	Same as above	Shipper must still use carrier in business upsurge.
Rate variable according to transit time	Carrier gets higher revenue if speeds shipment; lower revenue for slow service. Makes operating department accountable for service. Good carrier incentive system	Introduces uncertainty into charges. Good for high value goods shipments. Reduces costs of inventory in transit. Improves utilization of equipment owned or leased by shipper
Reduced rate for shipper finding backhaul shipment	Reduces empty car miles. Reduces costs of marketing and sales	Lower overall costs for car use
Carrier pays shipper an allowance, if it gets minimum percent of shipper routings.	Higher revenue	Lower overall transportation costs reduces options.
Higher charge for superior car	Carrier obtains higher revenue for more expensive car or vehicle	Shipper pays proportionately for standard versus special equipment. Higher charge might be less than car preparation of standard car.
Scheduled service	Requires tight operating control. Predictable operating needs meet shipper demands.	Can tighten inbound interplant and outbound move schedules. Can reduce inventory costs of goods in transit.
Special service	Obtain revenue from a service in addition to transportation (storage, en-route product handling, delivery inside customer facility, etc.)	Service might be more convenient than if provided by shipper firm. Service might be less costly than if provided by the firm.
Reduced rate for less free time	Improves equipment turnaround time	Requires fast loading and unloading. Provides incentive for this
Reduced rate for loss and damage allowance	Reduces loss and damage cost component	Shipper might be assured of lower loss and damage through efforts such as better packaging, bracing, etc., or the use of insurance.
Fee for assigned equipment	Receive revenue for cars assigned to shipper. Can enhance car utilization	Forces use of equipment or possible loss of it
Additional charge for fast return of shipper equipment	Incentive to move empty car of shipper on backhaul	Higher car utilization
Stable rate for long period	Problem if inflation high	Rate stability in purchasing or sales

reductions generally depend on the size and length of the volume commitment. Thus, many firms find it desirable to commit the total traffic of one or more product groups for entire divisions or regions, or even for the total system.

2. Less tariff support. Costly tariff support activities such as library maintenance, elaborate rate retrieval systems, and detailed rate audits can be reduced significantly or, perhaps, eliminated. Meanwhile, shippers and carriers can simplify rates for easy conversion to automated systems.

3. Better information exchange. A long-term contract helps to create a favorable environment for the installation of synchronized computer and EDI technology. The traffic manager can work with the carrier to develop and transmit information necessary to build customized reports that have useful details about linehaul performance, equipment utilization, freight expenditures, claims, and so forth. Further, the electronic transmission of bills of lading, freight bills, and claims can generate productivity gains by reducing paperwork and improving cash management.

4. More control. Contract agreements with relatively few carriers make performance surveillance easier. In addition, the shipper can devise contracts that offer protection against rate fluctuations and that include service guarantees for delivery and for the supply of specialized equipment. Greater rate and service stability coupled with better information exchange allows the traffic manager to have more confidence in the carrier and to make planning more effective. These guarantees have the potential to reduce inventory investment without sacrificing raw material, spare parts, or finished product availability. Guarantees also improve cash flows from payments tied to delivery dates, and they increase the productivity of shipper-owned equipment.

5. More confidentiality. Shippers see confidentiality as an important feature of contracting.[2] As discussed in Chapter 3, carriers do not have to divulge sensitive terms and conditions that might offend customers in competition with each other. Likewise, the shipper does not have to publicize any competitive advantage won through contracting.

6. More design flexibility. Contracting affords tremendous design flexibility. The terms and conditions can address virtually every aspect of transportation operations, documentation, information, financing, and so on. **Exhibit 8.4** lists a variety of contract-related innovations reported by shippers and railway companies.

For *both parties*, contracting paves the way for more cooperation. The shipper and the carrier shed adversary relationships to form partnerships. As partners, they are likely to create an atmosphere in which they will understand each other's operations, and thus, form a relationship that can benefit both parties. Ford

■■■■■■■■ *Exhibit 8.4* **Innovations in rail-shipper contracts**

- Tailor-made unit train and trainload operations
- Special credit and late payment terms
- Specialized demurrage and loading provisions
- Use of idle cars for temporary warehousing
- Elimination of claims below designated amounts
- Reload provisions to increase car utilization
- Special private car mileage payments
- Backhaul and triangular reload provisions
- Variable rates to respond to changing grain markets
- Preservation of abandoned service through intermodal contracts and volume commitments
- Investment risk sharing
- Guaranteed service and equipment
- Equipment use provisions to reduce cross hauling of empties
- Short-term "economy specials" to stimulate movements of certain commodities
- Allocation and better utilization of pier berthing space
- Load-ahead program for lumber
- Round-trip boxcar rates
- Market share incentives

Source: Interstate Commerce Commission, *Report on Railroad Contract Rates Authorized by Section 208 of the Staggers Rail Act of 1980* (Washington, D.C.: Office of Transportation Analysis, Section of Rail Services Planning, March 13, 1984), pp. 17–18.

Motor Company, for example, builds long-term relationships with transportation suppliers that actually contribute to Ford's logistics planning, as well as to technological and productivity improvements.[3]

Actually, the idea of partnership is well known to purchasing managers as a key element of *value engineering.* This concept, pioneered in the 1950s by General Electric, refers to a program that systematically searches for ways to improve existing designs instead of developing new ones. A vendor, for example, might discover some modifications to original contract specifications that allow certain component parts to be eliminated or simplified without altering product quality. Both parties agree to make the changes and share the benefits.

Similarly, in the contract carriage situation, a carrier might discover "design changes" that create freight consolidation, reloading, lane balancing, packaging, or other opportunities for improved productivity. In this light, computer and EDI applications can be viewed as value engineering.

Contracting has several disadvantages as well. Although contracting has tremendous potential in traffic management, the absence of competition or large traffic volume may quickly nullify much, if not all, of that potential.[4] Small volume shippers, as well as shippers that reside near the lines of large carriers, may have difficulty realizing some or all of the benefits already cited. They may find themselves at a disadvantage to larger competitors. These shippers must try to consolidate freight; use intermediaries such as freight forwarders, shipper associations, agents, or brokers; or look to other modes, including private carriage.

In addition, even when contracting is an option, it is likely to require a good deal of time and effort. Contracting generally is expensive and complex, and three to six months or longer may be required to gear up for the process. The contract program must be designed and staffed. Some reorganization may be required, for example, to create a centralized unit that can work with total company traffic to gain negotiating leverage. Further, legal resources must be drawn on for contract reviews.

The need for good negotiating, costing, and basic legal skills imposes additional training and educational requirements. Shippers cannot overlook contract administration of changes, contingency clauses, performance compliance, renewal, or termination, all of which begins *after* the contract is signed. Moreover, the cost and complexity of contracting means that nonperformance or default by the carrier generally will inflict greater penalties in contract than in noncontract carriage.

What contracting method? Once the decision to contract has been made, planners must include an evaluation of the two principal methods of contracting: competitive bidding and negotiation. Each method, of course, has variations. This section of the chapter examines each of these methods in terms of what it is and when it should be used. The discussion of other phases (solicitation, evaluation, and administration) reveals how each method is used.

Although *competitive bidding* is an established practice in purchasing management, it is relatively new to the field of traffic management. This approach requires the traffic manager to solicit competitive bids for well-defined transportation services and, all other things being equal, to make the award to the lowest bidder. Ordinarily, the bid represents a firm rate for services rendered. IBM and GTE are two highly visible companies that have implemented bid programs in the traffic function in recent years.

Competitive bidding produces the best results when *all* the following conditions are met:

1. Suitable traffic. The potential traffic revenue must establish sufficient incentive for the carrier to offer a competitive rate-service package. Other traffic characteristics, like balanced flows, also add incentives by cutting carrier operating costs. Nonetheless, the most important single requirement for a successful bid program is having traffic that permits the shipper to precisely define its needs. Bidding is most effective when transportation needs are clearly understood by the carrier. Carriers tend to inflate their bids to compensate for unspecified contingencies when the service requirements are vague. For similar reasons, service requirements that remain firm throughout the contract period make the bid method more effective. However, requirements should not be so restrictive that they limit the number of bidders to the point where competition suffers.

2. Adequate competition. In practice, a minimum of two carriers can provide adequate competition, but more bidders are preferable. Consideration of alternative routes or intermediate "bridge" carriers might enable the traffic manager to increase bid competition. The pool of potential bidders, however, comprises only the carriers that actually have the capability to meet the shipper's transportation service requirements.

3. Sufficient time to prepare. As already indicated, contracting frequently requires considerable background research and preparation. For initial programs, shippers may require five months or more to get started.

The *negotiation method* uses discussion, bargaining, and compromise to reach agreements of mutual benefit. The exact nature of the contract evolves from negotiations in which all aspects of rates and services are eligible for discussion. Thus negotiation offers a flexible approach in which rates may be revised, needs explored, and problems identified. Railroad contracts rely almost exclusively on the negotiation method.

This approach is generally preferable in the following situations:

1. Unsatisfactory bid conditions. If the shipper cannot satisfy all the bid conditions, then the negotiation method offers more promise than competitive bidding.

2. Traffic uncertainty. As in competitive bidding, the volume of traffic and the expected savings must justify the expense of formal negotiations. Where future traffic volumes may change significantly, say because the risk of economic recession and declining traffic volumes is high, the negotiation method is generally preferable. It permits the shipper and the carrier to examine and discuss economic adjustments and contingencies.

3. Significant change in carrier operations. If the carrier must alter its operations significantly to meet the shipper's needs, a thorough analysis and discussion of future relations and services is prudent. The negotiation method facilitates this type of analysis and discussion.

4. Available cost data. An experienced traffic manager, no doubt, will have a fair idea of what represents an acceptable range of rates. Nonetheless, cost data are necessary for reasonably accurate estimates of the carrier's unit costs and for effective evaluation of proposed rates. Consulting firms and other commercial vendors sell costing software complete with cost data (see Chapter 9).

Without these data—perhaps the data are unavailable or too costly to obtain—the traffic manager should rely on bid competition to obtain a reasonable rate.

Solicitation

During the solicitation phase, shippers invite bids or request negotiations for transportation services from the most qualified carriers. This phase involves two fundamental tasks: sourcing and notification.

Sourcing. The initial task is to survey the field and identify qualified carriers, that is, those having the capability and the commitment to meet the shipper's needs throughout the contract period. In purchasing and materials management, identifying and selecting qualified ''responsible'' sources of supply is called *sourcing;* it represents one of purchasing's most important tasks. Likewise, sourcing plays a key role in transportation contracting because the shipper will be searching for a group of potential partners.

After compiling a list of potential carriers, the shipper must screen the list to identify the most qualified sources of transportation supply. This process involves two key issues:

1. How to screen potential partners? Before making a long-term commitment of traffic, the traffic manager has to evaluate the strengths and weaknesses of carriers to ensure that only qualified carriers participate. Qualified carriers will have the operational and financial capability, as well as the business integrity, to meet the shipper's needs.

The level of traffic revenue, the complexity and duration of the contract, and the amount of knowledge or experience with carriers essentially determine the scope of a traffic manager's evaluation. More comprehensive analyses are necessary when the traffic revenues are large, the contract is complex or long term, and the shipper knows little of the carrier.

Generally speaking, the evaluation must consider three facets of the carrier's business. First, the traffic manager should investigate the carrier's *operational capability.*

- What is the carrier's record of, or reputation for, on-time delivery, loss and damage, strikes, and so on.
- Is there sufficient equipment, like special-purpose vehicles such as covered hoppers or refrigerated trailers, to ensure the necessary level of transportation supply?
- Does the carrier serve some or all of the shipper's traffic lanes?
- Does the carrier have computer data processing or EDI capabilities?

To get a clear idea of each carrier's operational capability, the traffic manager must ask these questions and others. Appendix 8.2 contains a comprehensive list of questions. (Chapter 2 amplifies that list with a discussion of the various transportation selection criteria.)

Second, the traffic manager will want to examine the carrier's *financial capability.* Financially weak carriers pose the greatest risks of nonperformance of contracted service standards. These risks imply higher administrative costs; for example, the firm will have to invoke penalty clauses or terminate the contract and begin negotiations with another carrier if the carrier fails to perform its duty. And recovery of freight left at terminal locations of bankrupt carriers may require extensive efforts. The greater risks here also imply a higher incidence of substandard service and higher total costs for the logistics system. Furthermore, if financially weak motor carriers are unable or unwilling to comply with the legal requirements for insurance, shippers may be exposing themselves to a serious risk (see Chapter 2).

Nevertheless, given a long-term commitment of sufficient revenues, an otherwise weak carrier may perform as well as any other qualified carrier. In these instances, the shipper may negotiate a rate-service combination commensurate with the higher risks.

An analysis of financial capability focuses on the carrier's income and balance sheets, which are published in quarterly or annual reports.[5] From these reports, American Trucking Association sources, or financial directories such as *Standard & Poor's Register,* the analyst can develop standard financial ratios that measure the health of the firm. Commercial consultants, of course, can perform the analysis for the shipper. Although not a financial expert, the traffic manager should be able to use this information intelligently.

Third, the evaluation should consider carrier *management capability and orientation.* Besides ability, carrier management should exhibit a clear record of good business integrity, especially if the carrier is an untried source. The traffic manager needs to have confidence in the carrier's degree of commitment to the letter and spirit of the contract. Moreover, if the traffic produces intense peak-period or seasonal demands, the carrier must demonstrate an ability to handle pressure.

It is also helpful to know if carrier management is service or operations oriented. As a general rule, operations-oriented carriers will try to secure traffic that meets *their* operating requirements, whereas service-oriented carriers will be more inclined to alter operations to meet *shipper's* needs.

The structure of management can reveal helpful insights into the carrier's capabilities, which are especially important when dealing with small- to medium-size companies. A high turnover rate among managers is one warning signal, for management instability is not conducive to the long-term development of productive shipper-carrier relations. A workforce dominated by one person is another warning signal, because the management of carrier operations and finances might become less effective in the absence of that person.

2. How many carriers? A study of transportation strategies for the 1980s commissioned by the Council of Logistics Management concluded that a promising strategy would be to concentrate total traffic with relatively few carriers to increase negotiation leverage and maximize volume discounts.[6] The specific policy adopted by the traffic function, however, generally depends on the nature of the traffic. Concentrating traffic with few carriers should work well for bulk or nonbulk commodities moved in volume quantities by rail, water, or truck transportation. The transport supply is stable for these commodities, and concentrating traffic with few carriers enhances leverage. The traffic manager, in fact, may have trouble finding more than a few rail carriers, for mergers have greatly expanded the scope of single-line service.

When contracted service calls for nonvolume shipment sizes, deciding how many carriers to use primarily depends on the firm's contracting objectives. Kaiser's goal, for example, is to have many strong carriers support its plants; thus, it spreads the traffic among many motor carriers. Likewise, IBM Canada Ltd. has established a competitive bid program that involves more than two-hundred motor carriers.[7]

In contrast, Ford assembles various inbound and outbound "contracting" packages that commit certain traffic to one or a few rail carriers for up to five years.[8] Ford abandoned the principle of using multiple carriers to ensure the supply of transportation services; instead, the company chose to establish long-term contracts with few carriers. Ford wants to establish an environment in which the carrier can handle freight more efficiently and will share the cost savings. Ford also wants to encourage innovations and better logistical tradeoffs. For similar reasons, the company accumulates and organizes its traffic to move over key traffic lanes and awards contracts through competitive bidding to a few carriers. For its domestic operations in two large rate territories, Ford reduced the number of truckload carriers from eleven to two, while six carriers handle almost all the nationwide less-than-truckload business.

Similarly, GTE has recently developed a bid program that concentrates about $25 million annually for two years in the hands of several air and motor carriers.[9] Previously, GTE used about seven hundred carriers, of which five hundred each handled less than $25,000 of traffic annually. The primary purpose was to gain purchasing leverage and to cut the cost of specifically tailored transportation services.

Notification. Invitation for bids (IFBs) or requests for negotiations (RFNs) may be sent to carriers through formal advertising, direct correspondence, or informal communication. All three methods of notification can be used in competitive bidding; however, the formal advertising approach, where the traffic manager gives public notice of IFBs in commercial or trade publications, helps promote competition. If enough qualified sources assuring competitive bids are already known, more personal forms of notification such as letters, telephone calls, or contacts with sales representatives may be desirable to encourage commitment and reduce costs. The negotiation method, too, generally relies on correspondence or informal communication to notify carriers of requests for negotiations or proposals.

Often prebid or prenegotiation conferences can make contracting more effective. Conferences, which may be held with individual carriers or groups of carriers, are especially helpful to spell out the details of service requirements or evaluation procedures in complicated contracts, thus avoiding bid mistakes. Conferences may also explore and identify precise needs. These meetings also place all interested carriers on equal footing in a competitive process. Uncomplicated contracts do not warrant the time and expense of these conferences.

Evaluation and Award

The evaluation and award phase of the contracting cycle involves the most critical and, often, the most complex tasks. To facilitate discussion, this phase is presented in the context of opening and closing activities.

Opening activities — Competitive bidding. In competitive bid programs, the shipper opens sealed bids for evaluation. The first task is to eliminate unresponsive bids or isolate them for corrective action. Responsive bids meet deadlines and specifications and are free of mistakes. Nonresponsive bids occur, for example, when the shipper requests bids for all traffic lanes and weight groups, and some carriers submit bids for only selected lanes or weights. GTE, for example, found that some bids for a common set of traffic lanes, which were to form the data base for evaluation, were incomplete. Other unresponsive bids introduced rates contingent on total contract traffic volume, even after GTE had held prebid conferences to explain its programs.[10]

If all carriers are equally qualified and all bids equally responsive, evaluation simply requires selecting the lowest bidder. Shippers, however, often find that carriers show a complicated mixture of strengths and weaknesses. The lowest bidder, therefore, may not produce the best results. Instead of concentrating on the lowest bid, the traffic manager must judge how alternative rate-service combinations will affect total logistic system performance.

Consequently, some form of objective analysis should help. As shown in **Exhibit 8.5**, one approach is to list specific service attributes and assign values that reflect their relative importance. The list, of course, ought to include the cost of service as an item to be ranked. An average ranking or some other scoring system can help reduce the number of bidders. Subjective business judgments, as well as more rigorous quantitative analyses, could then be applied to those that are similar in ranking.[11] To support the quantitative aspects of the evaluation, spreadsheet software appears to have a good deal of promise. Regardless

███████ *Exhibit 8.5* **Corporate criteria ranking**

1982 Air Freight Program Bid Criteria Evaluation*

21	Best rate for GTE
18	Immediate discount on face of each bill
16	Door-to-door-service
3	Released value liability
7	Monthly printout
9	Discount on all inbound and outbound
17	Discount coverage on minimum shipments
15	Discount allowance for multiple shipments
19	Volume-sensitive discount
11	Good-faith start
14	Airbill and bill of lading matchup
20	Service standards
4	Ninety-day claim settlement
2	Three-year claim filing period
1	Specified claim and limit
12	On-line tracing
8	No dimensional weight
13	Matching of all present and valid future rates
6	Fixed contract period with renewal options
5	Timely billing (within one week)
10	Negotiable rate escalator

*Most important valued 21; least important valued 1

Source: George A. Yarusavage, "Carrier Negotiation and Selection After Reregulation: One Shipper's Experience," unpublished paper submitted for partial fulfillment of the requirements for certified membership in American Society of Transportation and Logistics, 1983. This material is the property of the American Society of Transportation and Logistics, Inc., and is reproduced for educational purposes only by express permission. No other reproduction is allowed without express permission.

of the approach followed, the traffic manager should coordinate decisions with other departmental staff managers who will be affected.

Opening activities — Negotiation method. In contrast to competitive bidding, the negotiation method initially calls on the traffic manager to present the firm's transportation service requirements. A carrier may start the process with a presentation of its proposal for contract service. The opening presentation normally covers all the basic elements, including type of traffic and corresponding volumes by lane, expected service levels, equipment needs, and costs. Negotiations follow the opening presentation(s) until a preliminary handshake signals that the parties have reached general agreement on the key elements of the contract. One party will take responsibility for preparing the initial draft.

Responsiveness, as defined for competitive bidding, is not an issue in the negotiation method. The goal is to promote flexibility and encourage discussion of alternative solutions. Thus, to the extent that carriers express interest and negotiate in good faith, they are responsive.

Closing activities — Competitive bidding. Unlike the negotiation method, which involves extensive closing activity, the competitive bidding approach generally confines the closing to the legal review, signing, and award tasks. Boilerplate clauses already comprise elements of the service specifications that are often established unilaterally by the shipper before bidding. Regardless of the contracting method selected, however, the traffic manager may want to conduct a "preaward survey," whereby staff members, perhaps at various plants, visit the carrier's facilities to get a closer look at actual capabilities.

Closing activities — Negotiation method. Formal negotiations entail extensive closing activities. After studying the draft, the negotiators refine the terms agreed to in the opening and add, or "block in," myriad supplemental details. The shipper and carrier negotiate the numerous clauses, (see Appendix 8.1) called boilerplate, that address issues such as credit, billing, payment, reports, liability, claims, penalties, default, termination, or renewal. Interdepartmental coordination is an integral part of the entire planning process, and traffic staff often find it helpful to maintain communications with various department managers as the negotiators block in the details. This helps to eliminate potential pitfalls, which are discussed in the last section of this chapter.

The legal review, the award, and the signing follow closing negotiations. Lawyers for both parties examine the contract elements and language to ensure the contract follows sound legal practices and does not violate antitrust law. The shipper then awards the contract. It should be emphasized that the traffic manager also seeks competition when using the negotiation method and, when possible, makes the award to the carrier that ultimately offers the best package of rates and services.

Administration

The last phase of the contract cycle is administration, in which the traffic function handles changes or modifications to the contract. These actions are appropriate when carriers experience problems, such as labor strikes, or when changing economic conditions, such as rapid inflation, invoke contingency clauses or when value engineering applications appear. In addition, contract administration includes default, termination, and renewal actions. The legal nature of default and termination actions probably will mean added responsibility for the legal staff and more coordinating responsibilities for the traffic function.

The length of the contract period and the style of management generally determine the level of contract administration. Long-term agreements, as well as a formal management style, tend to produce thick contracts filled with rigorously defined clauses that trigger administrative activity. Some organizations, on the other hand, preferring a less conservative approach, will rely on informal working relationships. Experience with good performers, agreement on the particulars, and good faith offer enough protection for contract service. In these situations, little boilerplate is actually set down in writing.

Potential Pitfalls in Contracting

Because contracting can be complicated, the following material highlights potential problem areas that deserve special attention.

Traditional Carrier Duties

Shippers may assume wrongly that the traditional common carrier duties to which they are accustomed carry over to contract carriage. A comprehensive framework of common law, statutory law, tariff rules, and regulations guides these duties and offers the shipper considerable protection. For contract carriage, however, only what is written in the contract essentially affords protection.[12]

The issue of liability merits special scrutiny. Without express provisions in the contract, the relatively strict level of common carrier liability (discussed in Chapter 12) is not in force. Courts view contract carriage as a category of private transportation in which liability for loss or injury applies only when the carrier fails to exercise ordinary care. Although court actions for failure to perform—that is, for breach of contract—are still possible, such actions are likely to be limited to the issue of negligence.[13] By contrast, when performing common carriage, carriers remain liable for the full value of the goods unless an authorized exemption applies. The level of liability is sufficiently strict to say carriers are liable as insurers of the goods.

The traffic manager, therefore, must carefully negotiate and clearly specify the terms of liability. Prudent contracting dictates that the parties agree on equal protection from events beyond their control, that is, on the wording of the force majeure clause.[14] For example, a government quarantine may exempt the carrier from liability for delayed delivery. But equal treatment would exempt the shipper from liability where higher rates are incurred from not meeting contracted traffic volume requirements because of the quarantine.

Standard Tariff Provisions

Standard tariff provisions set forth minimum packaging rules for the safe transportation of property, as well as detailed rules for loading, unloading, routing, and virtually every other aspect of transportation rates and services. These rules, like traditional common carrier liability, do not automatically extend to contract carriage.

A convenient practice recommended by traffic professionals is to include specific tariff rules, say for packaging, or simply reference appropriate tariffs, such as the Uniform Freight Classification. This action makes the referenced matter a part of the contract. However, the traffic manager may want to simplify the typically cumbersome tariff language by rewriting the rules in plain English.

Contingencies

Potentially serious disputes can be avoided if shippers and carriers plan for contingencies by deciding whether adjustments or changes will be made unilaterally, bilaterally, or automatically. Contingency provisions for productivity improvements, rates, service performance levels, and minimum volume requirements are especially important when devising long-term contracts.

Shippers cannot afford to ignore probable improvements in carrier productivity, especially when dealing with railway companies. The savings from future cost reductions ought to be an item for negotiation. Otherwise, shippers can end up paying more than is necessary for transportation. Shippers without contracts or signing at later dates, moreover, will have a competitive advantage.[15]

Shippers must watch out for biased indexes that are used in rate escalation clauses. The rate escalation clause, a fairly standard feature in railroad contracts, authorizes the automatic adjustment of rates to offset rising costs. A potential pitfall in the use of escalation clauses lies in the selection of an index to measure rising costs. Critics contend, for example, that the regional or national indexes developed by the Association of American Railroads may seriously overstate the level of rising costs.[16]

In addition, these indexes reflect regional or national averages when the most accurate approach would tie the weights for key-cost components to specific routes. The carrier, however, may be less than enthusiastic about divulging that kind of proprietary information. Thus, the index weights that can be derived from the specific carrier's average costs, which are published in financial statements, appear to offer the best alternative. Unfortunately, designing these indexes is technical and complicated, and the effort will require knowledgeable staff.

Like the carriers that do not want to lock in rates over the long term when costs are likely to rise, shippers do not want to get caught with fixed rates when market competition induces discounting. Given adequate leverage, traffic managers may insist on the application of a "most favored shipper" clause. The clause specifies that contract rates cannot exceed the lowest applicable tariff rates that are legally published. For example, the carrier that cuts tariff rates below the contracted rate levels must also reduce the contract rates. GTE, which annually guarantees about $25 million in traffic revenues to a handful of winning bidders, has employed this clause successfully.

Reciprocity and Conflict of Interest

Traffic managers engage in reciprocity when they favor the carriers that buy the firm's products. The firm, in effect, trades purchases of transportation services for purchases of manufactured goods or some other type of products. An alert traffic manager will watch for the trading opportunities that serve the firm's interests and will coordinate action with marketing staff.

Shippers, however, should know that reciprocity occupies a gray area between legal and illegal behavior (see the section on antitrust in Chapter 3). To the extent that trading activities involve arm twisting or restraint of trade, they are illegal. Apparently, the threat of antitrust action has led many companies to purge "trade-relations staff" from their organization charts.[17]

When the traffic manager has a direct interest in a carrier that seeks the firm's traffic, such as stock ownership, there is a conflict of interest. The firm, of course, gets hurt when contract awards for significant transportation purchases are made to less desirable carriers. As buyers, traffic managers should abide by the same professional and ethical standards that purchasing managers follow.[18] Developing a policy manual that clearly explains the expected standards of conduct is sound management practice.

Summary

In the freer regulatory environment of the 1980s, contracting has become an increasingly important method of purchasing transportation. Contracting includes the tasks, skills, and activities necessary to conduct contractual relationships in the acquisition of transportation services.

Activities that make up the contracting process tend to pass through the following phases: (1) background research, (2) solicitation, (3) evaluation and award, and (4) administration. Altogether, these phases define the contracting cycle.

Background research of the firm's present and projected needs will determine whether contracting is appropriate and, if so, what contracting method to use. For contracting to work, each party to the agreement should gain something. Carriers usually expect a commitment of traffic, although other elements of freight traffic can contribute to efficient operations and profitability. Shippers benefit from reduced rates, less tariff support, better information exchange, more control, more confidentiality, and better design flexibility. Both the shipper and the carrier may benefit from applying the concept of value engineering to transportation contracting.

The absence of large volume traffic and effective competition can nullify the advantages of contracting. In addition, contracting programs generally are expensive and complex; they require staff to be trained in negotiating, costing, and contract law. Further, contract administration can be costly.

Competitive bidding and negotiation are the two principal types of contracting. The choice of method primarily depends on certain conditions that involve traffic characteristics, bid competition, planning lead time, carrier operations, and cost data.

In the solicitation phase of the contracting cycle, shippers invite bids or requests for negotiations. Both sourcing and notification tasks are accomplished in this phase. Two key issues are how to screen potential partners and how many carriers to use.

The evaluation and award of contract bids or proposals is generally the most critical and complex phase. During opening activities in the bid method, the shipper first eliminates or corrects unresponsive bids and then evaluates the bids using both objective and subjective criteria. In the negotiation method, a proposed course of action is presented initially and discussion of possible solutions ensues. Extensive closing activities occur after a preliminary agreement is reached. These activities include the ''blocking-in'' of details and the legal review. By contrast, the closing in the bid method essentially requires only the legal review.

Contract administration defines the last phase of the contracting cycle. Administrative tasks include contract changes, modifications, default, termination, and renewal.

Traffic managers need to stay alert for several pitfalls in contracting. First, managers may wrongly assume that traditional common carrier duties apply to contract carriage. However, only what is written in the contract affords protection to the shipper. Second, tariff rules, which set forth standard provisions for virtually all aspects of a shipment, do not automatically extend to contract carriage but must be referenced in the contract. Third, when negotiating contingency clauses, shippers need to examine inflation indexes carefully and to address the issue of bilateral or unilateral changes. Fourth, senior managers must watch for potential antitrust violations involving reciprocal purchase arrangements and for conflicts of interest by the staff that make transportation purchases.

Selected Topical Questions from Past AST&L Examinations

1. You are to draft a proposal for a contract rate in the following situation. Your objective is to minimize total logistics costs while satisfying the customer service goals of your firm. Therefore, your proposal should include specific provisions and any penalties that you think necessary if either the shipper (you) or the carrier do not perform to the terms of the contract.

   Situation: You are a manufacturer of a full line of small kitchen appliances (microwave ovens, toasters, food processors, and so on) from a single manufacturing plant in New England. You ship your products nationally and currently use public distribution centers in Atlanta, Chicago, and Salt Lake City. Your normal terms of sale are F.O.B. delivered, and your firm has experience in dealing with all modes of transportation.
   (a) Draft the above proposal.
   (b) Do you think the ability to make contract rates would favor the large or small shipper?
   (c) Would the ability to make contract rates have a different effect on the shipper of bulk commodities? (Fall 1981)
2. Present and analyze at least four key factors that should be considered by both shipper and railroad when establishing a contract rate. (Spring 1981)
3. Contract rates are said by some experts to be a major benefit of the 1980 Staggers Rail Act. Do you agree with that statement? If so, indicate clearly just what a contract rate is and how it differs from other rates. If you disagree, indicate what the potentially harmful effects might be. (Fall 1982)
4. Before 1980, J. C. Penney Co. used more than five hundred interstate motor carriers to deliver products to its stores. Today, fewer than a dozen trucking companies are used. Discuss why this trend has taken place. (Fall 1984)
5. The use of the "volume leverage" concept in transportation—the consolidation of an individual division's traffic into a single corporation negotiation package—has grown dramatically in the wake of recent congressional and administrative deregulatory actions. In theory, this concept obtains the lowest overall price level for all divisions regardless of individual size. In actual practice, however, company divisions that have pursued this strategy have been able to obtain even lower prices than the corporate parent, especially in air freight, LTL, and TL traffic. Explain how this result is possible and its consequences. (Spring 1985)

Notes

1. Douglas W. Seip, "How IBM's bidding program reduced rates by 20%," *Canadian Transportation and Distribution Management* (February 1983), pp. 41–42.
2. Interstate Commerce Commission, *Report on Railroad Contract Rates Authorized by Section 208 of*

the Staggers Rail Act of 1980 (Washington, D.C.: Office of Transportation Analysis, March 13, 1984).

3. Richard Haupt, "Profile of today's transportation user," *Traffic Quarterly* 40 no. 1 (January 1986), p. 59.

4. See the views of rail shippers as shown in Interstate Commerce Commission, *Report on Railroad Contract Rates*, pp. 4–5; Benjamin J. Allen, "The potential for discrimination with rail contracts—One point of view," *The Logistics and Transportation Review* 17 no. 4 (1981), pp. 371–385.

5. For a primer on this subject, see Edward J. Marien, "Measuring the financial condition of motor carriers," *The Private Carrier* (April 1984), pp. 16–25.

6. Temple, Barker, and Sloane, Inc., *Transportation Strategies for the Eighties* (Oak Brook, Illinois: National Council of Physical Distribution Management, 1982), p. 154.

7. See Patrick Gallagher, ed., "Stuck with overcapacity, truckers await the recovery," *Handling and Shipping Management* (January 1982), p. 29; Seip, "IBM's bidding program," p. 42.

8. Richard Haupt, Director of Transportation and Traffic, Ford Motor Co., "Carrier-shipper cooperation—A key to profits," speech presented before the Twenty-Ninth Annual L. L. Waters Indiana Transportation Conference (Bloomington, Indiana, April 7–8, 1982), as reported in *Traffic World* (April 19, 1982), pp. 39–41; Aden C. Adams and Carl W. Hoeberling, "Future of contract rates in rail transportation," *ICC Practitioners' Journal* 47 no. 6 (September–October), 1980, pp. 661–664; see also Temple, Barker, and Sloane, *Transportation Strategies*, p. 163.

9. George A. Yarusavage, "Carrier negotiation and selection after reregulation: One shipper's experience," unpublished paper submitted for partial fulfillment of the requirements for certified membership (Louisville, Kentucky, American Society of Transportation and Logistics, Rpt. R-5648, 1983), pp. 1–2; see also John J. Barry, "Transportation: Purchasing a service," *Inbound Transportation Guide* (January 1983), pp. 72–81; "Speakout: Deregulation, the consignee and inbound pricing," *Inbound Traffic Guide* (July 1983), p. 6.

10. Yarusavage, "Carrier negotiation and selection," p. 9.

11. For further reading about the carrier evaluation process, see John J. Barry, "Transportation: The purchasing of a service," *Inbound Traffic Guide* (January, 1983), pp. 72–85; Edward R. Bruning and Peter M. Lynagh, "Carrier evaluation in physical distribution management," *Journal of Business Logistics* 5 no. 2 (1984), pp. 30–47; Kent L. Granzin, George C. Jackson, and Clifford E. Young," The influence of organizational and personal factors on the transportation purchasing decision process," *Journal of Business Logistics* 7 no. 1 (1986), pp. 50–67.

12. For further discussion, see John W. Bagby, James R. Evans, and Wallace R. Wood, "Contracting for transportation," *Transportation Journal* 22 no. 1 (Winter 1982), pp. 63–73.

13. For further discussion, see Michael W. Uggen, "Railroad contract rates: A working analysis of section 10713," *ICC Practitioners' Journal* 48 no. 5 (July–August 1981), p. 534.

14. Robert S. Bernstein, Transportation Manager, Kennecott Copper, "Railroad contract rates," *Traffic World* (June 7, 1982), p. 101.

15. For further discussion, see George H. Borts, "Long-term rail contracts—Handle with care," *Transportation Journal* 25 no. 3 (Spring 1986), pp. 4–6.

16. For a detailed discussion of the indexing problem, see Richard B. Blackwell, "Pitfalls in rail contract rate escalation," *ICC Practitioners' Journal* 49 no. 5 (July–August 1982), pp. 486–502; David L. Anderson and Dean H. Wise, "Rail contract rate escalators: The hidden costs of shipper-carrier agreements," *Traffic World* (July 8, 1985), pp. 64–68. Borts, "Long-term rail contracts," pp. 6–10.

17. Martin, Farris, "Purchasing reciprocity and antitrust revisited," *Journal of Purchasing and Materials Management* (Summer 1981), p. 27; see also Edward McCreary Jr. and Walter Guzzardi Jr., "A customer is a company's best friend," *Fortune* (June 1985), pp. 180–182, 192, 194.

18. See "N.A.P.A. standards of conduct," *Guide to Purchasing*, National Association of Purchasing Agents, 1965.

Appendix 8.1 Checklist for Contracts Between Shippers and Railroads

This checklist is intended to serve only as a guide for consideration in the negotiation of contracts between shippers and railroads. It is not expected that every such contract will include provisions relating to every subject identified in the list and, of course, it is entirely possible that certain matters which should be included in a particular contract have been omitted here. No checklist of this type can pretend to be a substitute for the knowledge, ability, intelligence, and imagination of the negotiating parties.

The Examples referred to in and attached to the Checklist are intended only to illustrate the form of particular types of contract provisions. It is not suggested that any of the Examples is either ''standard'' or preferable to an almost infinite variety of other provisions which could readily be substituted.

  I. Nature of Movement
- A) Product(s)   [Note A]
- B) Origin(s)
- C) Destination(s)   [Note B]
- D) Packaging and/or transportation equipment required
- E) Routing (Example 3; where more than one railroad, consider rights and obligations of each throughout contract; joint or several liability)   [Note C]

 II. Term of Contract
- A) Beginning and end of term (Example 1)
- B) Is term ''keyed in'' to other contract (e.g., purchase or sale of goods)?
- C) Option(s) (whose?) to renew or extend; number of extensions; length of extensions; contract terms during extension period(s) (such terms could change, e.g., after amortization of equipment)

III. Volume
- A) Minimum volume required to be tendered by shipper, in pounds, packages, cars or trains; period(s) during which minimum to be tendered (week, month, year, etc.); minimum per shipment [Note D]
- B) Consequences of shipper's failure to meet minimum tender per

Source: Prepared by Stanley Hoffman for presentation at the National Industrial Traffic League program on ''Rail Contracts: Keys to Effective Negotiating,'' (New York, September 13, 1982). Copyright 1982 by Stanley Hoffman. All rights reserved. Reproduced with permission of Mr. Stanley Hoffman.

shipment or in specified period(s); liquidated damages (could be payment of "standard" tariff rates, or pay for deficit not shipped, or pay difference in carrier profit, etc.); opportunity to "make up" in subsequent shipment(s) or period(s); termination of contract (automatic or at carrier's option): railroad's alternate use of specifically designated equipment [Note D]

C) Maximum volume (per shipment or period, or both) required to be accepted and transported by railroad

D) Consequences of railroad's failure to accept or transport maximum volume; liquidated damages; opportunity to "make up" in subsequent shipment(s) or period(s); termination of contract (automatic or at shipper's option); shipper's right to use alternative transportation or alternative origin(s) and reduction of minimum tender; shipper's right to supply cars, locomotives, etc. (in lieu of carrier's supply)

E) Shipper's right to decrease minimum or increase maximum tender: When? How? Carrier's right to increase or decrease tender: When? How?

IV. Equipment (Cars)

A) Quantity

B) Type(s) (Specification)

C) Who supplies, maintains, repairs, cleans? (Shipper or railroad); mileage allowances (Example 6)

D) Dedication: possible use in connection with transportation other than contract (e.g., return haul)

E) When available? Where?

F) May equipment be used for shipper "advertising"?

G) Consequences of failure to supply or failure to supply adequately:
   (i) at inception of operation—see III(B) and (D), above
   (ii) due to damage, accident (derail, etc.)—maximum period for replacement
   (iii) due to failure to replace—see III(B) and (D) above

H) Right or obligation of equipment—supplying party to sell or lease equipment to other party upon termination, expiration, suspension, etc., of contract; right or obligation of non-supplying party to purchase or lease equipment from other party upon termination, etc., of contract; terms, including price, of purchase or lease

V. Service(s)

A) Definition of "transportation" [Note E]; terminal and accessorial services; transit (Example 17; See Note E); demurrage, penalty demurrage, etc. (Example 2)

B) Hours (times) of operation—switching; continuous train movement; 24 hours per day, 365 days or lesser periods; holidays; weather delays; consequences of force majeure occurrences (see III, above)

C) Notice(s) from railroad to shipper and shipper to railroad; in advance of arrival for loading, unloading

D) Shipper's (receiver's) obligation to load and unload in specified period; consequence of delays (see III, above)

E) Railroad's obligation to transport in particular train(s)

F) Operating conditions relative to safety: speed, train length, in-transit inspections, location of hazardous materials (also hazardous substances or wastes) cars in train

VI. Loading and Unloading Facilities

A) Description; location

B) Who supplies; When? Maintains? Replaces?

C) Consequences of breach of contract with respect to cost of facilities

VII. Force Majeure

A) Definition: include rail accidents? fire? requisition? confiscation? "bad-order" cars? weather delays?

B) Do strikes, etc., of shipper excuse railroad, and vice versa?

D) Consequence of occurrence: reduction in volume (minimum/maximum) (Example 15); include or exclude amortization of equipment? see III, IV, above

VIII. Rates, Charges; Payment

A) Statement of rates, charges for services [Note F]; overtime provisions; charges for additional crews, services, etc.; rate applicable to returned shipments [Note B]; return of empty private equipment (Example 7)

B) Escalation (and de-escalation); reference to index, labor costs, equipment costs, etc. (Example 8)

C) Escalation (and de-escalation) procedures; notices, right to terminate or revise contract; effective date of increase/decrease

D) Tariff rate lower than contract rate (Example 9)

E) Invoices: When, to whom, where? Payment: When, by whom, to whom, where? Discount for early payment? Penalty for late payment? Recourse to shipper/consignee?

F) Overcharge/undercharge: procedures; limitations period (Example 11)

IX. Termination

A) At will: by whom? when? procedure? consequences of early termination (Example 13); liquidated damages; disposition of equipment and facilities; termination on termination of purchase or sale contract

B) For fault (default): Definition of default; termination by whom? when? procedures; remedies for default

    X. Documentation
        A) Bills of lading (Example 4); receipts, notices, records of shipment;
           notations (Example 5); certifications
        B) Weight per package, car, shipment: agreed weights; scale weights;
           consequences of error; freight left in cars
   XI. Liability and Insurance
        A) Liability for employees of railroad, shipper, consignee
        B) Required insurances (public liability, contractual, etc.)
        C) Freight damage: agreed measure of damages (value per pound, ton,
           etc.; invoice value; etc.) (Example 10)
        D) Liabilities under sidetrack agreement(s); consistency with XI (A) and
           (B), above
  XII. Miscellaneous
        A) Assignment or non-assignment or transfer of contract (with or
           without written consent)
        B) Liens on cargo; prohibition against?
        C) Reports and inspections; examination, audit of books and records
        D) Patent indemnities (for equipment)
        E) Notices; where and how given?
        F) Modification of contract
        G) Guarantee by parent or affiliated corporation
        H) Environmental consideration; effect of government regulations?
        I) Confidentiality (Example 18)
 XIII. Tariff Provisions
        A) Reference to tariffs; ''frozen'' to prevent unilateral changes (see,
           Examples 4 and 17); not ''frozen'' (see, Examples 2 and 6)

Notes

Note A

Especially in contracts providing for a minimum percentage of certain traffic, it
is essential to identify the commodity with precision in order to avoid conflicts
as to whether or not the specified percentage of the intended traffic has
been met.

Note B

It is possible to provide for application of the rates specified in the contract from
or to points intermediate from or to a specified origin or destination, respectively; or for reconsignment in transit, or reshipment; or for return movement
to the origin of the returned shipment or another origin specified in the contract.

In such cases, however, where the contract is with a single railroad, care should be taken to avoid contract application via routes or to or from points on other railroads not party to the contract. Similarly, in cases of contract with two or more carriers, care should be taken to avoid single-line application under the contract, especially where such single-line movement could be handled by more than one of the contracting carriers, since such application could be considered to constitute collective ratemaking by the individual carriers. It should be possible, however, to provide for such single-line application by a separate contract between such line and shipper. See Example 12 for a "return shipment" provision, also providing for return after a car has reached a location intermediate to destination, which location is on the line of the second carrier in a two-carrier contract.

Note C

The possibility of subcontracting has been deleted since it does not appear that there are many cases where such an arrangement would be useful; in any event, it would appear to be unnecessary since other arrangements (such as per-car or other allowances to the shipper by one carrier party to a joint-line movement) may be made. Additionally, a subcontract may give rise to questions of applicability of the subcontractor's tariffs, since the shipper has no contract with such subcontractor.

Note D

Whenever a minimum volume is specified, the following possibilities should be considered:

(a) If a dollar amount is specified as payment for a deficit, will such amount be escalated; if so, on what basis? (See Example 14)

(b) If the payment amount is specified by reference to a tariff or specific rate or charge applicable to such deficit, how will changes in such tariff, etc., during the volume period be accommodated? (See Example 14)

(c) If the minimum volume will be adjusted for force-majeure occurrences, what is the formula for such adjustment? (See Example 15)

(d) If the contract is terminated prior to the end of the volume period, will the minimum volume be adjusted and, if so, how? (See Example 13)

(e) If the minimum volume is specified as a percentage of traffic, the traffic should be carefully defined. For example, where a shipper has a choice of more than one railroad route, it is possible to define the relevant traffic in terms of a percentage of rail traffic only. Similarly, if shipments of the same commodity are made in several types of packages or rail cars, the percentage specified should indicate whether it applies to all shipments of such commodity, regardless of form or packaging, or only to shipments of such commodity in a particular form or packaging.

(f) Where the minimum volume is stated in terms of cars, or shipments, or percentages of either, and such minimum is subject to adjustment in the volume period, the disposition of fractions should be provided for. Example: The minimum volume is 120 cars and the adjustment provision calls for a 6% reduction. Is the adjusted minimum 112 or 113 cars (94% × 120 = 112.8%)? (See Example 14)

(g) If a portion of the traffic relied on to make the volume minimum is not controlled by the contracting shipper (or consignee), will the minimum be adjusted if the controlling person requires routing by other than the contract route?

Note E

See Example 2, which is a provision that is useful where line-haul movements are covered under the contract and the rates therefore are specified in the contract itself, without reference to published tariffs. It should be adequate, in most cases, to protect against the "no duty" provision of 49 U.S.C. § 10713(h). For a similar provision which should be considered when the contract covers only a limited service, such as storage-in-transit (Example 17) or where contract movement will be subject to published tariff provisions, see Example 16 which, in addition to protecting against the "no-duty" provision, represents an effort to preserve the rights, obligations, etc., of the parties, despite the provisions of 49 U.S.C. § 10713(i), which provides that transportation under a contract is not subject to the provisions of Title 49 of the U.S. Code.

Example 16 is also useful where the contract is with one carrier but applies with respect to joint-line hauls (e.g., allowances paid to a shipper by an intermediate line-haul carrier). In such cases, it would seem essential to include a provision such as Example 16 because, without it, the status of transportation by the origin and destination carriers is questionable. Is the movement by such carriers (origin and destination) "transportation under such contract," even though they are not parties thereto? Thus, for example, if the contract specifies a minimum carload weight other than the minimum specified in the tariff, which minimum is applicable via the origin and destination carriers who are not parties to the contract?

Note F

Where the rates are set forth in the contract at alternating carload minima, a provision of the type below should be incorporated in the contract:

In no case shall the charges on any Shipment, computed at the actual weight thereof, exceed the charges on such Shipment computed at a higher carload minimum weight specified therefore in Section _____ of this Contract.

Examples

Example 1 — Term of Contract[1]

[Unless disapproved by the Interstate Commerce Commission, (hereinafter, "ICC"), this Contract shall be effective on the effective date specified in Tariff ICC___C___00___, and shall remain in effect for a period of one (1) year and,] thereafter, for successive periods of one (1) year each, unless any of the parties hereto terminate this Contract as of the last day of any such annual period by notice in writing of such termination given to each of the other parties hereto not less than ninety (90) days before the date of such termination.

Example 2 — Transportation and Related Services

Subject to the terms, conditions and provisions of this Contract, Railroads shall, from time to time during the term of this Contract, as, if and when requested by Shipper, accept for transportation hereunder and transport such shipments as may be tendered hereunder to Railroads or any of them, including the return of tank cars after unloading thereof, between the locations and via the routes specified in this Contract. Railroads shall have exclusive control of and be solely responsible for all such transportation.

As used in this Contract, the terms "transportation" and "transport" mean and include linehaul movement of loaded or empty tank cars and all switching, placement, terminal and other acts or operations necessary, convenient or desirable to safely and effectively, and with reasonable dispatch, perform such linehaul movement including, but not limited to, all acts or operations required

---

1. If it is intended to take advantage of the provisions of 49 CFR § 1039.2 (adopted effective June 27, 1983; see *Ex Parte* No. 387, Sub. No. 200, 48 F.R. 23824), which authorizes transportation under a contract to begin on the date of filing the contract, subject to specified conditions, the following may be substituted for the beginning portion of Example 1 shown above in brackets:

   Example 1 — Term of Contract

   This Contract shall become effective on the date when it is filed with the Interstate Commerce Commission (hereinafter, the "ICC") and transportation or service hereunder may begin on such date; provided, however, that performance under this Contract shall be subject to the conditions specified at 49 CFR § 1339.2, and provided, further, that in the event the ICC disapproves or rejects this Contract it shall be deemed null and void with respect to any transportation or service performed prior to such disapproval or rejection. This Contract shall remain in effect for a period of one (1) year and, . . .

by applicable governmental laws, rules or regulations with respect thereto. Shipments hereunder shall also be entitled to such other terminal services and privileges (including, but not limited to, reciprocal switching at origin or destination, with or without charges for line-haul movement) as may be provided in the tariffs of Railroads or in applicable contracts on file with the ICC, and shall be subject therefore to the charges, if any, and the rules, regulations and provisions (i) published in such tariffs as in effect on the date or dates when such services are provided or such privileges allowed, or (ii) contained in such contracts. If and to the extent applicable in accordance with the terms thereof, demurrage rules and charges, as published in Freight Tariff ICC _____, and such supplements thereto and successive issues thereof as may be or become effective from time to time during the term of this Contract shall be applicable to such shipments; provided, however, that without regard to the duration of any period during which any shipment hereunder may be held subject to such charges, the demurrage charge applicable hereunder for any chargeable day or fraction thereof during such period shall in no event whatsoever exceed the demurrage charge so published as applicable to the first day of such period.

Example 3 — Routing

Shipper shall route shipments hereunder via _____.
Except as otherwise provided in Section _____ of this Contract [Force Majeure], for any failure to adhere to such routing, Railroads shall be jointly and severally liable to Shipper for any resulting difference between the freight charges specified in this Contract and the freight charges applicable via the actual routing used.

Example 4 — Bill of Lading

Each shipment hereunder shall be tendered or shall be deemed to have been tendered on a uniform straight bill of lading substantially in the form published in the Uniform Freight Classification, Tariff ICC UFC 6000-A as in effect on the effective date of this Contract. The terms, conditions and provisions of such bill of lading shall be subject and subordinate to the terms, provisions and conditions of this Contract and, in the event of a conflict between the terms, conditions and provisions of such bill of lading and of this Contract, the terms, conditions and provisions of this Contract shall govern.

Example 5 — Bill of Lading Notation

Shipper shall insert on the bill of lading for each shipment hereunder the standard Transportation Commodity Code Number (STCC 00–000–00) of the Commodities and the Tariff number specified in Section 1 of this Contract, or any applicable successive reference, but the inadvertent omission of any such insertion on any one or more than one occasion shall in no event be deemed to be a breach of this Contract or to invalidate any of the terms, provisions and conditions hereof.

Example 6 — Equipment and Mileage

Shipments hereunder shall be tendered to and transported by Railroads in railroad tank cars furnished or caused to be furnished by Shipper. Each such tank car shall comply with the requirements for private rail cars in effect on the date of this Contract, as set forth in Circular OT-5 published by the Association of American Railroads, Operations and Maintenance Department, Operating–Transportation Division. Railroads shall pay mileage allowances to Shipper or to the owner of such tank cars, in accordance with the provisions of Mileage Tariff ICC PHJ 6007-G and such supplements thereto and successive issues thereof as may be or become effective from time to time during the term of this Contract. Nothing in this Contract shall affect or be construed to affect any arrangement, interchange agreement or other agreement between Shipper and Railroads with respect to the use, maintenance, operation or repair of, or payment for damage to or destruction of, such tank cars.

Example 7 — Rates and Charges

Except as provided in Section _____ of this Contract [Related Services], Shipper shall pay to _____, as full compensation to Railroads for each shipment hereunder consigned via the route specified in Section _____ of this Contract, charges computed at a rate of _____ dollars ($ _____) per one hundred (100) pounds of Commodities actually shipped, subject to a minimum weight of _____ thousand (_____) pounds and such charges shall include the return of tank cars after unloading, without payment of additional charges for such return.

Example 8 — Escalation

The rate specified in Section _____ of this Contract shall be increased or decreased, effective on the first day of each calendar quarter during the term of this Contract, in accordance with the procedures prescribed by the ICC pursuant to § 10707 a(a) (2) (B) of Title 49 of the United States Code and in effect as of each such effective date.

In calculating any increased or decreased rate pursuant to the provisions of this Section _____, (i) decimals shall be extended to four places only, and (ii) resulting fractions of less than a half cent shall be eliminated and resulting fractions equal to or more than a half cent shall be increased to the next whole cent.

Example 9 — Lower Tariff Rate

If, at any time during the term of this Contract, the charges published in any freight tariff on file with the ICC, and which would be applicable to any shipment hereunder in the absence of this Contract, are lower than the freight charges specified herein, the charges as published in such tariff shall be applicable hereunder to such shipment.

Example 10 — Liability for Damage to Freight

The provisions of Section 11707 of Title 49 of the United States Code, as the same may be amended or renumbered from time to time, shall be a part of this Contract as if set forth in full herein and shall be applied and interpreted in the same manner and shall have the same force and effect as if the said Section 11707 by its terms were expressly made applicable to railroads hereunder: provided, however, that notwithstanding the provisions of the said Section 11707, (i) it shall be a condition precedent to recovery hereunder that claims for loss, damage or delay of, or injury to, Commodities shall be filed with any of the Railroads within nine (9) months after delivery thereof or, in case of failure to make delivery, then within nine (9) months after a reasonable time for delivery has elapsed, (ii) an action at law for such loss, damage, delay or injury shall be instituted against any one or more Railroads only within one (1) year from the day when any of the Railroads has given written notice that Railroads have disallowed such claim or any part thereof specified in such notice, and (iii) no such claims with respect to any one shipment hereunder shall be filed or paid unless the amount payable therefore exceeds $_____, it being understood and agreed, however, that this provision shall not be construed as a deductible from any claim if the amount payable therefor exceeds the said amount.

Example 11 — Undercharges and Overcharges

Any action or proceeding by Railroads or any of them to recover charges alleged to be due hereunder, and any action or proceeding by Shipper to recover overcharges alleged to be due hereunder, shall be commenced not more than three (3) years after delivery or tender of delivery of the shipment with respect to which such charges or overcharges are claimed.

Railroads shall be jointly and severally liable for overcharges hereunder. To the extent permitted by applicable law, the expiration of the said three-year period shall be a complete and absolute defense to any such action or proceeding, without regard to any mitigating or extenuating circumstance or excuse whatsoever, unless the party named as defendant in any such action or proceeding has expressly agreed in writing to waive such defense in whole or in part.

Example 12 — Return Shipments

At any time after delivery but prior to substantial unloading of any Shipment at destination, or prior to arrival thereat, Shipper shall have the right to tender same to Railroads or either of them for transportation hereunder to any location specified as an origin in Appendix ___ (such tenders being hereinafter referred to, individually, as "Return Shipment" and, collectively, as "Return Shipments"), and Railroads or either of them shall accept and transport each Return Shipment to any such location designated by Shipper via the route specified therefore in Section _____ of this Contract, subject to the following terms and conditions:

(a) With respect to any Shipment tendered as a Return Shipment at the original destination thereof, charges on such Return Shipment shall be computed in accordance with Section _____ of this Contract as if such Return Shipment had been made (via the reverse routing) from the location designated by Shipper as the destination of such Return Shipment.

(b) With respect to any Shipment tendered as a Return Shipment at any location on _____ Railroad intermediate to the original destination of such Shipment, charges on such Shipment shall be computed in accordance with Section _____ of this Contract, and charges on such Return Shipment shall be computed in accordance with paragraph (a) of this Section _____ ; provided, however, that if the charges on any such Shipment or Return Shipment resulting from the application of any freight tariff or tariffs on file with the ICC (which would in the absence of this Contract be applicable to the actual movement of such Shipment or Return Shipment) are lower than the charges computed in accordance with the said Section _____ or the said paragraph (a), respectively, the charges resulting from the application of such tariff or tariffs shall be applicable hereunder to such Shipment or Return Shipment, or both, as the case may be.

(c) Except as provided in paragraphs (a) and (b) of this Section _____, all of the terms, provisions and conditions of this Contract, applicable with respect to Shipments, shall be applicable with respect to Return Shipments, except that (i) Return Shipments shall not be counted in determining whether the Minimum Annual Volume has been tendered during any annual period, and (ii) the provisions of Section _____ of this Contract [Force Majeure] shall be inapplicable with respect to Return Shipments.

Example 13—Cancellation or Termination

If, in any manner or for any reason, this Contract shall be terminated or cancelled before the end of an annual period, such annual period shall be deemed to terminate on the effective date of such termination or cancellation and the Volume Requirement (as defined in Section _____ of this Contract) shall be proportionately reduced.

Example 14—Minimum Annual Volume

Except as and to the extent otherwise provided in this Contract, Shipper shall, during each annual period while this Contract is in force and effect, tender for transportation hereunder not less than _____ percent (____%) of the aggregate number of carload shipments (in rail ____ cars) of one or more of the commodities specified in Appendix A, from the origins to the destinations specified in Appendix B, made by Shipper during such period (the said percentage, as adjusted from time to time pursuant to the provisions of this Contract, is hereinafter referred to as the "Minimum Annual Volume"). In calculating the Minimum Annual Volume for all purposes of this Contract, resulting fractions, if any, shall be eliminated.

If, during any such annual period, the Minimum Annual Volume is not tendered, charges shall be assessed in accordance with either paragraph (a) or (b), below, whichever is lower:

(a) Charges resulting from the application, to all Shipments actually made, of the provisions of any freight tariff or tariffs on file with the ICC which would, in the absence of this Contract, be applicable to such Shipments; or
(b) Charges computed on the Shipments actually made, at the rates specified in Appendix B, plus charges on each deficit Shipment computed at a rate of $ ____ per 100 pounds, subject to a carload weight per Shipment of _____ pounds. The number of deficit Shipments shall be calculated by deducting from the Minimum Annual Volume the number of Shipments actually made.

The rate specified in the foregoing paragraph (b) shall be increased or decreased in accordance with the procedures set forth in Section ____ of this Contract [Example 8] and, in the event that during any annual period the said rate shall be increased or decreased pursuant to such procedures, any charges which may be computed under the foregoing paragraph (b) shall be computed at the arithmetic average of all rates which shall have been in effect thereunder during such period, weighted so as to give effect to the number of days during such period on which each such rate was applicable.

Railroads shall invoice Shipper for the charges, if any, due under this Section _____ after the last day of each annual period or the effective date of termination or cancellation of this Contract, and, if correct, such invoice shall be paid not more than _____ (_____) days after receipt thereof by Shipper.

Example 15 — Disability

If, during any annual period, Shipments cannot be made via the route specified in this Contract from or to one or more points named in Appendix __ during a period of _____ (_____) or more consecutive days (including Saturdays, Sundays and holidays), due to or resulting from one or more of the following causes: Act of God, including but not limited to floods, storms, earthquakes, hurricanes, tornadoes, or other severe weather or climatic conditions; act of the public enemy, war, blockade, riot, insurrection, vandalism or sabotage; fire, accident, wreck, derailment, washout or explosion; mechanical breakdown in equipment or facilities of Shipper or Railroads; lawful or unlawful strikes, lockouts or labor disputes, governmental laws, ordinances, orders or regulations, whether valid or invalid and including, but not limited to, embargoes, priorities, requisitions, allocations or restrictions on facilities, equipment or operations, then each day of such period shall be deemed to be a disability day and the Minimum Annual Volume for such annual period shall be reduced by _____ for each such disability day.

Shipper shall send written notice to Railroads of each such disability within _____ (_____) days after the termination thereof and such notice shall specify the reasons for such disability.

Example 16 — Application of Laws, Tariffs, etc.

Except as and to the extent expressly provided in this Contract, (i) transportation under this Contract shall be and shall be deemed to be subject to all of the governmental statutes, laws, rules and regulations and to all of the terms and provisions of the uniform bill of lading and of the freight tariffs and schedules which would be applicable in the absence of this Contract, and (ii) the parties hereto shall have and shall be deemed to have the same rights, duties, obligations and liabilities which they would have in the absence of this Contract. If and to the extent that any such right, duty, obligation or liability which would, in the absence of this Contract, be enforceable in a proceeding before the ICC, is found to be unenforceable in such proceeding by reason of the existence of this Contract, such right, duty, obligation or liability shall be enforceable in an action at law or a proceeding in equity; provided, however, that nothing in this Section _____ shall be construed to modify or affect the terms and provisions of subsection (i) of Section 10713 of Title 49 of the United States Code, except that the term "transportation under such contract," as used in paragraph (1) of the said subsection (i), shall not be deemed to include all or any portion of the transportation of Shipments other than [storage-in-transit] as herein provided.

Example 17 — Storage-in-Transit

Carloads of _____ (Standard Transportation Commodity Code number _____), in railroad _____ cars, shipped by Shipper from _____, _____ to destinations in the United States (hereinafter referred to, individually, as "Shipment" and, collectively, as "Shipments") shall be entitled to all applicable storage-in-transit privileges on tracks owned or leased by Shipper and served by Railroad at _____, _____, as such privileges are defined in Tariff ICC _____, as in effect on the effective date of this Contract, and Shipments shall be subject to all of the terms, conditions and provisions of the said Tariff as in effect on the said date except that, in lieu of the charges stated in such Tariff for such privileges, Shipper shall pay to Railroad charges computed at a rate of _____ Dollars ($_____) for each Shipment accorded such privileges.

Example 18 — Confidentiality

Except as and to the extent required by law, the terms, conditions and provisions of this Contract shall be confidential and shall not be disclosed by any of the parties hereto to persons other than its employees and agents.

Appendix 8.2 **Background Research for Contracting**

What (product or commodity group characteristics)?
  Article description
  - Classification or tariff?
  - Alpha and numeric STCC?
  - Trade/brand names?
  - Solid, gas, liquid, or dry?
  - Bulk or packaged?

  Special characteristics
  - Hazardous materials?
  - Density, stowability, handling, susceptibility to damage?
  - Oversize or overweight?

  Quantity or volume
  - Per shipment?
  - Per period (day, week, etc.)?
  - Single or multiple vehicle lots?
  - Minimum contract volumes?
  - Maximum contract volumes?

Where (traffic lanes or routes)?
  Origins and destinations
  - Specific point-to-point hauls?
  - Single origin to multiple destinations?
  - Multiple origins to single destinations?

  Routes
  - Local?
  - Interline (rights and obligations of each carrier)?
  - Out-of-line or backhaul movements?
  - Diversions, stopoffs, transits?

How?
  Service
  - Transit performance?
      Transit time guarantees?
      Transit consistency guarantees?
  - Accessorial services?
  - Special handling?

Source: Adapted from Michael W. Uggen, ''Railroad Contract Rates: A Working Analysis of Section 10713,'' *ICC Practitioners' Journal* 48 no. 5 (July–August 1981), pp. 535–539. Copyright 1981 by Association of Transport Practitioners. Reprinted with permission.

- Frequency of dispatch?
    - Once, twice, etc., per day?
    - Continuous?
- Hours of operation?
- Choice of equipment or trains?
- Safety
    - Inspection requirements?
    - Hazardous material responsibilities?
    - Position of shipment in trains?
    - Operating speeds, train length?

Equipment
- Vehicle type, size, ownership?
- Specialized equipment?
- Quantities required?
    - Locations?
    - Timing?
    - Carrier supplies, repairs, maintains?
    - Shipper supplies, repairs, maintains?
    - (Purchase/lease additional capacity?)
    - (Cost sharing of equipment?)
    - (Mileage allowances?)
- Demurrage or detention provisions?
- Exclusive use?
- Loading and unloading requirements?
    - Blocking or dunnage?
    - Damage prevention devices (supply and return provisions)?
    - Where is it performed?
    - Who does it?
    - Whose facilities?
    - Arrival notice given?

Packaging specifications
- Bulk?
- Packaged?
    - Permissive or mandatory?
    - Who determines?
- Incorporation of published tariff rules or terms?

Billing, credit, payments
- Rate terms?
    - Units ($/cwt, $/vehicle, etc.)
    - Adjustments or modifications?
    - (Unilateral or bilateral?)
    - (Escalation formulas?)

        Most favored shipper (tariff rates apply when lower)?
- Credit terms?
    - Method of payment?
    - Credit period?
    - Net/discount basis?
    - Penalty for late payments?
- Billing/invoice procedures?
    - Single shipment freight bills?
    - Monthly statements?
    - Itemized summaries?
    - Documentation?
    - (Bills of lading?)
    - (Trip slips?)
    - (Inspection certificates?)
- Volume verification
    - Weight agreement?
    - Scale weight tickets? (whose scales?)

Liability, claims, and insurance
- Claims procedures?
    - Overcharge?
    - Undercharge?
    - Loss and damage?
    - Breach of contract?
- Filing requirements?
    - Time?
    - Burden of proof?
    - Documentation?
    - (Notice?)
    - (Reports and inspections?)
    - (Audits?)
- Measure of damages?
    - Released values?
    - Deductibles?
    - Special, liquidated, or consequential?
    - Required insurance coverages?
    - Mitigation?
- Settlement?
    - Payment schedules?
    - Compromise?
    - Offsets?
    - Arbitrations?
    - Mediation?

Litigation?
Attorney's fees or interest on claim amount?
- Liability for shipper or carrier employees?
- Liability for dock or sidetrack facilities?
- Force majeure
What elements are included (fire, accidents, weather delays, strikes, etc.)?
Consequences and remedies?

Term of contract
- Beginning and end?
- Renewal of extensions?
Automatic or negotiable?
Number of extensions?
Term of extensions?
- Related to sale of goods or other contracts?

Termination
- Automatic expiration at end of term?
- Carrier or shipper option to terminate at will?
Notice requirements?
Procedures?
Damages?
- Default or breach provisions?
Definition for both parties?
Procedures?
Damages and remedies?

Modification
- How, when, and by whom?
- Effect on remainder of contract?

Miscellaneous conditions
- Assignment of contract?
- Environmental regulations?
- Liens on cargo?
- Subcontracting?
- Performance bonds?
- Incorporation of published tariff provisions?
- Line abandonments?

Other contingencies (what if?)
Shipper does not meet specified contract volumes?
- Termination by carrier?
- Liquidated or consequential damages?
- Grace period?
- Alternative use of carrier equipment?
Carrier does not meet contracted equipment requirements?

- Contract termination by the shipper?
- Liquidated or consequential damages?
- Alternative equipment supply (private or other carrier)?
- Adjustment to minimum volume requirements?

Shipping patterns changes during the term of the contract?

An origin cannot ship or a destination cannot receive the goods?

The carrier uses a noncontract route?

- Absorption of extra charges via noncontract route?
- Refuse shipment?

Service standards are not met?

- Early arrivals and bunching of cars?
- Late arrival and operations are affected?

chapter 9

Costing and Negotiating

chapter objectives

After reading this chapter, you will understand:

The nature of shipment costing systems and procedures.

How to use shipment costing as a tool to support traffic management.

Some of the key points that traffic managers must keep in mind when planning and preparing for negotiations.

chapter outline

Introduction

Shipment Costing
The anatomy of an LTL
costing system
Costing truckload and
carload shipments

Shipment Costing as a Tool for
Traffic Management

Negotiating Rates and Services

Summary

Questions

Introduction

Costing and negotiating have become extremely important skills in today's traffic environment. Generally speaking, transportation cost analysis, or costing, refers to the task of determining how a carrier's costs of production react to changes in output. The competitive transportation environment of the 1980s has made carriers very sensitive to the cost of service, and many have developed sophisticated managerial costing systems to support decisions about pricing and operations. These systems enable carriers to calculate the cost of individual shipments and to assess the profitability of customer traffic moving on traffic lanes, that is, between specific origins and destinations.

This kind of information is also extremely useful to shippers. It is critical to negotiations about rates and services. In addition, shipment costing models support traffic planning and budgeting. Many companies now use costing software purchased from commercial sources and modified to suit their needs.

Although it was always possible to negotiate rates and elements of service, ICC regulation before 1980 stifled the ability of shippers to act like true purchasing managers. By contrast, contemporary traffic managers negotiate extensively with individual carriers and have had to learn new negotiating skills.

This chapter presents an overview of shipment costing systems and offers some insights into how traffic managers can use these tools to support decision making. The final portion of the chapter focuses on negotiation planning and tactics.

Shipment Costing

The process of costing individual shipments entails a set of procedures that transform expense and operating data into unit costs for each activity involved in the transportation of an individual shipment. A cost analyst describes the specific shipment characteristics to the system, and the system translates that information into units of work. Cost is computed for each activity by multiplying the work units by the cost per unit. The main elements of this kind of costing system are as follows:

1. Functional costs. Functional cost categories relate as closely as possible to unique operating and administrative activities required for transportation of a shipment. The objective is to allocate costs to the activities that create them. To accomplish this, a costing system compiles expenses organized in natural accounts, such as "fuel" and "materials," into functional cost categories, such as "linehaul," "pickup," and "delivery." The kind of transportation service of interest, say less-than-truckload (LTL) versus truckload (TL), will determine the functional cost categories.

2. Service units. Service units measure the work related to each functional cost category. Service activity is usually expressed in units of time, volume, or distance. For example, service units might be minutes per stop for pickup and delivery, minutes per hundredweight for dock handling, and vehicle-miles for linehaul movement.

3. Shipment and customer characteristics. Shipment characteristics, such as density, weight, handling, stowability, and packaging, affect the amount of work that the carrier must accomplish. Likewise, customer characteristics, such as the pickup or delivery location, the practice of tendering single or multiple shipments, and the methods used for packaging, affect carrier workloads. The cost analyst usually has to enter this information into the system.

4. Work equations. These equations translate shipment characteristics into service units. For example, dock time might be computed as handling minutes per cwt × shipment size (cwt). Detailed special studies that require extensive data may be needed to determine productivity relationships. Carrier operating practices and facilities, of course, affect these relationships.

   5. *Cost equations.*   Cost equations multiply the shipment-generated service units by unit costs. The cost of an individual shipment is determined by adding the results of each functional category.
   Carriers build costing systems with cost and productivity data developed from internal accounting information systems, engineering studies, and statistical models. Among other things, joint and common costs in transportation create difficulties for system designers.[1] *Joint costs* occur when the production of one activity unavoidably causes another activity. The problem is that the cost of each activity cannot be determined rationally. Managers must make some judgment about what constitutes an appropriate allocation of joint costs to multiple outputs. The fronthaul and backhaul situation in transportation presents a classic example. The act of shipping a load from point A to point B unavoidably causes a backhaul situation from B to A to arise. If the carrier can move a shipment from B to A, the problem is what costs apply to each leg of the trip.
   Common costs arise because certain transportation activities are common to two or more outputs. For example, the crew cost of a freight train is common to all commodities carried. The problem is how to trace common costs to each type of traffic or shipment. Managerial costing systems must arbitrarily assign common costs to individual shipments. For example, formulas may share the costs equally among all shipments, or they may allocate common costs in proportion to the size of the output, say, in tons.
   The costing systems developed or purchased by shippers also face similar design problems. In addition, shippers lack direct access to internal carrier cost or operating data. Traffic managers must use either (1) ICC costing systems and data or (2) internally developed (or commercially available) costing models calibrated with publicly available data. In either case, the data are not lane specific. They represent average figures for carrier networks, or they reflect regional averages. In most instances, though, ICC and commercial costing systems have the flexibility to incorporate more precise information as it becomes available. Although carriers are generally reluctant to disclose sensitive cost or operating information, traffic managers may try to obtain specific data from carriers during the negotiation process. In addition, traffic managers may establish actual service units for some activities (for example, pickup or delivery handling time) by direct observation.
   The ICC sells several automated costing programs to the public for a relatively small charge. The *Uniform Rail Costing System* (URCS), which became operational during the early 1980s, is perhaps best known to the transportation community.[2] URCS has a complex set of procedures that develop unit costs from a data base of about three thousand expense accounts and one thousand op-

erating statistics. It runs as an interactive computer program that permits cost analysts to develop variable costs and fully allocated costs of specific rail movements (see Exhibit 9.9 on page 251). Although this system is a substantial improvement over the previous costing procedure (Rail Form A), critics contend that URCS is conceptually flawed.[3]

In addition, the ICC has designed cost formulas to ascertain class I and II motor carrier costs. Highway Form A (know as the long formula) is used to determine territorial costs.[4] This costing system requires detailed data and the use of special studies. For those reasons, it is infrequently used. Highway Form B offers a simplified costing procedure to develop unit costs for single-line or interline shipments. The ICC sells an automated version known as the *Single and Interline Costing Program* (SICP).[5] Cost analysts who find it impractical to undertake special studies and who do not require extensive system refinements may find its relatively rough cost estimates helpful.

Although shipper, carrier, and ICC costing systems follow similar approaches, they are likely to produce considerably different results for the same set of shipment characteristics. The reason is that costing requires judgmental decisions about issues such as which data to use and which procedures to follow for allocating fixed, joint, and common costs. Those decisions can change ultimate cost determinations dramatically. In shipper-carrier negotiations, each party must be prepared to support its decisions on these issues with sound reasoning. In this context, costing is more of an art than a science despite the availability of sophisticated computer applications. Nevertheless, costing systems give both shippers and carriers far greater access to basic cost information today than in the past.

The Anatomy of an LTL Costing System

We can understand the costing process better, as well as can gain some insights into its use as a planning tool, by studying an application in a specific transportation setting. The discussion that follows illustrates the less-than-truckload (LTL) costing framework developed for the Regular Common Carrier Conference (RCCC).[6]

Before proceeding further, however, it is helpful to explore the nature of LTL operations. **Exhibits 9.1 and 9.2** illustrate two terminal networks that trucking companies have adopted for LTL traffic. A "cluster" arrangement is shown in Exhibit 9.1. In this system, the carrier picks up the LTL shipment and hauls it to one of several local terminals that cluster around a regional breakbulk center. Traffic typically moves in pickup trailers from these satellite terminals to a regional breakbulk center where the carrier strips incoming LTL shipments and reloads them for maximum capacity utilization of linehaul equipment. These loads then move to another regional breakbulk facility, where they are broken down and moved to local terminals for final delivery. Large interstate carriers

Exhibit 9.1 **Cluster network for LTL transportation**

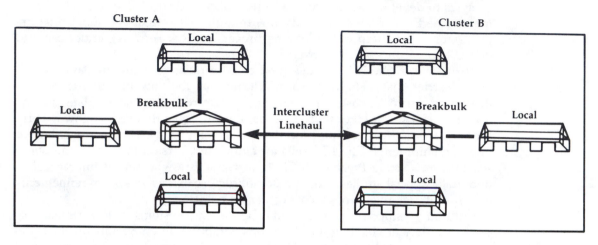

Exhibit 9.2 **Stand-alone breakbulk network for LTL transportation**

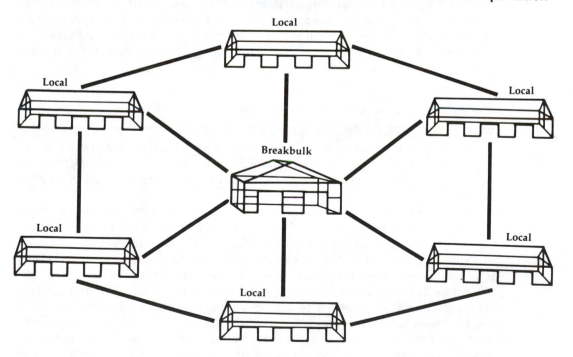

use cluster systems to make linehaul operations more efficient and to place terminals close to customers.[7]

Exhibit 9.2 shows a "stand-alone" system in which LTL shipments usually pass through a single breakbulk facility. Some LTL traffic, however, may move directly between local terminals. Regional carriers generally operate in this fashion.

In networks like these and others, carriers must perform certain operating and administrative activities. The RCCC costing system organizes these activities by functional cost categories as follows:

1. Pickup and delivery. Drivers normally complete their deliveries and then pick up shipments. Driver activities include terminal check-in or check-out, stem and peddle driving, and pickup or delivery of freight. At the beginning or end of each shift, drivers check in or out at the terminal, match freight against paperwork, and inspect equipment. *Stem* driving refers to the terminal-to-first-stop and the last-stop-to-terminal moves. *Peddle* driving includes all the between-stop moves accomplished during the daily run. Once at the stop, drivers perform activities such as opening and closing doors, locating the shipper or receiver, handling the freight, and checking the piece count against the bill of lading.

Carriers generally pay drivers by the hour for these activities, although some carriers pay by the hundredweight. The RCCC costing system measures pickup and delivery activities in minutes per stop and expresses unit costs in dollars per minute. Since the time spent making the stop involves a joint cost, the system determines average minutes per stop. Handling time at the stop depends on shipment characteristics such as weight and type of packaging. Moreover, in a multiple shipment pickup, the platform arrangement of the shipments may affect handling time. **Exhibit 9.3** illustrates these generalized cost relationships.

2. Dock and rehandling. Workers strip, load, stack, move, and check freight at terminals. Although the dock and rehandling functions involve similar tasks, "dock handling" concerns local terminal activity, whereas "rehandling" concerns intermediate breakbulk operations. The RCCC system assumes a relatively constant amount of time is required to perform certain activities such as checking piece counts or matching documents against freight. However, the size of the shipment and the number of pieces have a decided effect on the time it takes to move, stack, load, or otherwise handle freight. Handling time also depends on whether the shipper sorts and segregates multiple shipments, as it does on the type of packaging. Platform workers, for example, generally can use efficient material handling equipment for palletized (versus loose) shipments.[8]

Exhibit 9.4 illustrates general dock and rehandling cost relationships. Carriers usually realize higher productivity and lower unit costs (handling dollars

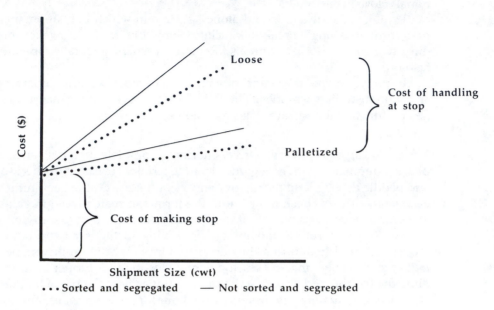

Exhibit 9.3 Generalized LTL pickup and delivery costs

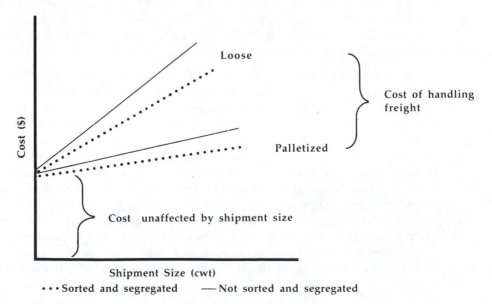

Exhibit 9.4 Generalized LTL dock and rehandling costs

per minute) for rehandling at breakbulk centers than for handling at local terminals, especially those in metropolitan areas.[9]

3. *Linehaul.* Linehaul service involves the terminal-to-terminal movement of many LTL shipments in the same vehicle. Thus linehaul costs must be apportioned to individual shipments. The RCCC system accomplishes this task by making "equivalent trailer miles" the service unit for each shipment. Using this approach, the system adjusts the cost of moving an average "lane-load" by the fraction of the trailer's capacity actually used by the LTL shipment.

The RCCC system measures trailer capacity in hundredweights. The weight of an individual shipment is divided by the average weight of a lane-load to determine the fraction of trailer capacity used. Designers of the RCCC system recommend allocating costs on the basis of cubic capacity, however, only when individual shipment density departs substantially from average lane-load density.[10] This method of computing linehaul costs is best illustrated by example. Suppose an LTL shipment weighs 6000 lb, has a density of 12 lb per cf, and travels 300 miles. It would occupy 500 cf of space (6000 lb/12 lb per cf). If the trailer capacity is 3000 cf, there are 50 equivalent trailer miles (500 cf/3000 cf × 300 miles) charged at the dollars-per-loaded-mile rate. **Exhibit 9.5** presents generalized linehaul costs for any specific traffic lane as a function of the density and weight of a shipment.

4. *Claims.* This component includes claims administration and loss and damage reimbursement. These costs derive from the carrier's basic duty to de-

Exhibit 9.5 **Generalized LTL linehaul costs for a specific traffic lane**

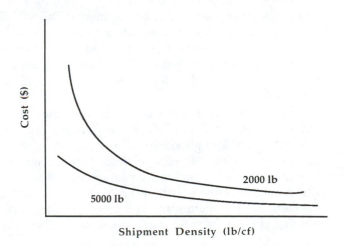

Shipment Density (lb/cf)

liver the goods in the same condition as received to the correct party and with reasonable dispatch (see Chapters 12 and 13). The service unit in the RCCC system is cwt. The unit cost is claims dollars per cwt. However, the system might be refined to reflect unit cost differences among commodity groups, major customers, or certain operations.

5. Clerical. Service units in the clerical function relate to shipment documentation. Activities include preparing manifests, determining charges, billing, and collecting (see Chapter 7). The basic system assumes clerical costs per shipment are the same for all LTL shipments. Changes in credit regulations, however, may justify clerical-dollars-per-shipment measurements broken down by major shipper accounts. Credit arrangements also affect a carrier's capital opportunity costs (as discussed in Chapter 7).

6. Sales and marketing. Sales and marketing activities include sales calls, promotion, bureau memberships, interline agreements, and administration. Some costs created by these activities are allocated equally to all LTL shipments. Other costs, such as sales force expenses, are allocated only to the noninterline shipments that originate in the carrier's own system.

7. Terminal overhead. The RCCC costing system allocates terminal costs in two ways. First, variable expenses related to the administration and operations of terminals are allocated equally to shipments. Second, fixed terminal costs such as facility costs, utilities, and maintenance are apportioned to shipments on a dollars-per-cwt basis.

8. General and administrative. Some expenses in this functional category (accounting, data processing, and general clerical tasks) are not directly traceable to individual shipments but vary with the volume of business. The formula equally allocates these costs to shipments. The RCCC system also assumes that indirect general and administrative costs (such as interest and bad debts) are directly related to the total cost of all the other functional cost categories. In other words, costs are allocated for these indirect G&A expenses by multiplying the unit cost (indirect G&A dollars per non-G&A expenses) by the total non-G&A expenses.

Exhibit 9.6 (pages 246–247) illustrates the entire process. The cost analyst enters the required set of shipment characteristics. In this illustration, a carrier will move 2000 lb of goods from Cincinnati to Detroit. The shipment comprises 10 pieces (each with a density of 10 lb/cf) tendered loose and requires local dock handling and break-bulk-center rehandling. The system determines individual-shipment costs from equations like those shown in Exhibit 9.6. The raw data and the formulas that determine unit costs, productivities, and cost-of-service equations constitute the main part of the system. The fully allocated (total) is $150.50

in this hypothetical illustration. The RCCC, of course, has tailored its costing program for the motor carrier's use. To use this software, shippers must be able to acquire the cost and operating data needed to make the program worthwhile.

The RCCC costing system can be modified by users to compute variable costs. For example, a contribution-to-burden model would compile only variable expense elements into the pickup and delivery, dock and rehandling, linehaul, claims, and clerical cost categories. The revenue from a shipment that exceeds the sum of these costs represents a contribution to overhead. The designers of the RCCC costing system, however, are cautious about this approach:

> *It should be recognized that determining which costs are fixed is very difficult and will depend on the time period involved, how management runs its operations, and the level of capacity utilization. A more accurate estimate of contribution, therefore, can be obtained only by studying . . . operations in substantial detail.*[11]

Costing Truckload and Carload Shipments

Costing systems for truckload (TL), railway carload (CL), multiple carload, and trainload shipments use accounting approaches similar to the LTL costing method. The individual components of the LTL, TL, and CL systems, however, are different because they reflect different types of transportation operations. Truckload operations, for instance, usually do not involve terminal operations. Primarily for this reason, TL costs and revenues are sensitive to the traffic and shipping practices of individual firms. Thus, TL costing systems relate service units and costs more closely to individual shipper accounts than do LTL systems. Shippers, of course, want their models to reflect the company's shipping practices and traffic.

Exhibit 9.7 illustrates the cost components of one TL shipment costing system in use today. Linehaul and pickup-delivery costs generate about 98 percent of total shipment costs for TL shipments weighing more than 20,000 lb.[12] Thus, this costing system concentrates on these two cost components. It also provides a finer breakdown of linehaul costs than the RCCC's LTL system.

Railway CL shipments involve a complex set of activities (see **Exhibit 9.8**).[13] After a shipper places an order for a certain type of freight car, railroad personnel attempt to locate an available car that will meet the shipper's requirements. An industrial switch engine will usually deliver the car to the shipper in one of two ways. First, it may "constructively" place the car on public team tracks where the shipper is given access to load the goods. Second, the car may be "actually" placed on the shipper's private loading sidings. A notable exception to these two types of delivery occurs in larger industrial plants that have complex track systems and switching requirements. An industrial switch engine usually delivers cars to a specified location, and a company-owned switch engine distributes the cars throughout the plant.

Exhibit 9.6 LTL costing

Shipment Characteristics

Origin	Cincinnati	Destination	Detroit	
Weight	2000 lb	Origin handling	Yes	
Density	10 lb/cf	Breakbulk (rehandling)	Yes	
Packaging	10 pieces loose	Destination handling	Yes	
		Linehaul miles	270	
		Interline-received shipment	No	

Unit Costs*

Pickup and delivery

Pickup $/minute	$0.50
Delivery $/minute	$0.50

Dock and rehandling

Origin dock $/minute	$0.40
Destination dock $/minute	$0.40
Rehandling $/minute	$0.30

Linehaul $/loaded-mile	$1.00
Claims $/cwt	$0.05
Clerical $/shipment	$1.00

Sales and marketing

$/Shipment	$0.50
$/Noninterline received shipment	$1.50

Terminal Overhead

$/shipment	$1.00
$/cwt	$0.40

General and administrative (G&A)

G&A $/shipment	$1.50
G&A $/non-G&A expenses	$0.05

Productivities*

Average minutes to make pickup	30.0
Average minutes to make delivery	40.0
Pickup minutes/cwt	1.2
Delivery minutes/cwt	1.5
Constant handling minutes/shipment	3.0
Origin handling minutes/cwt	2.2
Destination handling minutes/cwt	2.3
Rehandling minutes/cwt	2.0
Lane-load average capacity (cf)	3000

*All values are hypothetical and are for purposes of illustration only.

Exhibit 9.6 (continued)

Cost of Service

Functional cost	Cost equation	Amount
Pickup (P/U)	[30 minutes to make P/U + (1.2 P/U minutes/cwt × 20 cwt)] × $0.50/minute	$ 27.00
Delivery (DEL)	[40 minutes to make DEL + (1.5 DEL minutes/cwt × 20 cwt)] × $0.50/minute	$ 35.00
Origin dock	[3 constant minutes/shipment + (2.2 origin handling minutes/cwt × 20 cwt)] × $0.40/minute	$ 18.80
Destination dock	[3 constant minutes/shipment + (2.3 destination handling minutes/cwt × (20 cwt)] × $0.40/minute	$ 19.60
Rehandling	(2 rehandling minutes/cwt × 20 cwt) × $0.30/minute	$ 12.00
Linehaul	[(2000 lb ÷ 10 lb/cf)/3000 cf] equivalent-trailers × 270 miles × $1/loaded mile	$ 18.00
Claims	20 cwt × $0.05/cwt	$ 1.00
Clerical	$1/shipment	$ 1.00
Sales and marketing	($0.50/shipment) + ($1.50/noninterline-received shipments) × (0 interline-received shipments)	$ 0.50
Terminal overhead	$1/shipment + ($0.40/cwt × 20 cwt)	$ 9.00
General and administrative	$1.50/shipment + ($0.05/$ of non-G&A expenses × $141.90 of non-G&A expenses)	$ 8.60
Total cost of shipment		$150.50

Source: Based on the framework presented in *A Carrier's Handbook for Costing Individual Less-Than-Truckload Shipments*, (Washington, D.C.: Regular Common Carrier Conference, 1982).

■■■■■■■■■ *Exhibit 9.7* **Components of a truckload costing system**

User Input		System Output
Shipment characteristics	**Examples of input**	
Carrier	Common (regular or irregular route), independent contractor, private	Variable and fully allocated costs
Fleet		Per shipment
Equipment	Van, tank, flatbed, reefer	Per loaded-mile
Driver	Union, nonunion	Per total-mile
Linehaul miles	Oneway	Per cwt
Weight		
Region	Northeast, Midatlantic, Southeast, West, North Central, South Central	
Domicile state		
State involved in movement		
Cost index values per period	Fuel, wage, or inflation per year	

Functional Cost Categories	
Linehaul fuel	Licenses and permits
Linehaul driver compensation	Insurance
Linehaul equipment maintenance	Equipment ownership
Pickup and delivery	Operations, overhead, and miscellaneous
Highway use tax	Other capital asset ownership

Source: Based on the framework presented in *Microcomputer Software for Transportation Management — Truckload Cost Calculator User's Manual* (Washington, D.C.: Peat, Marwick, Mitchell and Co., 1983).

After constructive or actual placement, the shipper has a certain amount of time to load the car before demurrage charges accrue (see Chapter 11). When loading has been completed, the shipper notifies the railroad that the car is ready for pickup. An industrial switch engine may pull the car to an industry yard. Cars accumulate in this yard until a local train removes them or until enough of them have been gathered to make up a drag of cars that are moved to a larger yard. Instead of taking the car to an industry yard, the industrial switch engine may deliver it directly to a classification yard.

The classification yard will sort the car together with others into "blocks," or groups, of cars having a similar destination. These blocks are assembled into trains that may travel to intermediate yards, where the train may set off some cars, pick up others, or be completely resorted. It is possible, therefore, for a

Exhibit 9.8 Railway CL freight transportation process

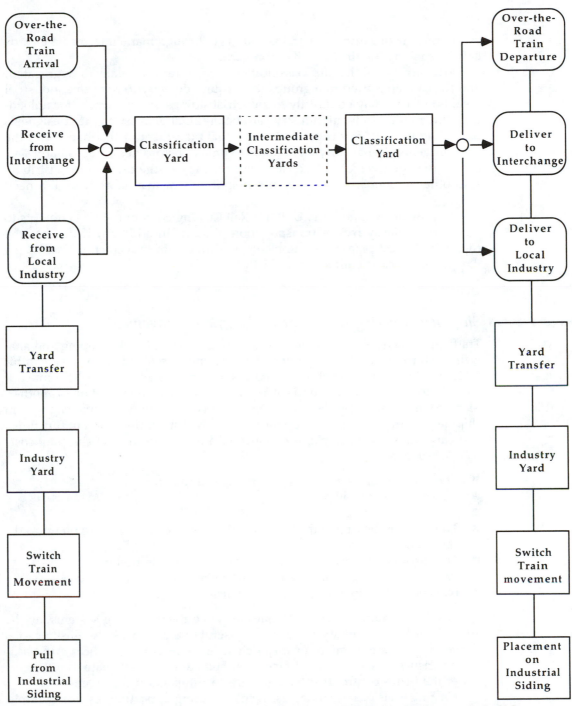

Source: S. J. Petracek et al., *Railroad Classification Yard Technology* (Washington, D.C.: U.S. Department of Transportation, Federal Railroad Administration, DOT-TSC-968, July 1976), p. 9.

car to travel with a number of trains and to go through many intermediate yards before reaching the final classification yard.

After arriving at the final classification yard, a train is usually disaggregated, and the cars are sorted into groups that require delivery to the same industrial area. Some cars may go directly to industrial sidings or team tracks for unloading. Others may go to an industry yard before final delivery to the consignee's unloading tracks. After constructive or actual placement of the loaded car, the receiver generally has forty-eight hours to unload before demurrage charges accrue. After the receiver unloads and releases the car, the carrier may distribute it to other locations requiring empty freight cars or may store it until a need arises.

As previously noted, the Uniform Rail Costing System (URCS) attempts to model this railway freight transportation process. In addition, URCS has the capability to cost piggyback shipments and unit train movements. **Exhibit 9.9** illustrates the basic components of URCS.

Shipment Costing as a Tool for Traffic Management

Traffic managers use costing models as powerful tools for developing and analyzing information critical to traffic planning and rate negotiations (see **Exhibit 9.10**). At the simplest level, these models answer the question of how much it will cost to ship so many pounds of a certain article from one location to another. More sophisticated applications address the issue of how specific changes in shipment characteristics or shipping practices will affect the different functional cost categories. For example, how much of a reduction in total shipping costs will be realized if

- The shipper sorts and segregates multiple shipments.
- Loose pieces are palletized.
- Fuel costs decrease.
- The traffic is moving in the backhaul direction and is helping to fill unused capacity.
- The carrier replaces a 45-foot trailer with 28-foot twin trailers.
- Privately-owned freight cars are used in place of railroad-owned cars.
- New packaging increases shipment density.

Besides providing answers to questions like these, costing systems lay the groundwork for sensitivity analysis. It is useful to know just how sensitive total shipment costs are to minor or major changes in shipping practices and traffic characteristics. The traffic manager can pinpoint and then evaluate the elements under the firm's control that have the greatest impact on carrier costs.

Costing is also useful for assessing cost-sharing opportunities. For example, one traffic manager using a small owner-operator truck firm determined that

■■■■■■ *Exhibit 9.9* **Components of the Uniform Rail Costing System (URCS)**

User Input		*System Output*
Shipment characteristics	**Examples of input**	
Region or carrier	Southern region, Western district, official territory, etc.; class I carriers	Variable and fully allocated costs:
Short-line miles		Per shipment
Movement type	Single line, interline, intraterminal	Per cwt
Car type	40-ft unequipped box, 50-ft equipped box, covered hopper, TOFC flat	
Number of cars		
Type of movement	Individual-car, multicar, unit train	
Car ownership	Railroad owned, privately owned	
Commodity type	Standard Transportation Commodity Code (STCC) identification	
Weight		
Special service parameters	Input related to TOFC, auto, lake transfer shipments	
Optional parameter changes	Circuity factors, number of industry switches, etc.	

Functional Cost Categories	
Running track	Freight car
Switching track	Specialized services
Road operations	General overhead and constant costs ratios
Yard operations	Loss and damage
Clerical and carload	

Source: Based on material presented in *Uniform Railroad Costing System* (Washington, D.C.: Interstate Commerce Commission, Bureau of Accounts, December 1982) and *Uniform Railroad Costing System Phase III Movement Costing Program User's Manual* (Washington, D.C.: Interstate Commerce Commission, Bureau of Accounts, April 1983).

new rigs could be bought by the owner-operator with the bank financing at 12 percent. The traffic manager's firm cosigned the loan causing it to change into a 9 percent loan. This 3 percent difference results in about $1300 a year less cost to the carrier, which must be shared with the traffic department each year.

The Xerox Corporation offers another illustration of the cost sharing concept (see **Exhibit 9.11**). Xerox established a National Dispatch Program in which it assumes the burden of scheduling trucks to pick up shipments among its na-

■■■■■■ *Exhibit 9.10* **Shipment costing at General Mills, Inc.**

Truck and railroad transportation costs exceed $300 million a year at General Mills, Inc. The transportation department uses several shipment costing systems for negotiation and planning purposes. Its irregular route truckload cost model is oriented toward long-haul truckload movements and permits GMI analysts to

- Specify thirty-six financial, operating, and equipment characteristics
- Conduct sensitivity analysis of elements such as fuel costs, annual mileage, driver wages, capital costs, and empty mileage
- Analyze private fleet costs
- Evaluate public policies that affect transportation (deregulation, truck size weights, user charges, energy impacts, and so on)

Although the truckload costing model has built-in default values for its functional cost components, General Mills often obtains cost information from the carriers it is negotiating with. The model's output includes operating cost per mile, cost per trip, cost per hundredweight, and cost per ton-mile. The program also computes and prints the sensitivity of trip costs to the amount of nonrevenue mileage assigned to a particular movement.

In 1977, General Mills purchased a Rail Form A Computer Costing Model. Because analysts had difficulties using this model, they frequently turned to the simpler ICC Carload Cost Scales for quick "ball-park" estimates of rail costs. General Mills now uses the Uniform Rail Costing System on a timeshare basis.

Source: Adapted from Jeffrey C. Kline, "Transportation Costing in Practice at General Mills, Inc.," *Proceedings — Seminar on Transportation Pricing, Costing, and User Charges* (Washington, D.C.: Transportation Research Forum, April 1982), pp. 29–30.

tional network of individual business units. In return for this service, Xerox receives reduced freight rates from participating carriers.

In a broader context, costing systems support various planning tasks. Some systems permit users to input cost index values and to forecast, say, the effects of rising or falling fuel prices on transportation costs. Forecasts of probable future shipping costs help shape transportation budgets (see Chapter 17) and other decisions such as plant and warehouse location. In addition, costing models are valuable tools for evaluating shipment consolidation programs (see Chapter 9).

■■■■■■ *Negotiating Rates and Services*

In contemporary traffic organizations, negotiating is regarded as an important management responsibility that demands new professional skills and knowledge. Today's traffic managers can create and initiate new ways to make "the buy" a profitable venture for the firm. The RCA Corporation, for example, estimates it has saved more than $100 million in its transportation budget with its new transportation purchasing initiatives.[14]

■■■■■■ *Exhibit 9.11* Xerox's National Dispatch Program

The National Dispatch Program has paid off handsomely for the Xerox Corporation. Generally, the program generates savings through "lane balancing," a process that matches truckload shipments between pairs of manufacturing, distribution, and supplier locations. It became obvious to some independent Xerox divisions that, although large volumes of product were being moved, there were a significant number of empty miles that prevented real savings in transportation costs. Xerox recognized the opportunity for cost reduction through integrating shipments from individual operating units to various destination points to balance flows better and reduce carrier empty miles. Essentially, the program combines a series of relatively short hauls into a continuous series of shipments that altogether form a far more productive long haul and assure revenue for up to twenty-eight straight days. In addition to the guaranteed business, carriers benefit from committed volumes leading to better vehicle utilization and from avoiding the task of finding freight to fill empty miles. In return for this service, Xerox negotiates significantly lower freight rates with selected carriers.

The key to the success of the National Dispatch Program has been the cooperation of the divisions and the carriers in integrating the requirements of each to the benefit of both. Early concerns were addressed through a series of workshops conducted to inform divisions and to get carrier input for program development. A nice surprise was that participating carriers have taken the initiative to extend the program to some third-party customers; these efforts provided additional economies resulting in further savings for all. Specifically, the first-year savings to Xerox have exceeded $2 million. The three-year projection is a $6 million reduction over costs that would have materialized without National Dispatch.

Source: "National Dispatch Saves Xerox $6 Million," based on a paper presented to the National Council of Physical Distribution Management (San Francisco, 1982). As reprinted by *Canadian Transportation and Distribution Management* (Canada's national physical distribution magazine, 1450 Don Mills Road, Ontario), November, 1983. Reprinted with permission.

As discussed in Chapter 8, both shippers and carriers gain something in successful negotiations. Nonetheless, benefits rarely are equally divided.[15] To come out ahead, traffic managers need to plan and prepare carefully, to know the subject matter being negotiated, to use sound negotiation tactics and good judgment, and to communicate effectively.

Skill in preparing and planning for negotiations is perhaps the most important single characteristic of successful negotiators.[16] The following discussion presents an overview of some key points that traffic managers need to keep in mind.

Set objectives. Establishing specific objectives is crucial to successful negotiations. Statements such as "We want reliable delivery service" or "The goal is to obtain a reduced rate for a volume guarantee" will not suffice, except as broad policy guidelines. Shippers must state their objectives in specific terms. For example, Nabisco Brands, Inc., which purchases about $15 million worth of intermodal transportation a year from shipping agents, established a general goal of having agents "become part of our business."[17] This broad statement was then translated into a specific five-year, volume-commitment objective. In addition, Nabisco established a "cost plus specific percentage markup" price objective and required agents to open their books to Nabisco.

Authorities recommend that managers use a "minimum-maximum-objective position" approach when setting objectives.[18] When formulating a *minimum position*, the shipper assumes all objectives will be met with a minimum cost. This position is something like making a wish list that assigns each component of transportation service an ideal rate and performance target. The *maximum position* assumes that all objectives will be met with a maximum cost. This position, for example, would indicate the highest rate a shipper is willing to pay. To determine the *objective position*, a shipper must estimate what the carrier's performance and price ought to be.

This approach to setting objectives requires detailed information. The traffic manager needs to specify an acceptable rate and service level for each position, which requires cost estimates for each functional cost category under consideration, as well as the rationale for the assumptions and methods used. Knowledge of the ground rules is also essential. The ground rules include antitrust violations, ICC regulations, legal purchasing fundamentals, and terms of sale.

A traffic manager planning for negotiation must estimate the carrier's probable minimum, maximum, and objective positions. The carrier's three positions will normally be higher than the shipper's. As negotiations proceed, the two sides will typically move toward their respective objective positions. The difference between these two positions represents the heart of negotiations.

Assess strengths and weaknesses. In the field of purchasing, assessing strengths and weaknesses is the most important responsibility of negotiated buying.[19] This task not only embraces the shipper's own internal evaluation but also requires an assessment of the carrier's needs, strengths, and weaknesses. An overview of the basic pricing situations that confront carriers will offer some insights into the strengths and weaknesses of a carrier's negotiating positions.

The pricing opportunities available to a carrier are first defined by whether it is operating at less than full capacity with or without excess demand. At less than full capacity, the carrier has spare equipment and labor resources available that could generally handle additional freight with only modest increases in costs. And additional business can be handled without further investment. In fact, if additional business is handled in existing runs and the carrier is operating at less than full capacity, the additional shipments often require only slight increases of fuel and handling costs. When the carrier is at full capacity with stable demand, a generally good match of equipment and labor is possible for movement economies. With excess demand, the carrier can price according to the relative demands and abilities of each form of traffic to pay. The carrier can also drop less profitable traffic to handle more profitable traffic, which did indeed occur in the common motor carrier and steamship industries during the early 1970s.

Most carriers face an excess capacity situation, such as rail lines that have only three or four trains per day, rail cars that sit idle for long periods, or trucks and terminals that operate at less than full potential. Even when an over-the-road truck is full, it sometimes must wait an additional day for the full load to be accumulated, thus, still representing a less-than-full-capacity, cost-revenue situation.

Obviously, a carrier that can price at, or above, its fully allocated costs is better off, because all variable and fixed costs, including any contribution to profit, are being recouped. With such a financial advantage, the firm will be able to replace equipment in the long run. A shipper paying the freight in this situation wants to investigate other available carrier or shipping options.

A price below fully allocated cost, but above variable cost, can result in a positive cash flow for the carrier, even though it will generally result in an accounting loss. If all revenue traffic is in this form, the carrier can often carry on operations until the fixed-cost assets require replacement, or a major financial commitment is due. Traffic rated below fully allocated costs and above variable costs can be handled as part of a carrier's overall mix over a long period. For example, a train operates between two points with adequate profit on all fifty cars. Ten more cars might be moved with no additional locomotive or crew requirements. The revenue from the ten extra cars will contribute to fixed costs and profits, even if they are charged a price below fully allocated cost but above variable costs.

If a carrier has high fixed costs or joint costs, the shipper may well be able to exploit this situation and negotiate a rate below fully allocated costs — especially in a competitive market. This form of pricing is the rationale behind many carrier contracts and pricing for proportionals. Carrier discounts for traffic above certain minimum volume levels are also based on this concept.

Pricing at variable cost will result in no additional cash contribution to the firm, and an accounting loss will be reported. At this price level, the carrier gains no benefit in volume. Pricing below variable cost for a move will result in an accounting loss and a cash loss. Operating at this level will so deplete the firm's liquid assets that it will cease operations.

The two price levels of at variable cost and below variable cost are not desirable for a carrier. They are presented here because some joint-cost, fronthaul-backhaul movements might be analyzed and seen in these situations. That is, through specific costing a fronthaul might produce sufficient revenues to cover a full round trip, but another analysis might show that the backhaul needs to cover one half the total trip costs. In a purely economic sense, the backhaul will provide favorable return to the firm with a revenue of one cent over and above the added cost of handling the backhaul. But with this analysis, an accounting loss below variable costs will be reported. Herein lies the need to understand how economic principles and accounting practices should be reconciled.

Rank issues. Ranking issues by order of importance will contribute to an effective agenda. Key areas of agreement, for example, can be set aside while the two sides concentrate on negotiating issues on which they disagree. As part of this effort, a shipper should list purchase criteria by order of importance. Rates might be ranked first, and ease of claims administration might be ranked last. This list identifies the importance of possible concessions, but, to keep some leverage, important concessions should be retained as long as possible. The shipper should also estimate how the carrier will rate these same criteria. If one criterion is rated high by the shipper but low by the carrier, the shipper should have no trouble convincing the carrier to yield. In the opposite case, the shipper can magnanimously yield to the carrier.

Consider order of presentation. The tradition in purchasing management is for sellers to make the opening presentation and for buyers to respond.[20] The buyer and the seller subsequently focus on areas of disagreement. By adopting this approach, a shipper can gain a tactical advantage; it creates an opportunity to place the carrier in a defensive position. During the initial response, for example, a traffic manager can request a point-by-point justification about cost breakdowns and assumptions.

In addition, most authorities recommend that both sides discuss issues in the order of their ease of solution.[21] The advantage of this method is that it fosters a positive, cooperative atmosphere before the two sides address the difficult issues.

Establish authority. It is essential for traffic managers to establish their authority to negotiate. Having strong support within the company will contribute greatly to a traffic manager's success. Whether negotiating alone or with a team of specialists, a traffic manager has to coordinate with other managers to serve their needs and interests. Team members should have clearly defined roles, which will prevent them from working at cross purposes or prematurely divulging the firm's objective position. Moreover, knowledge of a carrier's organization — that is, who has the legal power to make decisions — will enable the traffic manager to identify the carrier personnel that require attention. It is not unusual for trucking company sales persons to negotiate contracts without full knowledge of whether the operations department can actually perform certain services.[22]

Think about timing. The timing of concessions is important. As already indicated, traffic managers should save important concessions until last to retain negotiating leverage. A major concession, moreover, requires time for a detailed buildup. Time also affects negotiating leverage. Negotiating under the pressure of an early deadline dilutes leverage. The carrier simply has to wait until an approaching deadline forces a traffic manager to make concessions.

Make skillful use of questions and answers. Purchasing managers consider the skillful use of questions and answers one of the most important negotiating techniques.[23] On one hand, adroit questions that tactfully criticize the carrier help keep it in a defensive posture. On the other hand, answers should enhance the shipper's negotiating position. In this context, accuracy is not necessarily the best guide; a nonanswer or a vague answer may serve this purpose best.

Train and practice. It is difficult to understate the importance of training and practice to effective negotiations. Not surprisingly, some of the larger corporations with sophisticated logistic-traffic organizations have initiated formal buyer training programs for traffic managers as well as purchasing managers. For example, at the RCA buyer Training Seminars for Personal Development, held several times a year, key purchasing and transportation people learn negotiating skills and other purchasing techniques like evaluating costs, assuring supply, and identifying unreliable vendors.[24] Before 1980 and deregulation, RCA's fifty transportation managers might not have been included in negotiation training because they were involved in little real buying. RCA now has a three-pronged approach to help prepare transportation managers for the new business climate. In addition to the buyer training programs, RCA has initiated a team buying approach in which transportation managers from different divisions solicit bids for transportation and interview selected carriers. The company also has begun annual meetings with transportation people to keep them informed about deregulation and new purchasing methods.

Do not overlook location. The conventional sports wisdom that home teams have an advantage applies to negotiations as well. Technical support and data are nearby. In addition, physical and psychological fatigue will be less at home than on the road.

Summary

Costing is the task of determining how a carrier's costs of production vary with changes in output. The process of costing individual shipments entails a set of procedures to convert expense and operating data into unit costs for each major activity in the shipment process. After a cost analyst enters information about the shipment, the costing system computes the units of work generated by that shipment. Cost is then determined by multiplying work units by unit costs.

The main elements of these costing systems include (1) functional costs, (2) service units, (3) shipment and customer characteristics, (4) work equations, and (5) cost equations. In practice, joint costs and common costs cause difficulties for system designers.

The costing software purchased by shippers has similar design problems. These systems also lack direct access to the carrier's cost and operating data. Thus, they are usually calibrated with publicly available data. Besides commercially available costing software, shippers can purchase the ICC's Uniform Rail Costing System (URCS) and its Single and Interline Costing Program (SICP) for truck transportation.

Traffic managers use costing models for developing and analyzing information critical to traffic planning and rate negotiations. Besides estimating single shipment costs, traffic managers can evaluate how specific changes in shipment characteristics, or in shipping practices, will affect different cost components. Costing is also useful for conducting sensitivity analyses, assessing cost sharing opportunities, and supporting various planning tasks.

The capability to negotiate effectively is an important traffic management responsibility that demands new professional skills and knowledge. Traffic managers need to plan and prepare carefully, to know the subject matter being negotiated, to use sound negotiation tactics, and to communicate effectively.

Traffic managers need to keep several key points in mind when planning and preparing for negotiations. First, it is important to set specific objectives. Authorities recommend a "minimum-maximum-objective" approach to this task.

There is also a need to assess both shipper and carrier strengths and weaknesses. By understanding the basic pricing situations that confront carriers, shippers can gain some insights about the strength of a carrier's negotiating positions.

Ranking issues by order of importance helps establish an effective agenda. It also helps the traffic manager identify important concessions that should be retained as long as possible.

Shippers need to consider which side will make the opening presentation. In the interest of both sides is to discuss issues in the order of their ease of solution.

Traffic managers must establish their authority to negotiate. In a team situation, members should have clearly defined roles.

The timing of concessions is important. Important concessions should be saved until last to retain negotiating leverage. Managers should take time to build up important concessions and try to avoid negotiating under the pressure of a deadline.

Questions and answers should be skillfully used. Adroit questioning can keep the other side on the defensive. Answers should enhance the shipper's negotiating position; they need not be precise, especially when a nonanswer or a vague answer serves this purpose best.

Training and practice are critical to successful negotiations. Finally, location is important, for there is a distinct "home court" advantage.

Selected Topical Questions from Past AST&L Examinations

1. Discuss briefly the factors that are influential in negotiating with a carrier for a rate reduction. Be as specific as possible in your answer. (Fall 1980)
2. How can rail shippers use the Uniform Rail Costing System (URCS) recently developed by the ICC to their best interest? (Fall 1981)
3. The ability of a traffic manager to request and receive a rate change has been greatly expanded by the Motor Carrier Act of 1980 and the Staggers Rail Act of 1980. In such rate negotiations, what factors are considered by carriers in determining their rates? (Spring 1982)
4. Regulatory reform has lessened (to a large degree) the use of collectively set rates and has, instead, ushered in an era where shippers and carriers get together and, one on one, decide what rates will be charged by the carrier for providing transportation to the shipper. As a result, traffic and transportation managers have become proficient in the art of negotiation. Negotiation theory tells us certain principles must be followed to negotiate successfully. What are five of these principles? That is, what steps will take an effective negotiation cycle from preplanning to conclusion? (Fall 1982)
5. Describe and indicate the key concerns to shippers and carriers of the following cost elements that often enter into rate-service negotiation:
    (a) Stem or peddle runs
    (b) Terminal expense
    (c) Equipment costs
    (d) Cost or rate taper (Fall 1984)
6. A trucking firm has one truck available on Tuesday. The truck may be used to serve customer A or customer B, or it may stay idle. The cost in fuel, labor, wear and tear, and so forth, to serve either customer is $100. These costs are not incurred if the vehicle is idle. Customer A would pay $350 for the truck's services, whereas customer B would pay $525. The owner of the company asks a bookkeeper and an economist how much it costs to serve customer B. The bookkeeper answers $100, but the economist holds that the cost is $350. Explain their two positions. (Spring 1985)
7. Give three examples (from different modes) of how demand for transport service can systematically vary over time, that is, by hour of day, season of the year, and so on. What effect might this variation have on a carrier's equipment

utilization? What strategies might a carrier employ in this situation to enhance profits? (Spring 1985)

8. In today's newly regulated environment, carrier prices have become a negotiable variable in the carrier-shipper relationship. In constructing shipper specific prices, a carrier must analyze several characteristics of this relationship to ensure the price provides not only a profit to the carrier but also an incentive to the shipper to maintain a long-lasting relationship with the carrier. What are some of these characteristics to be analyzed in the price-making process? (Spring 1985)

Notes

1. For a discussion of other problems related to the accounting approach to costing, see W. G. Waters, II, "Statistical costing in transportation," *Transportation Journal* 15 no. 3 (Spring 1976), pp. 49–62; Wayne K. Talley, "Methodologies for transportation cost analysis: A survey," *Transportation Research Record* no. 828 (1981), pp. 1–3; David L. Shrock, "The functional approach to motor carrier costing: Application and limitations," *Proceedings — Twenty-seventh Annual Meeting Transportation Research Forum* 27 (1986), pp. 181–188.
2. Interstate Commerce Commission, *Uniform Rail Costing System* (Washington, D.C.: Bureau of Accounts, April 1983).
3. Mark E. McBride, "An evaluation of various methods of estimating railroad costs," *Logistics and Transportation Review* 19 no. 1 (1983), pp. 45–66; also "Economic costs, railway costs, and the URCS," *Proceedings — Twenty Third Annual Meeting Transportation Research Forum* 23 no. 1, pp. 375–381; Talley, "Methodologies of transportation cost analysis," pp. 1–3.
4. *Rules to Govern and Assembling and Presenting Cost Evidence* 337 ICC 298.
5. ICC Bureau of Accounts Stmt No. 2E1-82, *Highway Form B Motor Carrier Costing Program for Single and Interline Shipments*, March 1982.
6. Regular Common Carrier Conference, *A Carrier's Handbook for Costing Individual Less-Than-Truckload Shipments*, prepared by Temple, Barker, and Sloane, Inc. (Washington, D.C., February 1982); Arnold B. Maltz, "Know your carrier's costs? Factors influencing LTL motor carriers," *Proceedings — Twenty Sixth Annual Meeting Transportation Research Forum* 26 no. 1, pp. 483–492.
7. Robert M. Sutton, "Opportunity in LTL routing," unpublished report (Scotch Plains, New Jersey: Distribution Management Associates, 1982), pp. 3–4; see also "Know thy carrier," *Handling and Shipping Management* (June 1982), p. 46.
8. See Alan D. Schuster, "Statistical cost analysis: Engineering approach," *Logistics and Transportation Review* 14 no. 2 (1979), pp. 151–174; Maltz, "Know your carrier's costs?", pp. 483–484.
9. Sutton, "Opportunity in LTL routing," p. 4.
10. Regular Common Carrier Conference, *A Carrier's Handbook*, p. 127.
11. Regular Common Carrier Conference, *A Carrier's Handbook*, p. 120.
12. Richard D. Armstrong, "The relative importance of linehaul and terminal costs for individual motor carrier shipments," *Proceedings — Twenty Sixth Annual Meeting Transportation Research Forum* 26 no. 1, pp. 459–463.
13. The following description of the railway freight transportation process is adapted from S. J. Petracek et al., *Railroad Classification Yard Technology* (Washington, D.C.: U.S. Dept. of Transportation, Federal Railroad Administration, DOT-TSC-968, July 1976), pp. 7–9; see also Robert C. Dart Jr., *Contracting for Coal Transportation* (New York: McGraw-Hill, 1982).
14. Jean Crichton, "RCA recipe for transportation purchasing," *Inbound Traffic* (January, 1985), p. 19; see also Richard Haupt, "Profile of today's transportation user," *Traffic Quarterly* 40 no. 1 (January 1986), pp. 54–64.
15. See Chester L. Karrass, *The Negotiation Game* (New York: World Publishing, 1970), p. 4.

16. See Howard Raiffa, *The Art and Science of Negotiation* (Cambridge, Massachusetts: Belknap Press, 1982), pp. 120–121.
17. Allen Wastler, ''In search of the win/win situation: New transport negotiating strategies,'' *Traffic World* (October 14, 1985), pp. 27–31.
18. See Roger Fisher and William Ury, *Getting to YES* (Boston: Houghton Mifflin, 1981); Raiffa, *The Art and Science of Negotiation*, p. 38; Dobler et al., *Purchasing and Materials Management* (New York: McGraw-Hill, 1984), pp. 220–221.
19. Dobler et al., *Purchasing and Materials Management*, p. 220.
20. Dobler et al., *Purchasing and Materials Management*, p. 223.
21. See Dobler et al., *Purchasing and Materials Management*, p. 223.
22. See Kurt Hoffman, ''Trucking,'' *Distribution* (December 1984), p. 61.
23. Dobler et al., *Purchasing and Materials Management*, p. 225.
24. Crichton, ''RCA recipe for transportation purchasing,'' pp. 19–24.

chapter | 10

Shipment Consolidation and Load Planning

chapter objectives

After reading this chapter, you will understand:
The general forms of shipment consolidation.
The purpose and nature of a consolidation program.
The cost and service tradeoffs involved in shipment consolidation.
The techniques and practices used to accomplish shipment consolidation.

chapter outline

Introduction

General Forms of
Shipment Consolidation

Planning Framework
Cost and service tradeoffs
Techniques and practices

Summary

Questions

Introduction

Logistics planning includes two basic methods of product consolidation: inventory and shipment.[1] Inventory consolidation concentrates on the number, type, and location of storage points. Shipment consolidation involves the practice of aggregating two or more small shipments into a single large shipment.

Small shipments are a large part of most businesses' total traffic. Generally speaking, small shipments weigh less than 10,000 lb. The majority of them, however, weigh less than 1000 lb. The trucking industry, which is well suited to haul small shipments, dominates this segment of the transportation market. Air freight carriers and intermediaries such as freight forwarders, shipper cooperatives, shipper's associations, shipper's agents, transportation property brokers, and warehouse operators also play important roles.

Shipment consolidation enables the firm to reduce the cost of transportation. As shipment size increases, transportation costs per hundredweight generally decline. Such economies of density result from the effects that shipment size, weight, and pieces have on carrier operating costs and on the rate structures that embody such effects.

Discount programs, especially in the trucking industry, have increased the payoff of shipment consolidation programs. Many trucking firms have used discounts to encourage shippers to consolidate freight into the larger LTL weight groups above 5000 lb that permit more efficient operations. Similarly, air freight companies have structured rates to encourage shipments of 1000 lb or more. Thus, consolidation programs do not have to concentrate primarily on building truckload-sized shipments but can be successful when consolidated shipments weigh as little as 5000 to 6000 lb.

Consolidating shipments, moreover, can improve transportation service. Deregulation has fostered a climate in which carriers and traditional intermediaries offer innovative shipment consolidation services that not only reduce costs but also provide faster, more dependable deliveries. Traffic managers now have considerable flexibility to negotiate published or contract rates. Negotiations can focus directly on the cost reductions that carriers realize when shipments are consolidated. And shippers can negotiate specific arrangements with one carrier for dedicated service that includes all assembly and distribution activities. Previously, shippers had to rely mainly on the availability of third parties, such as public warehouse operators or local cartage companies, to perform assembly and distribution.

This chapter introduces the general forms of shipment consolidation and presents the planning framework for developing and managing a consolidation program. Although the concept of shipment consolidation extends to bulk commodities and volume movements, the primary focus is on small shipments.

General Forms of Shipment Consolidation

Virtually all forms of shipment consolidation involve the aggregation of customer orders across time or place or both. Aggregation across *time* occurs when the shipper holds orders or delays purchases to consolidate shipments. Rather than ship each order immediately, the shipping plan might schedule the release of shipments every second, third, or fourth day.

Aggregation across *place* involves the consolidation of shipments to different destinations within the same general area. For example, the traffic manager might "pool" orders to create larger LTL, TL, CL, or TOFC shipments. The pooled freight is sent as a single consignment to a breakbulk facility such as a warehouse, distribution center, or carrier terminal. At this facility, the LTL carrier, warehouse operator, or local cartage company splits up the shipment, or breaks bulk, and delivers the individual consignments to each customer (see **Exhibit 10.1**). Breakbulk and delivery is often referred to as "distribution" service.

This form of consolidation does not hinder the normal flow of traffic. Shippers, however, probably will have difficulty generating the traffic flows neces-

Exhibit 10.1 **Shipment consolidation — Aggregation across place**

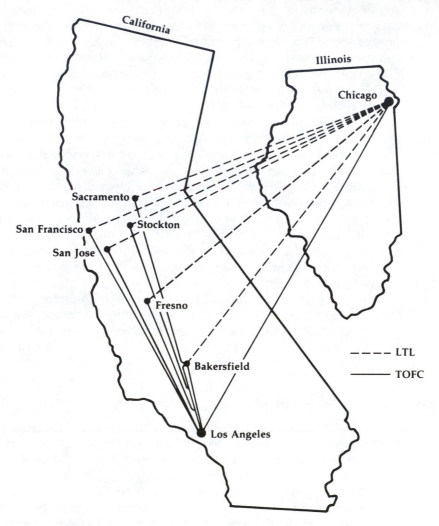

**TOFC consolidation versus LTL shipments
from Chicago to California customers**

Source: Walter L. Weart, ''The Techniques for Freight Consolidation,'' *Traffic World* (March 19, 1984), p. 20. Reprinted with permission of the magazine and the author.

sary to sustain this practice. Nonetheless, the shipper can increase the opportunities to aggregate across place in several ways.

One way is to expand the market area to increase the potential number of shipments eligible for consolidation. Expansion can be accomplished by using an intermediate breakbulk facility. Another way is to use stopoff services. The shipper consolidates orders at the plant for several customers and has the carrier stop along the way for partial unloading. The shipper, moreover, might combine the stopoffs with breakbulk and distribution services (see **Exhibit 10.2**).

A third option is to pool the traffic of different shippers. This is normally accomplished by third parties such as special property brokers, agents, shipper associations, and freight forwarders — all of which are especially important to firms that cannot provide the volume of small shipments necessary to develop effective consolidation programs. Seeking new alternatives, some shippers have initiated co-loading arrangements in which two or more shippers combine their LTL freight to form full truckload shipments.

Exhibit 10.2 **Stopoff consolidation with breakbulk distribution**

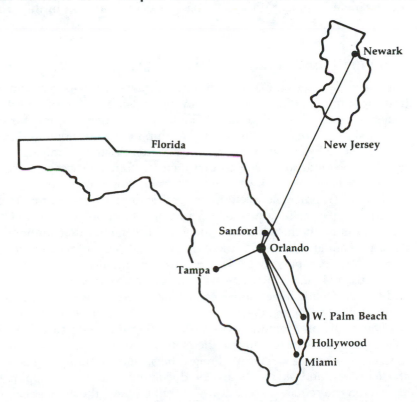

Stopoff at distributor at Orlando. Final destination Tampa.
Distribution to Sanford, Hollywood. Miami, and W. Palm Beach

Source: Walter L. Weart, ''The Techniques for Freight Consolidation,'' *Traffic World* (March 19, 1984), p. 51. Reprinted with permission of the magazine and the author.

The aggregation of orders across both *time and place* allows for maximum consolidation opportunities. This form of consolidation requires shippers to hold and accumulate orders made by customers located in the same general market area. Again, the traffic manager can expand consolidation opportunities by using intermediate breakbulk centers, stopoffs, or third-party intermediaries.

Thus far, the discussion has focused only on the forms of consolidation for outbound shipments. For inbound traffic flows, the process simply reverses itself. For example, the shipper might require different vendors to send their small shipments to the breakbulk facility for assembly into one large load. The consolidated shipment would then go directly to the plant. This reverse logic applies to the other consolidation methods as well.

Shipment consolidation also includes product mixing. Tariff rules, or separately negotiated contract arrangements such as ''single-factor rating programs,'' define the rates applicable when shippers combine different articles into a single shipment. Product mixing is examined later in this chapter.

Planning Framework

The traffic manager must identify and evaluate alternative assembly, distribution, or stopoff configurations within the logistical network. **Exhibit 10.3** illustrates one possible configuration. The traffic manager also must investigate the potential payoff of the various forms of shipment consolidation. These tasks are accomplished by analyzing traffic flows to identify the promising configurations and the consolidation methods that meet customer service requirements.

Analysis of traffic flows requires a breakdown of product groups by lane, mode, and shipment characteristics. Traffic departments normally define lanes in terms of zip code zones, standard point location codes (SPLCs), cities, or states. Shipment characteristics include all elements that impinge on consolidation potential, such as shipment frequency, direction, weight, cube, handling, stowability, susceptibility to damage, and seasonality.

Traffic managers in the firms that conduct periodic transportation system audits will already have the necessary data base. In the firms that do not make these audits, traffic managers must initiate data collection and make it part of the consolidation program. When completing this task, care must be taken to collect data for representative periods and to avoid periods that reflect unusual market or seasonal variations. Freight bills, bills of lading, sales invoices, purchase orders, and payments are the documents that most support this effort.

Given the required breakdown of traffic flows, the analyst can explore various routing and consolidation arrangements. The shipping patterns of some products, or the limitations of some facilities, will probably preclude the use of some forms of consolidation. The absence of industrial rail sidings, for example, clearly eliminates all-rail stopoff service. Similarly, shipment frequency and size might allow a maximum shipping weight of only 10,000 lb and, thus, eliminate

■■■■■ *Exhibit 10.3* **A consolidation configuration**

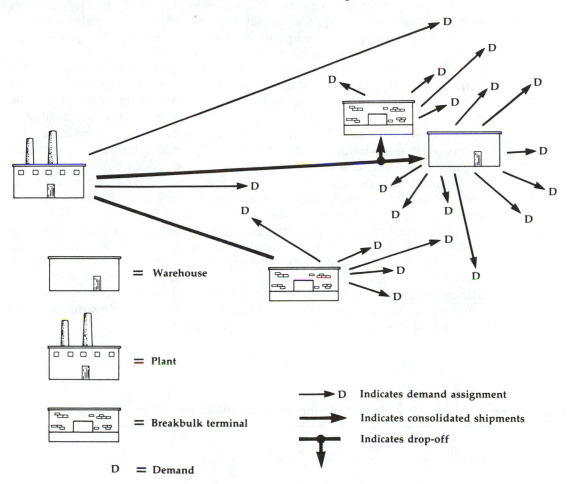

= Warehouse

= Plant

= Breakbulk terminal

D = Demand

→ D Indicates demand assignment

→ Indicates consolidated shipments

Indicates drop-off

Source: Based on ''Assign2 Facilities Location Model,'' Distribution Models, A Division of Edplan Corp., New York, NY 1983.

TL or CL arrangements. Experience and judgment should help the traffic manager isolate the more promising arrangements.

Optimization of complex networks, however, will require sophisticated computer-supported models. Large firms with capable staff and computer resources may choose to build their own models. Other firms with less capability can use the services of various commercial sources. Numerous vendors offer software packages that have the flexibility to fit a variety of network configurations.[2]

Regardless of the approach used to identify consolidation opportunities, the traffic manager must not overlook the service dimension. Marketing and production needs affect the frequency and size of orders. Furthermore, customer service and production requirements set delivery standards; they also set needs for special services such as exclusive use of vehicles, heating, refrigeration, expedited handling, or other similar accessorial elements.

Service requirements, perhaps, represent the most critical elements of the traffic flow analysis. Coordinating the development of those service standards is especially important, because they have an impact on the scope of consolidation opportunities. As the hypothetical situation in **Exhibit 10.4** illustrates, consolidation is sensitive to delivery-time standards. In this example, holding orders for five service days might allow the shipper to aggregate shipments for five delivery points and to increase the shipping weight from 5000 lb to 30,000 lb. Yet marketing managers may consider five service days an intolerably long time. Marketing or production requirements for expedited deliveries or other special services also limit the scope of opportunity.[3]

Thus, the traffic manager must carefully evaluate the net contribution of consolidation alternatives, which requires assessing cost and service tradeoffs. When the evaluation is completed, the traffic staff can work closely with other departments to implement promising options. That the success of any consolidation program mainly rests on interdepartmental coordination and cooperation cannot be emphasized too much. **Exhibit 10.5** profiles the successful program established at National Starch and Chemical Corp.

■■■■■■ *Exhibit 10.4* **Consolidation opportunities and holding time**

BB = Breakbulk facility   D = customer demand points

Service Days Allowed	Consolidation Weight	Market Area
1	5,000 lb	D1,D5
2	8,000 lb	D1,D5,D4
3	15,000 lb	D1,D5,D4,D3
4	20,000 lb	D1,D5,D4,D3
5	30,000 lb	D1,D5,D4,D3,D2

Source: Based on ''Assign2 Facilities Location Model,'' Distribution Models, a Division of Edplan Corp., New York, NY, 1983.

███████ *Exhibit 10.5* **National Starch & Chemical Corp. saves millions on small shipments**

National Starch & Chemical Corp. cut 10 percent off its LTL freight bill in 1984 and saved the company millions of dollars. The National Small Shipments Traffic Conference and *Distribution Magazine* recognized National Starch as "Shipper of the Year" in 1985 for this achievement.

The company, headquartered in Bridgewater, New Jersey, manufactures starch, specialty chemicals, resin adhesives, flavorings, and seasonings. Its four plants—in Chicago; Plainfield, New Jersey; Indianapolis, Indiana; and Island Falls, Maine—also serve as LTL consolidation points. For example, the company consolidates in New Jersey, ships to Cincinnati, and breaks the shipment up and peddles from there. It also has a pool truck operation that goes to Baltimore, Cleveland, Cincinnati, and the St. Louis area.

From these locations, National Starch lets its "freight consolidation coordinators" do their stuff. These traffic people work in the field with production, customer service, and corporate traffic to coordinate small shipments and slash shipping costs.

An open-order ship list lets the National Starch freight consolidation coordinators know what's in the hopper for the next three days. From that list, they can plan their consolidations accordingly. The coordinators have to think about more than transportation scheduling, though. Rescheduling a shipment may mean that manufacturing may have to move production up or back, or that a customer may have to accept a shipment a day earlier or a day later. Traffic managers emphasize, however, that consolidations can actually improve service. They insist a consolidation program can cut some transit times in half.

National Starch uses a hand-prepared routing guide at each of its facilities. The guide, which is updated constantly by four or five people, is a directory of which carrier to use to any point in the country. No one within National Starch is allowed to vary from the guide.

Source: Adapted from Denis Davis, "Reining in Runaway LTL Costs," *Distribution* (September 1985), pp. 36–46. Reprinted with permission from *Distribution Magazine*, Chilton Co., Radnor, Pa.

Cost and Service Tradeoffs

Like everything else in the realm of logistics, consolidation involves tradeoffs. Benefits are derived primarily from lower transportation costs and better transportation service. On the other hand, consolidation sometimes lengthens the order cycle, thus adversely affecting customer service. It also imposes increased administrative costs for program planning and management. The following discussion examines these fundamental tradeoffs and several other miscellaneous considerations.

1. Transportation costs. Economies of density in transportation operations make it possible for carriers to haul larger shipments at lower rates per hundredweight. For the most part, shipment consolidation favorably affects the carrier's pickup, delivery, and dock handling costs (see Chapter 9). A truckload shipment, for example, minimizes the number of stops. The carrier normally makes one stop at origin for pickup and one stop at destination for delivery. There is no dock handling at origin, destination, or intermediate breakbulk terminals. Furthermore, the shipper normally loads the freight, and the receiver unloads it.

By contrast, small shipments require the carrier to make more stops for pickup or delivery, or both. A linehaul tractor-trailer, for example, may stop at several different locations to pick up a few large LTL shipments destined to the same general area. The shipments are transported directly in the linehaul vehicle to destination. There, the trucking company delivers each LTL consignment to different customers or, possibly, to the same party. The carrier avoids terminal (dock) handling, but the added stops increase the cost per hundredweight.

In other operations, shipments may be so small, and pickup or delivery points so dispersed, that it is not economical for the carrier to use linehaul equipment. The carrier must send out specialized pickup and delivery trucks. At origin, they stop at various pickup points and bring many small shipments to the breakbulk terminal. Platform crews sort the shipments by destination and load linehaul equipment as fully as possible. The freight then moves to destination, where the process is reversed.

In this type of breakbulk operation, pickup, delivery, and dock handling costs increase dramatically. Carriers make more stops, yet they pickup or deliver less weight per stop. In addition, during pickup or delivery, as well as during dock handling, more pieces inflate the handling minutes per hundredweight. Smaller, lighter pieces similarly inflate handling costs.

In 1982, the Surface Transportation Assistance Act permitted motor carriers to use 28-foot twin trailers. These "twins" have added significant flexibility to LTL operations. Each twin trailer can be treated as a separate full trailer for operational purposes; that is, it does not have to go through breakbulk consolidation once it is filled with freight bound for a single destination or a set of destinations. Further, these trailers reach full weight and full visible capacity when the load density is near the average general commodity level of 12.5 pounds per cubic foot. By contrast, a standard 40-foot trailer requires an average traffic density of about 17 pounds per cubic foot for full loads (to "cube out").[4]

The emergence of trucking companies that offer innovative assembly and distribution (A&D) services dedicated to specific company logistics systems has been another significant development. These A&D carriers offer a total consolidation service package that enables them to bypass traditional third-party intermediaries. Furthermore, because they can provide fast and reliable linehaul transportation from distant sources, some shippers are using them as key components of just-in-time logistics systems with long supply lines.[5]

2. Transportation service. From the tradeoff perspective, faster and more consistent transit times allow inventories to be reduced without changing customer service standards. In part, inventories, or safety stock, are kept to protect against stockouts caused by late deliveries — that is, when the goods arrive one, two, or more days late. When reliable transit performance reduces that variation, it also allows the firm to reduce safety stock without risking more stockouts.

Faster transit times decrease the cost of inventory in transit. Capital remains tied up in the consignment for a shorter time. Moreover, faster deliveries may generate earlier payments and speed cash flow. This benefit would occur, for example, when the terms of sale indicate that the sales invoice is due and payable on the delivery date. Large consolidated shipments made to the same customer, however, require additional storage space and increase average inventory levels.

3. Customer service. Consolidating shipments generally requires customers to accept slower and, sometimes, less consistent service. This relationship represents a key tradeoff. Order cycle length and consistency are two of the most important elements of logistical customer service. The order cycle refers to the length of time between the placement of the order and the delivery of the goods. Holding orders for shipment consolidation generally will increase the length the order cycle and may also increase order cycle variability.[6] As a general rule, the delays from holding orders offset the faster, more reliable transit times realized by many consolidated shipments. As already suggested, dedicated A&D carriers, coupled with carefully scheduled orders from multiple sources, may be the exception to this rule.

Customers accustomed to delivered prices or freight collect terms (see Chapter 7) will probably have to accept new billing arrangements. Such action is necessary because carriers usually require prepayment of charges for consolidated shipments. The new arrangements will require the shipper to prepay and charge back to each customer the pro rata share of the cost of the consolidated shipment.

4. Planning and control. To build the information system support necessary for a consolidation program may require a good deal of effort. The design and development of suitable data bases for traffic flow analysis, planning, and control can involve extensive preparations. When key data elements are contained in various files spread throughout different departments, like accounting, order processing, or comptroller, the task is especially difficult. The traffic manager probably will have to work with computer system analysts to develop needed data bases. In other firms, such as large industrial concerns, extensive data bases already will be in place, and extracting the data required for analysis should involve only a small incremental effort.

In both the development and the administration of consolidation programs, the traffic manager will have to devote much time and effort in coordinating plans with other departments, especially marketing, production, and order processing. Managers must work together to establish guidelines for issues such as:

- The acceptable holding time for shipment scheduling
- The procedures to ensure that orders made by the same customer, or vendor, are cross-referenced
- The method of notification for the new marketing programs that might alter the timing and size of customer or purchase orders
- The need to change terms of sale in sales contracts or purchase orders to gain control over the shipments

5. Other considerations. By performing services normally conducted at the carrier's terminal, the traffic department takes on additional managerial responsibilities and freight handling costs. Shippers, for example, may have to set up a central shipping or receiving area to sort traffic, assemble loads, break bulk, and perform other activities related to terminal operations.

The cost of processing shipping documents is also likely to increase. Because consolidated shipments generally require prepayment, the shipper has to calculate the pro rata share of the lower total freight costs and bill each customer separately.

Consolidation does, however, reduce the expense of tracing or monitoring shipments. Traffic personnel monitor one consolidated movement instead of many small shipments.

Techniques and Practices

In evaluating tradeoffs, traffic management is confronted with many different techniques and practices. Some of these involve straightforward applications; others are more technical. Likewise, the way in which techniques are pieced together can form simple or highly complex applications.

The following discussion concentrates on fundamental consolidation techniques and practices. In addition, **Exhibit 10.6** identifies ten shipment consolidation opportunities. With experience and some ingenuity, the traffic manager can conceive of many different ways to apply these techniques—either alone or in combination.

Weight breaks. Like most other nontransportation businesses, carriers offer quantity discounts. A weight break identifies a break-even quantity. At the weight break, the cost of shipping a smaller quantity and paying a higher rate

Exhibit 10.6 Ten shipment consolidation opportunities

1. *Single customer, multiple-order* — Created when different plants or divisions in close proximity to one another order several times from the same vendor through decentralized "open buying" or "blanket" orders. Back orders create additional freight that could add to the consolidation mix. Carriers often offer "deficit" weight-break considerations for this type of consolidation.
2. *SKU-unitized shipment* — By bundling, palletizing, shrink wrapping, stretch wrapping, using slipsheets or corrugated shipping containers, "unitized" shipments can be assembled. Many carriers, such as UPS and specialized truckload carriers, have "piece-type" charges or allowances that offer good savings opportunities for this type of consolidation.
3. *Postponement of SKU packaging and labeling* — With many distributors now having private labels, and many manufacturers using multiple-brand strategies to gain market share, savings can be realized by consolidating unlabeled and different SKU-packaged products into larger shipments, which are then moved to more concentrated markets.
4. *Multiple consignee, unit load* — Using this technique, freight is trapped by the shipper or carrier for aggregate movement to destination areas such as three-digit ZIP sectional areas. This results in cost savings and often improves customer service.
5. *Consignor sequence-loading of multiple-consignee aggregate tenders* — Through shipper sequence-loading of freight for the carrier, both parties can gain service and cost efficiencies.
6. *Multiple-consignor shipment* — By using shippers' agents, shippers' associations, freight transportation property brokers, and other third parties, freight can be aggregated into larger volume loads, which can favorably affect cost and customer service.
7. *Balanced inbound and outbound shipments* — In combining LTL and TL shipments for single or multiple parties, carrier efficiencies — and thus shipper and receiver benefits — can be gained.
8. *In-transit inventory storage and handling for redistribution to local markets* — Trading off inventory-carrying costs for freight savings is an important area for analyzing the benefits of consolidation. Low freight rates in relation to inventory-carrying costs per unit have increased the velocities of inventories through distribution systems.
9. *Freight billing* — Freight and carrier billing terms of sale, how and when carriers "pro-batch" their bills, and how and when shippers and receivers "pro-batch" carrier payments are all subtle ways of consolidating activities between shippers, carriers, and receivers. They all have tremendous potential impact on shipper and carrier cash flows and working capital and profits. Another consolidation of this type is "master billing," which allows for the grouping of small shipments for billing purposes into large stopoffs of LTL freight. This is common for intrastate traffic in large states such as California and Pennsylvania.
10. *Small parcel and courier traffic* — Various ways exist for consolidating the tremendous volume of small parcel communications between shippers, carriers, and other business parties. Most notably, these include the new forms of electronic mail in active or passive dialogue, facsimile reproductions, and electronic data interchange (EDI) of business transactions between parties.

Source: Walter L. Weart and Edward J. Marien, "Everybody Out of the Pool," *Distribution* (May 1985), p. 54. Reprinted with permission from *Distribution Magazine*, Chilton Co., Radnor, Pa.

is the same as the cost of shipping the minimum quantity that authorizes the discount rate.

Exhibit 10.7 illustrates the weight break concept for an LTL rate of $9.68 per cwt for shipments falling into the 2000–4999 lb (2M) weight group, and another LTL rate of $7.28 for the 5000–10,000 lb (5M) weight group. At first glance, the rates appear to indicate that all shipments in the 2000–4999 lb range take the $9.68 rate. If this were true, rates would not accurately reflect increases in weight. For example, a 4000 lb shipment would cost $387.20, whereas a 5000 lb one would cost $364.00. By adding 1000 lb of rocks to the shipment, the shipper could save $23.20.

Fortunately, carriers apply the principle that charges should not be higher for shipping less. This principle is the heart of the maximum charge rules found in tariffs. Generally, the rules state that the charge for a shipment in the lower weight group should not exceed the charge for the minimum-sized shipment in the higher weight group.[7]

Referring to Exhibit 10.7, the charge for a shipment in the 2M weight group cannot exceed $364. That is, the weight-break quantity (Q) produces a charge

Exhibit 10.7 **Weight break concept**

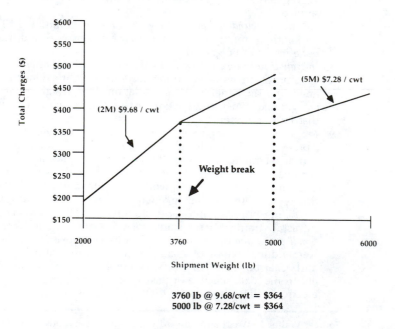

3760 lb @ 9.68/cwt = **$364**
5000 lb @ 7.28/cwt = **$364**

of $364 for either the 2M group or the minimum-sized (5000 lb) shipment in the next higher 5M group. Here Q is determined as follows:

$9.68 \times Q = \$364$ or $Q = \$364/\$9.68$ per cwt $= 37.6$ cwt $= 3760$ lb.

When shipping 3760 lb, the shipper is indifferent to the spread between 2M and 5M rates because they both produce the same total charge ($364). In practice, the following rules would apply:

1. For all shipments weighing less than the weight break (3760 lb), use the higher ($9.68) rate and the actual shipping weight.
2. For weights greater than the weight break but less than the minimum weight of the next weight group, use the lower rate and the minimum weight. In the previous example, a 4000-lb shipment would show the billing weight and rate as follows: 4000 lb billed as 5000 lb at $7.28 per cwt.
3. When the actual weight exceeds the minimum weight, the actual weight times the lower ($7.28) rate is the correct way to compute charges.

As shown in Exhibit 10.7, the 2M and 5M rate structures create a cost plateau of $364 between the weight break (3760 lb) and the minimum shipment size (5000 lb) of the next weight group. Such plateaus have two important ramifications for shipment consolidation.

The *first* is that it is possible to consolidate freight and not realize any savings. For example, this outcome would occur when the consolidated shipment weight (say, 2000 lb combined with 1000 lb) still does not exceed the breakpoint. In fact, the charges for breakbulk, split delivery, or other distribution services may actually increase total transportation costs.

The *second* point is that, under the right circumstances, the shipper can benefit greatly from the rate zone defined by the cost plateau. Any additional freight that increases the shipping weight above the weight break up to the minimum weight does not incur extra linehaul freight charges. The shipper, in effect, has already paid for shipping the minimum weight. When making a 4000-lb shipment, the shipper has already paid for 5000 lb. With an annual volume of 200,000 lb, the shipper could save $3640 a year by adding another 1000 lb (see **Exhibit 10.8**).

If consolidation should involve only one customer, however, the $3640 saved must be evaluated against the additional cost of the larger average inventories, including in-transit inventory, that the additional lot sizes create in the new system.

The weight-break concept also extends to comparable weight-rate alternatives. For example, shippers can compute weight breaks for the various weight groups and for alternating CL and TL rates (see Chapter 5). The generalized formula for the weight break is:

$$\text{Regular rate} \times Q = \text{Discount rate} \times \text{Minimum weight} \tag{1}$$

or

$$Q = \frac{\text{Discount rate} \times \text{Minimum weight}}{\text{Regular rate}} \tag{2}$$

Exhibit 10.9 illustrates the application of this formula to motor carrier weight groups.

Pool service. Pooling freight represents a particular form of shipment consolidation. Specifically, pooling is the process of forming larger shipments for linehaul transportation by grouping together small shipments going to different locations in the same general area. For outbound shipments, the service combines the linehaul shipment of pooled freight with distribution, or split delivery service, from the destination breakbulk point. For inbound traffic, the consolidated linehaul load is assembled from the small shipments of various suppliers that are sent to the assembly facility.

Traditional tariff rules for "pool truck" or "pool car" service require one bill of lading, one shipper, and one receiver. These requirements usually mean a third party must participate. Third parties include local cartage companies, agents, shipper associations, public warehouse operators, or other similar intermediaries.

For inbound traffic, an intermediary usually takes responsibility for assembly operations and acts as the shipper. The consolidated load moves to the plant or firm, which is the single receiver. For outbound traffic, the shipping department normally prepares a single pooled truckload or carload consignment for delivery to a third party, who acts as the receiver and performs the breakbulk and distribution service.

Exhibit 10.8 **Annual savings from reduction of deficit weight**

Item	Old Method	New Method
Annual company purchases	200,000 lb	200,000 lb
Shipment size	4,000 lb	5,000 lb
Shipments per year	50	40
Cost per shipment	$364	$364
How shipped	4,000 lb as 5,000 lb	5,000 lb
Annual freight bill	$18,200	$14,560
Savings		$ 3,640

Exhibit 10.9 **Weight breaks for LTL and TL groups**

Class	Weight Group	Weight Range (lb)	Rate ($/cwt)	Maximum Charge($)	Weight Break (lb)
100	L5C	<500	17.29	67.65	391
100	5C	500–999	13.53	106.70	789
100	1M	1,000–1,999	10.67	193.60	1,814
100	2M	2,000–4,999	9.68	364.00	3,760
100	5M	5,000–9,999	7.28	672.00	9,231
100	10M	10,000–19,999	6.72	940.00	13,988
100	20M	20,000–29,999	4.70	1,170.00	24,894
70	30M	30,000–39,999	3.90	1,216.00	31,179
70	40M	>40,000	3.04		

Example: Weight break for 5C and 1M weight groups

$$\text{Weight break} = 789 \text{ lb} = 7.89 \text{ cwt} = \frac{\$10.67/\text{cwt} \times 10 \text{ cwt}}{\$13.53/\text{cwt}}$$

The growing number of specialized agents and intermodal transportation companies, a freer and more innovative tariff pricing environment, and the ability to contract for transportation services, together allow firms to go beyond the traditional approaches to fashion pool service arrangements. Some motor carrier tariffs, for example, require only 10,000 lb as the minimum weight for pooled shipments. Other motor carriers have initiated "one stop shopping" in which the carrier performs the entire service.

The total transportation cost for pooled service is the sum of the linehaul charge, the fee for breaking bulk (or assembly), and the delivery charge. As shown in **Exhibit 10.10**, pool service saves transportation dollars when the total cost is less than the sum of the charges for direct shipments of each order. This outcome primarily depends on the following elements:

1. *Volume and regularity of orders.* The effectiveness of shipment consolidation is sensitive to the volume and regularity of customer orders.[8] As the number of orders a day increases, the shipper has more opportunities to pool the orders to create effective shipment sizes. Stable flows of pooled shipments, moreover, allow carriers to efficiently utilize their equipment. This result, in turn, helps create additional negotiating leverage. For these reasons, pool service is generally confined to the major markets that attract large and stable traffic flows.

Exhibit 10.10 **Pool consolidation**

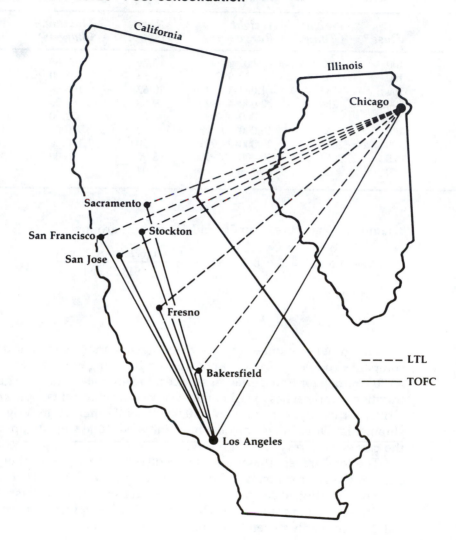

(a) TOFC consolidation versus LTL shipments from Chicago to California customers

(continued)

Exhibit 10.10 *(continued)*

	Direct LTL Charges ($)			Pool Truck Charges ($)			
Customer location	Shipment weight (lb)	Rate ($)	Total	TOFC*	Han- dling†	Local‡	Total
Sacramento	3,750	22.54	302.49	237.38	37.50	236.50	511.38
San Francisco	6,500	17.72	1151.80	411.45	65.00	307.45	783.90
San Jose	560	27.26	152.66	35.45	5.60	47.30	88.35
Fresno	2,850	20.10	572.85	180.41	28.50	191.50	400.41
Stockton	2,100	19.56	410.76	132.93	21.00	189.06	342.99
Bakersfield	5,100	17.23	878.73	322.83	51.00	163.00	536.83
Los Angeles§	9,140	15.25	1525.00	578.56		85.00	663.56
Total	30,000		4994.29	1899.99	208.60	1219.81	3327.42

All shipments assumed to be Class 70 LTL—No discount RMB 583 Sec 1

*TOFC charges prorated against each shipper's weight at 56.33 cwt including drayage ($1,900/30,000 lb)

†Handling at distributor 1.00 cwt

‡Rates from distribution point to final delivery

§Nose load no handling—spot rate. All rates from Tully Trucking, Los Angeles, CA

LTL Rates not discounted

(b) Dollar savings from outbound consolidation

Source: Walter L. Weart, "The Techniques for Freight Consolidation," *Traffic World* (March 19, 1984). Reprinted with permission of the magazine and the author.

2. Linehaul distance. The spread between the linehaul cost of small and large shipments generally increases with distance. In other words, the linehaul savings from freight consolidation typically increases with the length of the haul. Nevertheless, recent evidence suggests that the total cost advantage of new A&D trucking service compared to direct LTL service is not very sensitive to distance or shipment size.[9] Apparently A&D carriers gain their advantage over direct LTL shipments primarily from more efficient small shipment pickup and delivery.

3. Pickup and delivery. For pooling to be feasible, the rule of thumb states that vendors or customers should not stray far from the breakbulk centers located in major markets. Delivery charges tend to escalate rapidly as delivery distance increases or delivery points become more dispersed. High delivery charges can quickly consume any savings in linehaul costs from consolidation. During the 1980s, competitive market forces have eroded the rate structures, preserved in the past by regulation, that artificially restricted delivery distances. Now negotiations concentrate more on the carrier's cost of service. As illustrated in Exhibit 10.10, the carrier's cost of service may justify departures from that rule of thumb.

4. Assembly or breakbulk operations. The fee that some carriers and third parties charge for these operations is additional to the transportation cost for pool service. By adding this cost to the other pooled shipment costs, the shipper can assess the impact of the fee on the feasibility of pooling freight.

The arithmetic becomes complicated, however, when the shipper has to establish shipping or receiving departments at central points to assemble and disassemble small shipments. The cost of this activity will vary from firm to firm. Some firms will have to reorganize traffic operations and, perhaps, even build new facilities. Others will require only minor changes in operating procedures.

Furthermore, the effect of shipment consolidation on centralized operations, once set up, may vary. Interrupting the natural flow of goods to consolidate shipments might produce uneven workloads and less efficient platform operations for some firms. On the other hand, because most shippers do not operate their dock facilities at peak efficiency, consolidation programs may actually improve dock productivity.[10] Thus, consolidation programs, which foster standardized handling procedures and routine supervision, can create productivity increases that absorb almost all extra handling costs. In addition, computerized scheduling of orders will help reduce dock congestion.

5. Shipment release time. How soon orders are released for shipment affects shipment size and customer service. There are three fundamental release policies. First, the *scheduled release* policy requires shipment dispatch at regular intervals, every four days, for instance. Second, the *early release* policy, likewise, accumulates orders over a fixed period, but pooled shipments are released early if their weight exceeds a certain threshold, say 10,000 lb. Third, the *unscheduled release* policy authorizes shipment release only when the weight reaches a designated threshold.

The unscheduled release policy should produce the lowest cost of transportation because shipments are always held until the required weight thresholds have been reached. But this policy also will have the most adverse effect on order cycle length and variability.

The scheduled release policy is more likely to produce longer order cycles than an early release policy.[11] The impact of scheduled release or early release on order cycle variability, however, is not so clear. Several simulation studies indicate that order cycle variability does not necessarily increase when consolidation occurs.[12]

Stopoff service. Many railway and motor carriers authorize stopping in transit for partial unloading or to complete loading. Other modes of domestic transportation do not generally offer this service.

In stopoff service, shipments are subject to volume rates and minimum weights. Moreover, various rules and regulations govern the application of stopoff service. Railway companies usually publish rules in separate stopoff tariffs.

By contrast, trucking firms include stopoff provisions in separate rules tariffs and in specific rate tariffs. Although the shipper must carefully examine the rules that apply to specific stopoff services, the limitations are generally as follows:

1. The entire shipment requires only one bill of lading. The shipper must provide the information necessary for the carrier to perform the stopoff, such as the identity of the points and parties and whether the carrier should perform partial loading or unloading.
2. Like other CL or TL shipments, the stopoff shipment requires shippers to load and receivers to unload.
3. Stopoff points must be on routes authorized for specific CL or TL rates. The authority to switch cars to public or private sidings at these points is also required for railway shipments. Normally, stopoff points are "directly intermediate"; that is, they are at locations directly on the route from origin to destination. For competitive reasons, carriers sometimes authorize "constructively intermediate" stopoffs at off-line points. In **Exhibit 10.11**, for example, Railway A must authorize a stopoff at X to retain traffic; otherwise, the traffic would have to move by Railway B. If constructively intermediate stopoffs are not authorized, off-line points are excluded, or the carrier may impose out-of-route penalty charges.
4. Tariff rules limit the number of stopoffs, the use of stopoffs for both loading and unloading, and the application of stopoffs to certain types of traffic such as export-import freight.

The total charge for stopoff service is computed from the linehaul rate and the service charge. The legally applicable linehaul rate multiplied by the correct weight produces the linehaul charge. The correct weight is the actual weight or the minimum weight, whichever is greater. The actual weight is defined as the weight of the complete load—that is, after partial loading or before partial unloading.

■■■■■■■ *Exhibit 10.11* Constructively intermediate stopoff point

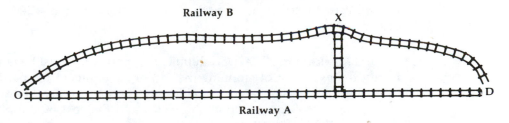

Railway A permits constructively intermediate stopoff
at X en route from O to D to retain traffic.

Tariff rules typically require shippers to pay the highest charge applicable to any portion of the itinerary. Otherwise, a shipment from Chicago to Indianapolis with a stopoff in Los Angeles would penalize the carrier. The distinction between "charge" and "rate" is important. Carriers apply the highest charge, rather than the highest rate, because a lower rate coupled with a higher minimum weight may earn more revenue than a higher rate with a lower minimum weight. For a shipment having an actual weight of 60,000 lb, the principle can be demonstrated as follows:

Rate	Minimum weight	Charge
$3.00	60,000 lb	$1800
$2.80	80,000 lb	$2240

The extra service charge usually takes the form of a flat fee for each stopoff and vehicle. In railway service, moreover, tariff rules may call for additional switching charges.

In the past, and even now to some extent, the method of identifying the applicable through rate has led rate analysts to search for unusual rate structures that would permit more favorable rates to apply to stopoff shipments. For example, although rates generally increase with distance, an analyst might discover that a carrier has published a competitive commodity rate to a distant city that is lower than a class rate for a nearby customer. The analyst can save money, therefore, by consolidating the intermediate stopoff traffic with the traffic of the distant customer, instead of the nearby customer. In today's environment, however, traffic departments have come to rely more on estimating the carrier's cost and negotiating rates than on cleverly applying tariffs.

Exhibit 10.12 demonstrates how the shipper can use regular and pooled stopoff service to reduce transportation costs. Regular service involves partial loading or unloading of LTL or LCL freight only. Pooled stopoff service, on the other hand, involves partial loading or loading of pooled shipments, either alone or in combination with straight LTL/LCL freight.

Besides the pool service considerations already studied and apart from tariff or contractual restrictions, the following additional factors affect the feasibility of using stopoffs:

1. Shipment size. LTL/LCL shipments ("partials") must be large enough to justify the extra cost of stopping the carrier's equipment for partial unloading or loading. Minimum weights significantly affect the tradeoff between the total charge for consolidated stopoff shipments and the cost of shipping small lots direct. As **Exhibit 10.13** demonstrates, when the combined weight of two or more partials is considerably below the minimum weight, stopoff service does not pay.

2. *Customer service.* Customers at downline locations sometimes must endure the delays in switching and loading or unloading that occur at intermediate stopoff sites. For railway stopoff service, intermediate customers may have to absorb the cost of extra handling, as well as blocking or bracing to secure the balance of the load. The shipper may partition and secure loads in support of customer service. This effort, of course, adds cost.

Other practices. Other important consolidation practices involve overflow shipments, mixed freight, multiple shipment pickup and delivery arrangements, and negotiated "single factor" rating programs. In many instances,

Exhibit 10.12 **Pooled stopoff service**

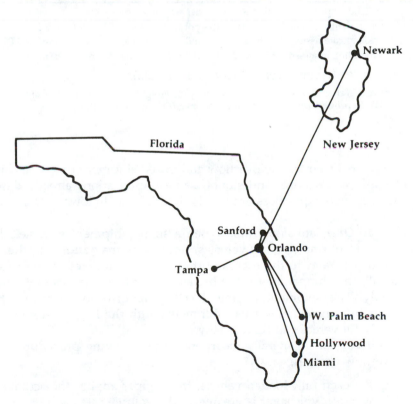

Stopoff at distributor at Orlando. Final destination Tampa.
Distribution to Sanford, Hollywood. Miami, and W. Palm Beach

(a) Example of Stopoff Savings Opportunities

(continued)

Exhibit 10.12 (continued)

	Direct LTL Charges ($)			*Pool Truck Charges ($)*				
Location	Shipment Weight (lb)	Total	Linehaul	Stopoff*	Handling	Local Distribution	Total
Tampa (final destination)	14,500	1522.00	1084.60				1084.60
Miami (SO)	1,248	243.48	93.35	11.48		82.37	187.20
Sanford (SO)	1,320	236.15	98.74	12.14		47.78	158.66
Hollywood (SO)	2,640	423.46	197.47	24.29		141.77	363.53
W. Palm Beach (SO)	2,937	445.24	219.68	27.02		131.87	378.57
Total	22,645	2870.33	1693.84	75.00		403.79	2172.36

*A $75.00 stopoff at local distributor carrier was assessed.

All shipments assumed to be class 70—not discounted—SMC 530 Sec. 1

Truckload rate assumed to be $1.50 per mile—1129 miles; stopoff charge: $75.00

No handling assessed—by distribution (Florida Terminal & Truck) Orlando, FL

(b) Dollar Savings from Stopoff Consolidation

Source: Walter L. Weart, "The Techniques for Freight Consolidation," *Traffic World* (March 19, 1984). Reprinted with permission of the magazine and the author.

shippers can integrate overflow and mixed shipment practices with the weight-break, pool service, and stopoff service applications already described. These miscellaneous practices are briefly described as follows:

1. Overflow shipments. When a single shipment occupies the full visible capacity of one or more vehicles, tariff rules may authorize the application of TL or CL rates to the excess portion of the shipment. If the excess portion is small, the shipper avoids the higher LTL or LCL charges that would otherwise apply. The shipment planner has the opportunity to reduce transportation costs by consolidating small shipments with the large customer orders that already fill vehicles to full capacity.

Tariff rules generally incorporate the following procedures for computing overflow charges:

1. For each fully loaded vehicle, the shipper applies the actual or minimum weight, whichever is greater, to the volume rate.
2. For the vehicle containing the excess portion, the shipper computes the charge by multiplying the volume rate by either the actual weight or the minimum weight for the partial lot.

■■■■ *Exhibit 10.13* **Shipment size and stopoff feasibility**

	Origin	Stopoff	Destination
Route:	O	———→ S	———————→ D

Rate ($/cwt):	O–S	S–D	O–D
LCL/LTL	1.00	1.00	1.50
CL*/TL*	.70	.90	1.00

*Minimum weight = 36,000 lb

Stopoff charge = $75 per stop per vehicle

Customers at S and D order 10,000 lb each

Consolidated shipment with stopoff

O–D:	20,000 lb as 36,000 lb @ $1.00/cwt =	$360
Stopoff charge:		75
		$435

Direct LCL/LTL

O–S:	10,000 @ $1.00/cwt	= $100
O–D:	10,000 @ $1.50/cwt	= $150
		$250

In the trucking industry, "capacity load" rules typically attach a 5000 lb minimum weight to the excess portion. That figure, however, may vary with the size, or cubic capacity, of the vehicle. In the railroad industry, the minimum weight for the excess freight depends on the car type and size. For example, standard-sized flatcars (open equipment), boxcars (closed equipment), and specially equipped cars carry 4000 lb, 10,000 lb, and 15,000 lb minimum weights, respectively.[13]

2. Mixed shipments. By expanding the pool of shipments eligible for consolidation to include different articles, rather than a single product or product group, the shipment planner may dramatically increase consolidation opportunities. In addition, strategically located product mixing facilities allow shippers to create customized assortments to enhance customer service.

Published tariffs provide three basic methods for mixing freight into a single shipment. The first method uses all-container and FAK (freight, all kinds) rates. These rates, though, are relatively high. Consolidating low-rated articles, or mixing low- and high-rated articles, and shipping them FAK probably will not save money.

The second method uses the article descriptions that authorize specific commodity mixtures. Such descriptions are mostly found in motor carrier tariffs. Sometimes, the published rate applies to a list of commodities. Frequently, the list identifies articles related to a larger commodity group, such as "boots or shoes, NOI." Occasionally, item descriptions permit other separately described and rated items to form a certain percentage of the total weight.

The mixed shipment rules found in rail and motor carrier tariffs establish the third method.[14] When the shipper consolidates two or more separately rated articles, these tariff rules set forth the procedure for computing charges. The rail classification (UFC) version for volume shipments requires the shipper to:

- Multiply the actual weight of each article by its straight (that is, non-FAK or nonmixture) rate.
- Use the highest minimum weight among the articles.
- When the minimum weight is higher than the actual weight of all the articles taken together, apply the highest rate to the deficit weight.

The key advantage of this formula, unlike the motor carrier (NMFC) version, is that the carrier does not apply the highest rate and minimum weight published for any article in the mixed shipment. Most motor carriers, however, participate in separate rules or rate tariffs that include a mixed shipment rule that follows the rail formula and supersedes the classification rule. In some cases, the motor carrier version will even authorize the use of the lowest rate for any deficit weight. Thus, by loading only one pound of a low-rated article along with other higher rated articles, the shipper then can reduce the charge for any deficit weight.

These rules, moreover, allow the computation of charges in terms of alternative combinations of LTL and TL (or LCL and CL) shipments to create the lowest charge. These optional calculations do not require the shipper to physically alter the load in the vehicle in any way. For example, given articles A, B, and C loaded in the same vehicle, one option is to apply the formula to all three articles. Another option is to compute charges for A and B as a mixed truckload (and apply the formula) and C as a separate less-than-truckload shipment. There are other combinations, of course, from which the shipper will select the lowest charge.

To benefit fully from these formulas, the shipper will need to review specific details of mixed shipment and other related rules that have important implications for load planning. For example, rules list certain commodities that do not qualify. They also stipulate other restrictions that make the use of mixed shipment formulas contingent on the way in which tariffs describe articles and shippers package and load freight. **Exhibit 10.14** illustrates these contingencies and shows how related classification rules fit together.

■■■■■ *Exhibit 10.14* **UFC rules governing mixed shipments**

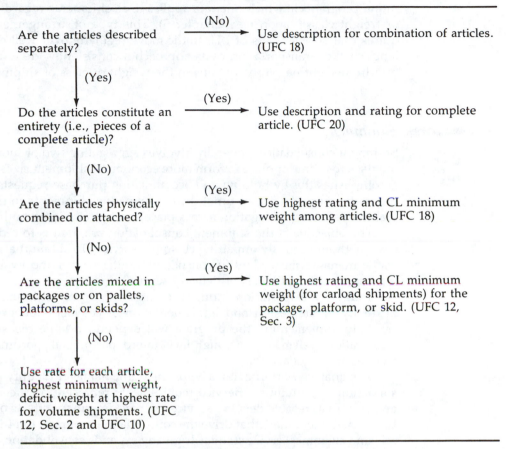

Are the articles described separately? — (No) → Use description for combination of articles. (UFC 18)

(Yes) ↓

Do the articles constitute an entirety (i.e., pieces of a complete article)? — (Yes) → Use description and rating for complete article. (UFC 20)

(No) ↓

Are the articles physically combined or attached? — (Yes) → Use highest rating and CL minimum weight among articles. (UFC 18)

(No) ↓

Are the articles mixed in packages or on pallets, platforms, or skids? — (Yes) → Use highest rating and CL minimum weight (for carload shipments) for the package, platform, or skid. (UFC 12, Sec. 3)

(No) ↓

Use rate for each article, highest minimum weight, deficit weight at highest rate for volume shipments. (UFC 12, Sec. 2 and UFC 10)

3. Multiple shipment pickup and delivery. Simply by consolidating multiple shipments for single pickup, the shipper often can reduce transportation costs. As discussed in Chapter 5, motor carriers often publish "aggregate tender" discounts for shippers who make two or more shipments available at one time for one pickup. Although consolidated for pickup, the shipments remain independent. Each has a separate bill of lading and may go to any authorized destination. With less frequency, shippers may find similar discounts applicable to delivery.

These discounts are a product of the high cost of handling small shipments. Often, they allow 20 percent off the class rate and require a minimum weight of 5000 lb. Their purpose is to give firms incentives to set up dock and scheduling procedures to promote more productive pickup or delivery operations.

4. Single-factor rating programs. Some shippers have installed single-factor rating programs in which one rate applies to all shipments regardless of weight or freight classification (see Chapter 5). This type of arrangement, which requires that sizable flows of LTL traffic move regularly over well-defined traffic lanes, makes transportation costs approach a consistently identifiable cost per hundredweight despite variations in the weight or class of shipments.

Summary

Shipment consolidation generally involves aggregating two or more small shipments across time or place to form more economical shipment sizes. The shipper accomplishes this by holding and accumulating purchase requests or customer orders (aggregation across time), by consolidating shipments to the same general market area (aggregation across place), or by combining both methods.

The objective of the shipment consolidation program is to reduce logistical costs without seriously impairing customer service. To obtain that objective, the traffic manager must identify consolidation alternatives and evaluate them in relation to different levels of delivery service.

Today's transportation industry environment gives shippers tremendous flexibility in the design and implementation of consolidation programs. Successfully implementing the program will depend, in large measure, on good information system support, high-level interdepartmental coordination, and effective rate negotiations.

The analysis of traffic flows helps the traffic manager identify potential consolidation opportunities. Service requirements represent a crucial part of that analysis. Fast, reliable delivery service can have important effects on sales—that is, the customer orders that drive the consolidation program. On the other hand, relaxing customer service standards greatly expands consolidation opportunities and, thus, potential savings. The traffic manager needs to evaluate cost and service tradeoffs for transportation, customer service, planning and control, and other miscellaneous elements that might tip the balance.

In the analysis of tradeoffs, traffic management must assess the merit of alternative consolidation techniques, including weight breaks, pool service, and stopoff service.

Selected Topical Questions from Past AST&L Examinations

1. With the increased price competition and rapidly changing transportation rates, many traffic managers are not calculating the weight breaks because ''by the

time they do, they change." What is weight break, and why would a traffic manager want to calculate it? (Spring 1982)

2. What major tradeoffs must the traffic manager consider when building a consolidation program?
3. What kind of information is helpful for analyzing traffic flows?
4. What is the most critical element of traffic flow analysis? Why?

Notes

1. For further discussion of how consolidation fits into logistics planning, see Donald J. Bowersox, *Logistical Management*, 2d ed. (New York: Macmillan, 1981), chaps. 5 and 7.
2. For various listings, see Thomas C. Dulaney, "Shape up, ship out," *Business Computer Systems* (August 1984), pp. 71–80; Colin Barrett, "Software review," column in *Traffic World*.
3. See Andrew W. Lai and Bernard J. LaLonde, "A computer simulation study of freight consolidation alternatives at the distribution center level," *Proceedings, ORSA/TIMS Joint National Meeting* (November 1976).
4. See Charles A. Taff, *Commercial Motor Transportation*, 7th ed. (Centreville, Maryland: Cornell Maritime Press, 1986), p. 93.
5. David L. Anderson and Robert J. Quinn, "The role of transportation in long supply line just-in-time logistics channels," *Journal of Business Logistics* 7 no. 1 (1986), pp. 68–87.
6. See Martha C. Cooper, "Freight consolidation and warehouse location strategies in physical distribution systems," *Journal of Business Logistics* 4 no. 2 (1983), pp. 53–73; James M. Masters, "The effects of freight consolidation on customer service," *Journal of Business Logistics* 2 no. 1 (1980), pp. 55–74; and George C. Jackson, "Evaluating order consolidation strategies using simulation," *Journal of Business Logistics* 2 no. 2 (1981), pp. 110–138.
7. See NMFC Rule 595 and UFC Rule 15.
8. See Jackson, "Evaluating order consolidation," pp. 120–122; Cooper, "Freight consolidation," p. 61.
9. Anderson and Quinn, "Transportation in long supply line just-in-time logistics channels," pp. 77–80.
10. Malcolm J. Newbourne, *A Guide to Freight Consolidation for Shippers* (Washington, D.C.: Traffic Service Corp., 1976).
11. Jackson, "Evaluating order consolidation," p. 128.
12. Cooper, "Freight consolidation," p. 63.
13. See UFC Rule 24.
14. See UFC Rule 10 and NMFC Rule 645; also check the same item numbers in rules and rate tariffs for modified versions.

chapter | 11

Accessorial Services and Charges

chapter objectives

After reading this chapter, you will understand:
The basic nature of accessorial services.
The role of accessorial services in the transportation
purchase decision.

293

Introduction

Accessorial Linehaul Services
Transit privileges
Reconsignment and diversion
Stopoffs
Pool service
Protective service

Accessorial Terminal Services
Pickup and delivery

Demurrage and detention
Storage
Weighing
Switching

Summary

Questions

Introduction

Freight transportation entails many services besides the actual linehaul move. Accessorial services are incidental to the linehaul or part of terminal operations; tariffs contain the rules and charges for these services. Nonrail carriers, especially trucking companies, use all-encompassing "rules tariffs" to define the conditions and charges for terminal and special services. Railway companies,

Exhibit 11.1 **Accessorial services**

Linehaul	*Terminal*
Transit privileges	Pickup and delivery
Reconsignment and diversion	Demurrage and detention
Stopoffs	Storage
Pooling service	Weighing
Protective service	Switching

294

on the other hand, prefer to publish separate tariffs devoted exclusively to certain types of accessorial services, such as transit privileges or stopoffs.

This chapter surveys the linehaul and terminal accessorial services shown in **Exhibit 11.1.** Private (shipper-furnished) railcar service is not shown in Exhibit 11.1 because it is discussed with fleet management in Chapter 14.

Accessorial Linehaul Services

Transit Privileges

Historically, railway companies (and a limited number of trucking companies) have offered shippers transit privileges that allow for an interruption in the through movement so that manufacturing or merchandising activities can take place at an intermediate transit point. These privileges include a broad range of processes or operations that add time, place, or form utility to the goods in transit. Although storage in transit remains the most popular application, railway carriers have authorized transit privileges for milling, fabrication, cleaning, sorting, mixing, repackaging, and similar activities. After processing the goods at the transit point, the shipper reforwards the identical inbound articles or their by-products to the marketplace. The rates charged by the carrier assume a single through movement rather than two short hauls. In other words, the shipper pays the less costly origin-to-marketplace through rate plus a small charge for the service, instead of the separately established combination of rates for shipments to and from the transit point.

Railway companies introduced the transit privilege more than a century ago as a device for stimulating competition and traffic. For example, as shown in **Exhibit 11.2**, the privilege effectively neutralizes the transportation cost advantage of favorably situated producers or consumers by equalizing the freight charges among competing firms. In this illustration, the Minneapolis Construction Company has the option to purchase steel roofing from two suppliers, one located in Chicago and the other in Milwaukee. Both suppliers fabricate steel roofing from sheet steel manufactured in Chicago. The Chicago fabricator, however, enjoys a transportation cost advantage of 25 cents/cwt, which allows it to capture the market and Railway A to haul all the traffic. By offering the privilege of fabrication-in-transit to the Milwaukee supplier, Railway B helps the supplier compete on an equal footing with the Chicago fabricator and stimulates some steel roofing traffic for itself.

The cost of both the linehaul service and the transit privilege is the total transportation cost to the shipper. Most railway tariffs use the *balance-of-the-through-rate* approach to compute linehaul charges. This approach requires the shipper to pay for the initial inbound movement. When reforwarding the goods from the transit point, the shipper subtracts the inbound rate from the applicable

■■■■■■ *Exhibit 11.2* **Transit privileges**

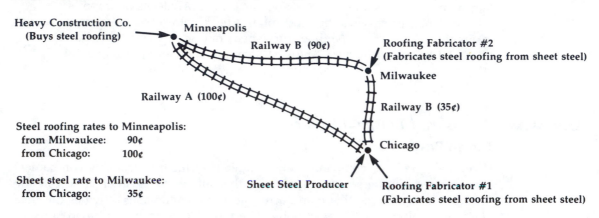

Both fabricators make steel roofing from sheet steel produced in Chicago and sell it to Heavy Construction Co.

Fabricator 1 Cost (¢/cwt)			*Fabricator 2 Cost (¢/cwt)*	
			Without transit	**With transit**
	Through rate		Combination rate	Through rate
		Sheet steel— Chicago to Milwaukee	35	35
Steel roofing— Chicago to Minneapolis	100	Steel roofing— Milwaukee to Minneapolis	90 125	65 100
		Transit change	0	1
Total	100	Total	125	101

through rate and uses the difference (the balance of the through rate) to compute charges. The shipper also pays a relatively small amount for the transit privilege. For example, in Exhibit 11.2, Fabricator 2 pays an inbound rate of 35 cents and the balance of the through rate of 65 cents (100 cents − 35 cents), and a transit charge of 1 cent for a total of 101 to make a through rate of 100 cents.

Other tariffs rely on *claims* to settle the freight bill. This method requires the shipper to pay the separately established rates applicable to each leg of the complete trip and then to file a claim with the carrier for a refund. The difference between the total linehaul charges already paid and the charges calculated with the through rate is the amount of the refund. In Exhibit 11.2, Fabricator 2 pays the combination rate of 125 cents/cwt for the through movement and then files a claim for the difference between the through rate (100 cents/cwt) and the combination rate (125 cents/cwt). The transit charge is an extra cost.

Both methods, therefore, give the shipper the benefit of the through rate. The origin and the final destination ordinarily determine this rate as long as it applies to the route via the transit point. Nonetheless, shipments to or from the transit point may have the longest haul or may show the highest rate, or both. To address this situation, carriers publish a rule, unofficially known as the "three-way" rule, that requires the shipper to examine the applicable rates from

- The origin to the transit facility
- The origin to the final destination
- The transit facility to the final destination

The rate that yields the highest *charge* among these alternatives is the through rate. **Exhibit 11.3** illustrates an application of the three-way rule.

Transit rules also prescribe conditions for the following other elements that affect the determination of charges:

- Weight loss or gain during processing
- The inbound weight (actual or carload minimum) entitled to transit privileges
- The application of alternating rates and minimum weights
- Nontransit tonnage reforwarded with transit tonnage
- Added or absorbed switching charges
- The option to use transit tonnage that has originated at various points to make up a single outbound shipment from the transit point
- Time limitations
- The option to conduct transit operations at off-line (constructively) intermediate points

Not surprisingly, the implementation of this service requires extensive paperwork, policing, and record keeping. Traffic managers must apply in writing to the carrier for authority to act as transit operators. In addition, carriers demand that transit operators keep complete records and submit reports periodically. Such documentation contains detailed information about the commodities that pass through the transit station, including a breakdown of inbound and outbound shipments by categories such as commodity, origin, mode of transportation, and weight. Further, transit operators must agree to inspections and policing by the railroad industry's weighing and inspection bureaus.

Although transit-type services are still important marketing tools, today's pricing environment has eroded the historic pattern of transit ratemaking. Deregulation has allowed railway companies to cancel transit privileges and to re-

███████ *Exhibit 11.3* **Three-way rule**

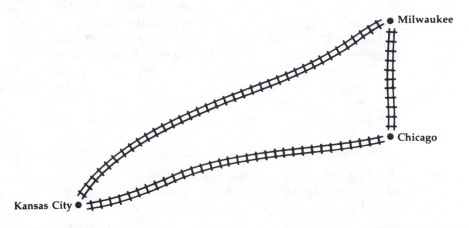

Situation	Origin: Chicago Destination: Kansas City Route: Chicago to Milwaukee to Kansas City
Rates*	Chicago to Milwaukee 35¢/cwt Milwaukee to Kansas City 100¢/cwt Chicago to Kansas City 90¢/cwt
Three-way rule	The through rate to use in computing linehaul transit privilege charges is the rate that produces the highest charge.
Through rate	100¢/cwt

*It is assumed that all rates have the same minimum weight.

structure rates to meet the needs of specific transit operators. By substituting competitive combination-rate structures for transit privileges, some carriers have eliminated costly record keeping and policing responsibilities. In 1981, for example, Conrail introduced a "no frill" grain tariff that quotes distinct rates for grain and its by-products arriving at or departing from intermediate transit stations. The tariff contains thirty-seven pages and replaces four thousand pages of tariffs, special arrangements, transit provisions, and routing guides.[1]

Furthermore, contract ratemaking has diminished the importance of transit tariffs. Many long-time users of transit privileges have abandoned traditional published tariff arrangements to forge specific contractual agreements with railway carriers.

Reconsignment and Diversion

Reconsignment or diversion service allows shippers to amend the shipping contract by altering the billing instructions or creating an additional movement.

Specifically, this service consists of a change in (1) the name of the consignee or consignor, (2) the route, or (3) the destination.

Historically, tariffs distinguished between diversion and reconsignment. Diversion referred to a change in the billed destination of goods en route. Reconsignment, on the other hand, meant any change made after the arrival of the goods at the billed destination but before the actual delivery. This distinction, though, has disappeared, and practitioners and tariffs now treat the terms synonymously.

For competitive reasons, the railway industry has traditionally allowed liberal use of reconsignment. Shippers have come to view railway reconsignment service as a commercial necessity and an obligation of common carriage. This view, however, does not extend to motor carriage; trucking companies publish rules stating that the companies agree only to try to follow a request for reconsignment. In other words, motor carriers refuse to accept responsibility for situations in which they are unable to perform reconsignment.

Shippers use reconsignment services primarily for the following reasons:

1. Marketing-distribution tool. Most railway reconsignments involve sales-in-transit purchases. Such arrangements add flexibility to the distribution process by permitting traffic managers to substitute in-transit inventories for warehouse facilities and field inventories. This process generally works as follows: In anticipation of sales, sellers consign shipments to themselves at locations close to distant markets. As the goods move toward the market, buyers place orders. Then, to fill the orders, sellers reconsign the shipments.

Such flexible distribution is often essential to market many types of products successfully. For example, to minimize spoilage, firms selling fresh fruits and vegetables strive to reduce the length of time between harvesting and consumption. Reconsignment service, coupled with sale-in-transit agreements, enables these sellers to use low-cost but relatively slow railway carriage to reach longhaul markets. That is, rather than wait until buyers place orders before making shipments, sellers ship the goods immediately and then find buyers for the shipments already en route.

Western lumber mills have similarly used reconsignment service to achieve delivery times that are competitive with southern mills more favorably situated. By "rolling" freight cars full of lumber toward midwestern and eastern markets, western producers shorten the time it takes to deliver shipments once the buyer places an order.

The flexibility of reconsignment service also helps sellers promptly respond to changing market conditions. When the price of commodities, such as grain or fresh produce, changes frequently in different market areas, sellers can respond by diverting shipments to capture the highest price. Likewise, brokers, agents, or other intermediaries in the distribution channel can purchase goods, head them in the direction of several contiguous markets, and then resell them as opportunities arise. These examples of sales-in-transit applications do not, of course, exhaust the ways shippers use reconsignment service as a marketing-distribution tool.

From the railway carrier's perspective, such sales-in-transit applications facilitate the flow of volume traffic. To assist this process further, carriers establish "hold points" that serve as the nominal billed destinations from which cars will be reconsigned to buyers. That is, the shipper forwards the car to one of these points with instructions to "hold for reconsignment." Railway companies also realize that this service has a negative side. Freight cars often stand idle for a week or more awaiting forwarding instructions, which adversely affects car utilization.

2. Contingencies. Shippers also use reconsignment to resolve problems related to unexpected contingencies or emergencies. That is, business conditions may necessitate reconsignment action in the following situations:

- Buyers cancel orders or go bankrupt
- Critical shortages of supplies appear unexpectedly at certain plants
- Strikes, production breakdowns, or natural catastrophes make delivery impossible or undesirable

3. Avoid cost of reshipment. Reconsignment service often reduces the cost of shipping by permitting traffic managers to avoid the more expensive process of reshipment. When a shipment reaches its billed destination, the traffic manager always has the right to reship the goods to another location. Each move, though, ordinarily requires a separate bill of lading, and the shipper pays the separately established charges for transportation to and from the reshipping point. By contrast, reconsignment service assumes one continuous movement as long as the carrier retains control of the vehicle or rail car. Thus, carriers generally charge the single-factor through rate rather than the combination of rates via the reconsignment point.

Trucking firms, however, often restrict the through rate benefit. Some motor carrier rules apply the highest rate for any segment of the route; others allow the through rate but charge for circuity. Further, both rail and motor tariffs make this service subject to numerous conditions, including payment, backhauls, inspections, shipment sizes, and many similar items.

Generally, the amount paid for the reconsignment service increases with the level of effort required. The location of the shipment when the shipper issues reconsignment instructions, of course, primarily determines the carrier's incremental effort.

Although shippers may use phone calls or other electronic means to expedite the processing of reconsignment instructions, the requests must be confirmed in writing. Any party making the request must also satisfy the carrier that he or she has the authority to do so. The party that has title to the goods in transit normally has the right to reconsign. Tariff rules give the details and prescribe other numerous requirements.

Stopoffs

Shippers use stopoff service to interrupt the linehaul movement for partial unloading or to complete loading. This service permits shippers to pay the volume through rate rather than the rates for less-than-volume shipments loaded or unloaded at stopoff points. Stopoffs play an important role in shipment consolidation (as discussed in Chapter 10).

Pool Service

When pool service is available, shippers can aggregate separate shipments headed toward the same market into a single large shipment. The pooled freight moves to destination terminals where carriers or third parties perform breakbulk and split-delivery tasks. (Pool service is also discussed fully in Chapter 10.)

Protective Service

Mechanical refrigeration and heating are two frequently used methods of protecting shipments. Obviously, refrigeration can be vitally important for the successful transportation of fresh foods, dairy products, and many other perishable commodities. Still other products may require heating.

As an extra service, carriers may supply special-purpose equipment to protect such commodities. For example, railway companies supply insulated boxcars. In addition, the interior of some boxcars is coated with white epoxy to protect food products from contamination. Either the linehaul rate includes the use of special-purpose equipment, or the carrier establishes a separate charge for it. Nonetheless, protective services are available only when authorized by tariff. The charges in tariffs generally depend on the specific type of protective service requested and on the commodity.

Accessorial Terminal Services

Pickup and Delivery

Linehaul rates for less-than-truckload shipments normally include pickup and delivery of freight. Likewise, the linehaul rates of freight forwarders and air freight companies frequently include pickup and delivery service. Nonetheless, shippers pay additional charges when pickup and delivery service requirements exceed certain size, weight, distance, or handling conditions. Most motor carriers, for example, charge shippers extra for picking up single shipments of 500 lb or less; for service on Saturdays, Sundays or holidays; and for service in congested inner-city points. Further, shippers must assume the expense for tasks like sorting, packing, assembling, or inspecting freight, or they must pay the carrier to accomplish these tasks. For truckload (or carload) shipments, carriers also expect shippers and receivers to complete loading and unloading or to pay the carrier to perform these tasks.

Demurrage and Detention

Demurrage refers to the delay or detention of carrier's equipment. Originally shipowners established penalty charges for detaining vessels beyond the time authorized for loading or unloading. But now demurrage includes charges for delays to other types of transportation equipment—railcars, trucks, and aircraft. The basic purpose of demurrage is twofold: (1) to compensate carriers for the cost of equipment, and (2) to impose an additional penalty element to encourage the prompt release of equipment.

Railway demurrage. Railway demurrage encompasses delays related to the accessorial services undertaken at intermediate points, as well as to the loading, unloading, or other special services provided at origins and destinations. For example, railway companies charge for the delays that involve equipment at intermediate reconsignment-holding, stopoff, transloading and transit points. Carriers, however, exempt some types of equipment from demurrage charges, principally, private cars on private sidings.

The railroad industry has traditionally established uniform charges and rules for demurrage in one governing tariff.[2] These charges and rules apply to the international, interstate, and intrastate traffic of all participating carriers.

The following discussion presents the long-standing demurrage plans and practices of the railway industry. The ICC, however, may soon deregulate demurrage charges and simultaneously remove the railway industry's antitrust immunity to set demurrage charges collectively.[3] This action, of course, would free shippers (or carriers) to negotiate charges and terms.

The *basic demurrage plan* permits shippers one day (twenty-four hours) to load or hold cars for reconsignment, and two days (forty-eight hours) to unload. Loading includes furnishing forwarding directions for outbound cars. Similarly, unloading includes notification that the car is either empty or partly unloaded and available to the railroad. After the expiration of this "free time," shippers must pay for each day of demurrage.

The demurrage day begins at 7 A.M., but the exact date that the clock starts depends on several conditions—the type of placement, the notice of arrival, and the day itself (that is, whether it is a Saturday, a Sunday, or a holiday). Carriers must notify shippers of placement to start the clock. At times, though, the carrier cannot place the car on designated delivery sidings because of a condition attributable to the consignee. Instead, the carrier must "constructively place" the freight car on public delivery tracks or at other holding points on the railroad's line. Although actual placement constitutes notice, constructive placement usually requires notification by mail or telephone.

Furthermore, Saturdays, Sundays, and holidays do not count as demurrage days unless they follow the second chargeable day. A graduated rate schedule determines the charge per car per day. The current schedule, for example, charges $20 per car for days 1–4, $30 for days 5–6, and $60 for day 7 and beyond. **Exhibit 11.4** demonstrates the accounting procedure.

Occasionally, events beyond the shipper's control cause delays. The demurrage tariff grants relief from charges in the following special situations:

- Strike interference
- Bunching— placing cars in accumulated numbers in excess of scheduled car orders or daily shipments
- Run around— carrier places new arrivals at receiver's siding before placing cars already awaiting placement
- Weather interference
- Railroad error

Shippers have the option of subscribing to an alternative *average-agreement* demurrage plan, which applies only to the loading and unloading operations not part of interplant service. Such an agreement enables shippers to use unexpired free time to offset some demurrage charges. The accounting procedure works as follows: A credit is earned for each car released within the first twenty-four hours of free time. Conversely, a debit is charged for each day after the expiration of free time. The charge per debit day escalates in the same manner as the charge in the basic plan. At the end of each month, carriers deduct the total number of credits from the total number of debits and bill shippers for the remainder. The credits, however, cancel only the debits that accrue during the first four chargeable days. Further, loading credits can cancel only loading debits, and vice versa. Like the basic plan, the average agreement provides allowances for delays arising from exceptional circumstances; however, it places more restrictions on these allowances than the basic plan does.

Although the majority of shippers favor the average agreement, the railroad industry is concerned about the workability of the plan in its present form. This concern stems from the relationship between per diem and demurrage charges. *Per diem* is the charge one carrier must pay another for the use of foreign cars on its lines (for example, a Conrail car on Chessie System lines). This charge has a time-mileage structure established for the different types of freight cars.

███████████ *Exhibit 11.4* **Basic demurrage plan accounting procedure**

Day	Status	Charge ($)
Monday	Free	—
Tuesday	Free	—
Wednesday	Charge	20
Thursday	Charge	20
Friday	Charge	20
Saturday	Charge	20
Sunday	Charge	30
Monday	Charge	30
Tuesday	Charge	60
•	•	•
•	•	•
•	•	•

(a) Example 1 — Unloading

Day	Status	Charge ($)
Wednesday	Free	—
Thursday	Free	—
Friday	Charge	20
Saturday	Excluded	—
Sunday	Excluded	—
Monday	Charge	20
Tuesday	Charge	20
Wednesday	Charge	20
Thursday	Charge	30
Friday	Charge	30
Saturday	Charge	60
•	•	•
•	•	•
•	•	•

(b) Example 2 — Unloading

The average-agreement plan, however, was originally designed to simplify accounting procedures when per diem was the same for all cars. Now a boxcar costing $2.40 per day for per diem can earn a credit that cancels a debit of a trilevel (autorack) car costing $40 per day.[4] The deregulation of demurrage charges will, no doubt, permit the railway industry to rationalize its accounting procedures and charges.

The volume of rail traffic will determine the extent to which traffic managers should monitor and control demurrage. Occasional users of rail service should educate employees about the importance of timely loading and unloading to prevent demurrage. Firms like Ford Motor Company or General Motors — which have tens of thousands of freight cars serving a vast network of suppliers, plants, distribution centers, and dealers — need to monitor and control the activities that affect demurrage. Setting the proper pace of departures, for example, is very important. When ordering empty cars for loading, traffic personnel must schedule shipments so that the rate of empty-car placement does not exceed the work capacity of loading crews at origin. Likewise, when planning outbound shipments, schedulers must stagger departures to help prevent bottlenecks at destinations.

Other demurrage control measures include a periodic review of facilities and equipment and the development of management reports. Among other things, reviews ought to include the length of sidings, the type of material handling equipment available, and the number and timing of crew shifts.

In addition, demurrage plays a role in long-range transportation planning. It is an important consideration in the design of plant and distribution center facilities, in the purchase of material handling equipment, and in the use of privately owned or railroad-owned cars.

Ford Motor Company, for example, has an elaborate computer-based system for monitoring all loaded and empty railcars in North American via the Supplier On-Line Management Information System (SOLMIS). Once a supplier ships a railcar to a plant, the outbound movement information is input into SOLMIS, and a history of the movement is initiated. Railway companies continually update the system with the latest railcar movements from various interchange points.

Exhibits 11.5–11.8 depict how Ford monitors the flow of railcars from supplier to destination and how the company calculates its projected demurrage liability:

1. *Exhibit 11.5 — Notice of Constructive Placement.* After receiving the notice of constructive placement, plant traffic inputs this information into SOLMIS for the update of the file.
2. *Exhibit 11.6 — Plant Switch Order.* Plant material handling, or part control, prepares the switch order, and traffic forwards it to the railroad. The switch order advises the railway company of the railcars to be pulled out empty and the placement of required railcars by track and sequence of "spot order." Plant traffic again will update SOLMIS with the information about railcars being switched into and out of the plant.
3. *Exhibit 11.7 — SOLMIS Reports.* Users can generate complete sets, as well as group or individual reports, at any time. For example, Report 1 shows the loaded railcars en route. Details include the car identification (car number) and car code (type of car), origin, scheduled transit days (FT DA), switchout date (month, day, and hour), and estimated time of

Exhibit 11.5 Sample notice of constructive placement

THE CHESAPEAKE AND OHIO RAILWAY CO.

NOTICE OF CONSTRUCTIVE PLACEMEN

CHESSIE SYSTEM FORM DEM I-C
REV 12-69
MADE IN U.S.A

Thur.
DATE
4-4-85

STATION

7439 Wayne M,

TO

Ford Motor Co.
P.O. Box   5
Wayne, M,

You are hereby notified that the following cars consigned to, or ordered to or by you, cannot be delivered on account of your inability to receive, or because of other conditions attributable to you, and tender of same is hereby made. These cars are subject to the demurrage rules published in the tariffs lawfully on file, and charges in accordance therewith will be made for detention beyond the free time therein provided.

| CAR | | ORIGINAL CAR (IF LADING HAS BEEN TRANSFERRED) | | CONTENTS | POINT OF SHIPMENT | ARRIVAL | | REMARKS |
INITIALS	NUMBER	INITIALS	NUMBER			DATE	HR.	
IHRC	10006			A PTS	Connersville	4-3	930PM	X GRB3
Sou	33904			—	Culpeper	—	—	—
WP	4003			—	Athens	—	—	—
NW	86096			—	Chgo Hgts	4-4	1230A	X TL82
CR	243417			—	—	—	—	—
Milw	4791			—	—	—	—	—
MP	271062			—	—	—	—	—
MP	271223			—	—	—	—	—

J. Hagle

Agent

ORIGINAL COPY OF THIS NOTICE RECEIVED

AT 1:30 A M 4-4 19 85 BY Ron Santure PER

(Hour) (Date) (Consignor or Consignee)

(If Consignor or Consignee refuses to acknowledge receipt of this notice as provided above, the person delivering same should fill out, date and sign the following certificate on the copy kept on file by the agent.)

I hereby certify that the copy of this notice was delivered by me to

Mr. _____
representing the Consignor or Consignee named herein.

At _____M. on _____19_____

_____
(Signature of Person Delivering Notice)

(If this notice is not delivered personally by the agent or his representative, it must be served by mail, and the following certificate on the copy kept on file by the agent should be filled out, dated and signed.)

I hereby certify that the copy of this notice was mailed to the

Shipper or Consignee named hereon, at _____ M. on _____

19_____

_____
Agent

Source: Ford Motor Company. Reprinted by permission.

Exhibit 11.6 Sample switch order

SWITCH ORDER

Total Empty ___6___

Spot #1 begins at bumper.

Date __4-3-85__

⊙ This indicates where railroad actually placed railcar and is to be filled out after the tracks are set.

CARS ON SPOT				CARS TO BE SPOTTED		
Track #2 4-3-85 Time: 10³⁰pm				**Track #2** 4-3-85 Time: 10³pm		
Car Number	Code	Spot	Release	Car Number	Code	Spot
CPAA·205942	CL ᔆ/ᴮ	1	1			1
CR·237608	CL	2	MT CR	NW·861088	CL	2
MP·27157 S	CG	3	MT CO			3
DT·J·26012	WH	4	MT CR			4
GTW·306066	WH	5	MT CR			5
NW·355027	BU	6	MT CR			6
ATSF·36856	BU	7	MT CR	DT·I·26675	DB	7
GTW·305860	DB ᔆ/ᴮ	8	8			8

				MID-SHIFT		
						1
						2
						3
						4
						5
						6
						7
						8

Track #3		Time:		Track #3		Time:
		1				1
		2				2
		3				3
		4				4
		5				5
		6				6
		7				7
		8				8

Source: Ford Motor Company. Reprinted by permission.

Exhibit 11.7 **Supplier on-line management system (SOLMIS)**

REPORT #1

4/4/85   7:23.82 AM   LOADED AUTOMATIC ETA REPORT

RAILCAR NUMBER	CAR CODE	ORIG	FT DA	SWITCHOUT MO-DA	HR	TFC ETA	MTL ETA	R.R. NAME	TRAIN	S C	LOCA- TION	AS OF MM/DD	HR
CR 297647	AC	A753F	5	04/03	17	06/13	07/03	CR	JEIN3	A	BIGFO	04/03	23
CR 295447	BP	B821A	5	04/03	15	08/03	08/03	CR	DICCS	W	PHIMI	04/03	22
CR 297861	BP	B821A	5	04/02	15	07/05	07/13	CR	CAP13	P	HARRI	04/04	06
IHRC010014	PF	CC05A	6	03/30	03	05/04	05/13	NW	3WJ03	A	HOMES	04/03	13
NW 861387	PF	CC05A	6	04/03	03	07/08	07/13	NW	4FB02	A	BELLE	04/04	05
GMO 056712	176	D726F	5	04/03	16	08/03	08/03	UP		W	COLUM	04/03	17
DTI 026184	FDE	EF03A	2	04/03	15	05/03	05/03						

REPORT #2

4/4/85   7:22.26 AM   AVAILABLE TO THIS PLANT REPORT

RAILCAR NUMBER	ORIGIN CODE	SO MO-DA	HR	CAR CODE	ARRIVAL MO-DA HR	CP MO-DA	PEN MO-DA	DEMURRAGE DB/CR$	PEN$
NW 861083				NCL	03/02 03			**NO CP DATE**	
CR 237498	A753F	04/01	17	AC	04/04 06	04/04	04/12	20-	1
CR 297718	A753F	03/29	17	AC	04/01 04	04/01	04/07	40	1
CR 297879	A753F	03/29	17	AC	04/01 04	04/01	04/07	40	1
GMO 056743	D726F	03/28	16	176	04/02 06	04/02	04/08	20	1

REPORT #3

4/4/85   7:21.09 AM   SET AT THIS PLANT REPORT-AP 16A

RAILCAR NUMBER	ORIGIN CODE	SO MO-DA	HR	CAR CODE	ARRIVAL MO-DA HR	CP MO-DA	PEN MO-DA	SET MO-DA	HR	DEMMURRAGE DB/CR$	PEN$
IHRC010006	CC05A	03/29	03	PF	04/03 10	04/04	04/12	04/04	07	20-	0
DTI 026160	EF03A	04/02	06	FDE	04/03 03	04/03	04/09	04/03	18		1
CR 293410	FD04A	03/26	03	UT	03/27 13	03/29	04/05	04/04	06	80	1
LN 180903	KODAK	03/28	16		04/01 04	04/01	04/07	04/02	06	40	0
SBD 314541	KODAK	03/28	16		04/01 04	04/01	04/07	04/02	06	40	0
SBD 325412	KODAK	03/28	16		04/01 04	04/01	04/07	04/02	06	40	0

Source: Ford Motor Company. Reprinted by permission.

arrivals. This report also provides information about the latest status of these cars, including the railroad and train, car status, the junction point, and the date. Report 2 identifies the available railcars on hand at a certain plant. This report also shows arrival (ARR), constructive placement (CP), and penalty-charge (PEN) dates, as well as debit, credit, and penalty amounts. Report 3 contains the same information as Report 2, but lists the cars set at the plant for unloading.

4. *Exhibit 11.8—Demurrage Report.* The demurrage report, which is run at the end of each month, calculates the projected demurrage liability. This report is used to audit the monthly demurrage billing. For each car, the report provides information about the railroad (car owner), origin, load-empty status (LE), arrival date, constructive placement (CP) date, penalty date, release date, credits (CR), debits (DB), late penalty charges (P1 and P2), and total demurrage charges (also summarized by railroad at the bottom of the report). Most of the plants have an average-agreement demurrage plan with the serving railway companies.

Motor carrier detention. Motor carriers assess detention charges against shippers as a penalty for holding a carrier's drivers or trailers beyond a stated period. The charges and conditions for detention are found in the rules section of "rate" tariffs or in governing "terminal and special services" tariffs. Since the ICC abandoned its requirements for uniform detention provisions in 1982, motor carriers have published a variety of rules. The following discussion outlines some of these provisions.

Both the amount of free time for loading or unloading and the level of charges depend on three conditions: (1) whether vehicles with "power" (the tractor) remain with the trailer, (2) whether the driver stays, and (3) how much the shipment weighs. Carriers permit the longest free time (perhaps twenty-four hours) for the truckload deliveries that do not require power units or drivers to remain. On the other hand, less-than-truckload deliveries, in which both the power unit and the driver remain, receive the shortest amount of free time. Here, charges usually accrue during fifteen-minute intervals. These intervals expand, however, as the total weight of all the shipments made available for delivery (or pickup) increases.

Free time begins when the driver notifies the firm of the arrival of the vehicle for loading or unloading; it ends with the task of loading or unloading. The shipper has not completed loading until the carrier has the documentation necessary to forward the shipment, that is, the bill of lading. Similarly, unloading includes the payment of charges, if required before delivery, as well as the receipt for delivery. Carriers permit adjustments to free time in certain instances. For example, detention rules may exclude weekends and holidays from the free-

4/4/85   6:28.88 AM   DEMURRAGE FOR APRIL

RAILCAR NUMBER	RAIL ROAD	ORGIN	L E	ARRIVAL DATE	CP DATE	PENALTY DATE	RELEASE DATE	CR 20	DB 20	P1 30	P2 60	DEM CHRG
CR 238184*	CR	A753F	L	03/31 02	04/01	04/07	04/03 17		1			20
CR 237071*	CR	EF03A	L	03/30 02	04/01	04/07	04/03 17		1			20
CR 293443*	CR	FD04A	L	03/29 11	04/01	04/07	04/04 04		1			20
CR 297841*	CR	FD04A	L	03/24 13	03/25	03/31	04/03 17		4	2	2	260
SCL 637005*	COBO	F593J	L	04/02 06	04/03	04/09	04/03 17	1				20-
MPA 03L094*	COBO	G597F	L	04/02 23	04/09	04/09	04/04 04	1				20-
SBD 141952*	COBO	KODAK	L	03/26 21	03/27	04/02	04/04 04		4	2		140
LN 187659*	COBO	KODAK	L	03/28 23	03/29	04/06	04/03 17		2			40
LN 521359*	COBO	KODAK	L	03/29 23	03/29	04/06	04/03 17		2			40
SBD 321848*	COBO	KODAK	L	04/01 04	04/01	04/07	04/03 17		1			20
ATSF036856*	CR	MS01A	L	03/30 14	04/01	04/07	04/03 22		1			20
CO 493116*	CR	MS01A	L	04/01 04	04/01	04/07	04/03 11		1			20
CR 237605*	CR	MS01A	L	04/01 04	04/01	04/07	04/03 11		1			20
CR 237629*	CR	MS01A	L	04/01 04	04/01	04/07	04/03 11		1			20
NW 355027*	CR	MS01A	L	03/28 02	03/29	04/06	04/03 22		2			40
MP 271575*	COBO	MS01A	L	03/31 01	04/01	04/07	04/03 22		1			20
NW 861074*	COBO	MS02A	L	03/31 23	04/01	04/07	04/03 11		1			20
CPAA205942*	CR	MS03A	L	03/31 02	04/01	04/07	04/04		1			20
CR 237366*	CR	MS03A	L	03/28 02	03/29	04/06	04/03 17		2			40
CR 237608*	CR	MS03A	L	04/01 04	04/01	04/07	04/03 22		1			20
NW 861088*	CR	MS03A	L	04/01 04	04/01	04/07	04/04 04		1			20
SOU 042597*	CR	MS03A	L	03/27 03	04/27	04/02	04/03 17		4	2		140
CO 493230*	CR	MS03A	L	04/03 03	04/03	04/09	04/03 17	1				20-
CR 237717*	CR	MS03A	L	04/03 03	04/03	04/09	04/03 17	1				20-
DTI 026675*	CR	MS05A	L	04/03 03	04/03	04/09	04/04 04	1				20-
GTW 305860*	CR	MS05A	L	04/03 03	04/03	04/09	04/04 04	1				20-
NW 860667*	CR	MS05A	L	04/03 03	04/03	04/09	04/03 11	1				20-
DTI 026012*	CR	MS09A	L	04/03 03	04/03	04/09	04/03 22	1				20-
GTW 306066*	CR	MS09A	L	03/28 02	03/29	04/06	04/03 22		2			40
NW 861089*	CR	MS09A	L	04/03 03	04/03	04/09	04/03 11	1				20-
DTI 026758*	CR	PP03A	L	03/29 06	03/29	04/06	04/04 04		2			40
NW 861548*	CR	PP03A	L	03/31 02	04/01	04/07	04/03 17		1			20
CR 281042*	CR	TC03A	L	03/31 02	04/01	04/07	04/03 17		1			20

SERVING RAILROAD	RELEASE COUNT	DEMURRAGE COUNT	CREDIT $20	DEBIT $20	1ST-PEN $30	2ND-PEN $60	$DEMURRAGE CHARGES
COBO	8	8	2	11	2		$240
CR	26	26	7	29	4	2	$680
TOTAL	34	34	9	40	6	2	$920

END OF REPORT

Source: Ford Motor Company. Reprinted by permission.

time computation for simple trailer "drops" of truckload shipments. Further, carriers may authorize adjustments for delays caused by labor strikes.

The management of truck detention, like railcar demurrage, requires careful scheduling, periodic reviews of facilities, performance reports, and employee education.

Storage

Carriers consider freight stored when it is held in their possession after the free time for loading or unloading has expired. If through no fault of their own, carriers must involuntarily store undelivered shipments or shipments awaiting instructions for linehaul transportation, shippers are required to pay storage charges. These charges, like demurrage and detention, are penalties that increase with time to encourage prompt action by shippers and receivers. In the railway industry, storage charges generally apply only to less-than-carload freight, unless the lading consists of hazardous materials.

When a carrier stores the goods, it assumes the role of a warehouse operator instead of a common carrier. This transition is important because the liability of a warehouse operator is much less than that of a common carrier (see Chapter 12).

Weighing

Transportation companies quote freight rates for units of weight such as 100 lb (cwt), 2000 lb (short tons), or 2240 lb (long tons). Obviously, obtaining accurate shipping weights is important for both shippers and carriers. The shipping weight includes packaging materials and containers and excludes a certain amount of the *dunnage* (such as blocking, bracing, and flooring materials) that shippers use to secure loads. Railway tariffs, for example, allow 2000 lb of dunnage without charge, although motor carrier tariffs generally place no limits on the amount of dunnage. In addition, carriers frequently permit shippers to exclude the weight of pallets, platforms, or skids.

Transportation companies face two choices for weighing freight: (1) assume the task themselves, or (2) accept the shipper's evidence of actual weights. If carriers choose to weigh shipments to assess freight charges, they do not charge for the weighing service because it is considered a necessary part of the linehaul. Should shippers wish to double-check carrier's weights, however, extra charges may be levied. The difference between the results for the initial weighing and those of the reweighing determines whether the carrier will adjust the billing, as well as who will pay for the weighing. That is, when the difference exceeds a certain tolerance (for example, 1 to 1.5 percent of the initial weight), carriers customarily adjust the billing to reflect the lowest weight, and they absorb the cost of reweighing.

There are good reasons for making frequent requests to reweigh inbound shipments. For example, human error (say in typing or transcribing figures), vice (theft or pilferage), mechanical failures in the scale itself, and weather (ice, snow, or heat) often produce inaccurate weights.

The railway industry makes the most extensive use of weighing the freight themselves. Since few firms have carload shipment volumes to justify their own track scales, the railway carriers themselves, or the industry's weighing and inspection bureaus, accomplish most of the actual weighing. On the other hand, motor carriers, water carriers, airlines, and freight forwarders generally accept shippers' weights. Motor carriers, however, must frequently weigh their trucks at roadside scales for compliance with highway load limits. This requirement gives carriers in the truckload market the chance to verify the shipper's weights and to reweigh shipments when the divergence is great.

By using shipper-certified weights, transportation companies, including railway carriers, eliminate the costs of weighing, as well as other incidental costs such as extra switching. And shippers gain the advantage of faster, more reliable service by using shipper-certified weights.

Railway companies (and other modes of transportation) implement the second option of using the shippers' weights through actual- and average-weight agreements. The firms or the industry-sponsored organizations that maintain private scales use the *actual-weight agreement*. The carrier agrees to accept the shipper's actual weight inserted on the bill of lading. The shipper, in turn, agrees to (1) keep complete records, (2) make them available for inspection, (3) maintain the scales in good working order, and (4) pay for any undercharges promptly.

Carriers and shippers prefer to use the *average-weight agreement* whenever possible. This agreement takes advantage of the nearly constant weight of standardized articles and packing, like canned goods packed in cases of twenty-four units. The two parties conduct tests jointly to determine the average weight of the standard shipping unit. Carriers then authorize shippers to calculate the weight of each shipment by multiplying the average weight per unit times the number of units. Shippers, in turn, agree to keep records and to make them available for inspection.

The railroad industry's weighing and inspection bureaus administer these agreements. In addition, the Association of American Railroads promulgates track-scale specifications and rules for scale operations. Motor carriers have their own weighing organizations that perform like the railway bureaus.

Carriers also authorize the use of *estimated weights* that do not involve agreements. Instead, shippers compute estimated weights from formulas cited in tariffs. These formulas are derived from well-established weight-to-volume properties of certain commodities, such as petroleum, oil, lubricants, and other bulk liquids. For example, the weight of a tank car full of gasoline is computed by multiplying the number of gallons times 6.6 lb/gal.

Switching

Switching service refers to the railway transportation of freight cars within the yard limits of a single station or within an industrial switching district that comprises several stations. Competitive conditions determine the size of the switching limits rather than the actual operating yard space.

Railway companies switch cars in conjunction with linehaul service and as part of independent terminal activity. When providing linehaul service, railroads usually do not charge shippers for the origin and destination switches on either private or public sidings. When the road movement involves interline routing, a "switch" carrier may switch freight cars at the origin location or destination location, or both. Such a carrier serves as an agent of the "road" carrier, but the road carrier retains responsibility for billing and claims. Further, like the motor carriers that provide free pickup or delivery for competitive reasons, road carriers often absorb switching charges.

Reciprocal switching arrangements are another common practice used for interline movements. These arrangements call for the participating carriers to perform switching operations for each other at a common station without charge to the shipper. At no extra cost, traffic managers can route shipments over the lines of any participating carrier serving the station, regardless of the carrier that has the physical connection to the firm's private siding. Lacking reciprocal agreements, the switch carrier usually demands a share of the linehaul revenue, and the road carrier must absorb this cost or collect it from the shipper. The Staggers Rail Act, however, provides railway companies with considerable freedom to cancel joint rates and switching agreements or to impose surcharges on interline traffic. This freedom, together with the mergers and consolidations in the industry, has led some carriers (especially Conrail and the Chessie System) to engage in cancellation and retaliation. Not surprisingly, shippers consider this behavior an undesirable by-product of deregulation and an important consideration when dealing with carriers.

Sometimes interline routes require *intermediate switching* to "bridge" two road carriers, which do not have a direct connection. Third-party switch companies, which are separately or jointly owned by many railway carriers, perform this bridge service. The road carriers frequently absorb the cost but may pass it on to the shipper.

As mentioned earlier, railroads also perform switching service as an independent terminal activity. For example, they switch cars among contiguous plants in an industrial area and make *intraplant switches* for large industrial firms.

Summary

Transportation companies offer shippers many services related to linehaul and terminal operations.

Transit privileges permit an interruption in the through movement to undertake some manufacturing or merchandising activity at a transit point but protect the through rate. In the 1980s, however, pricing innovations and contracting have replaced most traditional tariff applications of these privileges.

Shippers use reconsignment and diversion service to amend the routing or delivery instructions shown on the bill of lading. Buyers and sellers frequently use this service with sales-in-transit purchase agreements. In addition to giving shippers a flexible marketing-distribution tool, reconsignment and diversion is useful in emergencies. Furthermore, the cost of reconsigning goods is usually less than the cost of reshipping them.

With stopoff service, shippers can partly load or unload freight en route and still apply the through rate. Stopoffs and pool service are key elements in many freight consolidation programs.

Many commodities require refrigeration or heating. Protective service includes providing special-purpose equipment for such commodities, as well as the necessary additional labor.

Although freight rates for less-than-truckload shipments usually include pickup and delivery, shippers may have to pay additional charges when pickup and delivery requirements exceed certain size, weight, distance, or handling limits. Normally, shippers must load volume shipments, and receivers must unload.

Companies pay demurrage or detention when they delay carrier's equipment beyond the time allotted for loading or unloading shipments. The railway industry offers two options: the basic plan and the average plan. Unlike railway demurrage rules, motor carrier detention rules lack uniformity. For many firms, demurrage and detention are important cost elements and, therefore, require considerable planning and control.

When carriers must involuntarily hold goods after free time for loading or unloading has expired, they charge for storage. A carrier then assumes the role of a warehouse operator, which is significant because the carrier's liability for loss and damage is much reduced.

Obtaining accurate shipping weights is an important responsibility for both shippers and carriers. Shipping weights include packaging and packing materials but exclude dunnage up to certain limits. The option to reweigh shipments and the chance to establish weight agreements present opportunities to control cost.

Switching involves the railway transportation of freight cars within yards or industrial switching districts. Carriers switch cars as either a linehaul service or an independent terminal activity.

Selected Topical Questions from Past AST&L Examinations

1. The terms "diversion" and "reconsignment" are often used interchangeably but are technically different. Define diversion and reconsignment, and distinguish between them. Explain how their use could be advantageous to shippers. (Spring 1979)

2. Define transit privilege, and explain its benefit to the shipper. (Spring 1980)

3. The interpretation of demurrage and detention rules has created many conflicts between shippers and carriers.
   (a) Define demurrage and detention.
   (b) What are the intended purposes of each?
   (c) Why are there the conflicts? (Fall 1980)

4. For many traffic managers, one of the most complex activities involves administering and using transit privileges. Transit privileges were developed to negate the effect of tapering rates. Explain. (Spring 1982)

5. Define reciprocal switching. What advantages does it offer a shipper and a railroad? (Spring 1982)

Notes

1. "Non-traditional ratemaking," *Modern Railroads* (June 1982), p. 37.
2. Freight Tariff PHJ 6004-N (H. J. Positano, agent).
3. Ex Parte 462, decided January 30, 1986 and reversed June 18, 1986.
4. George Stern, "Innovative pricing to improve utilization of rail plant and equipment," *Annual Conference Proceedings of the National Council of Physical Distribution Management* (1980), pp. 323–330.

chapter | 12

Carrier Liability for Loss and Damage

chapter objectives

After reading this chapter, you will understand:

The sources, scope, basis, and limitations of liability for overland common carriage.

The principal discrepancies between overland, air, and water carrier liability regimes.

Liability issues related to contract and exempt carriage.

chapter outline

Introduction   *Summary*

Liability Regimes   *Questions*
Domestic overland common carriage
Water common carriage
Air common carriage
Exempt and contract carriage

Introduction

Freight loss and damage causes shippers and carriers to lose billions of dollars each year. The nature and limitations of cargo liability directly affect the ability of shippers to recover damages for lost or damaged freight. During the past ten years, deregulation, new transportation legislation, and ICC administrative reforms have created uncertainty and confusion in the field of cargo liability. Liability provisions for overland, water, and air transportation are less uniform today than in the past. Even within the overland freight transportation category, it is not always clear in today's unregulated market when or if the historic standards of cargo liability apply to railway or motor carriage. As a result, today's shippers find the traditional protections afforded to them have been weakened. These changes make it imperative for traffic managers to stay abreast of the changing legal rules for cargo liability.

This chapter introduces the fundamentals of carrier liability law. Although the primary focus is on domestic overland common carriage, the chapter highlights the principal discrepancies between overland, air, and water carrier liability regimes. The chapter finishes with a discussion of several important issues related to contract and exempt carriage.

Liability Regimes

The sources, scope, basis, and limitations of liability identify the key elements of a regime. Though the following discussion attempts to present the legal guidelines and concepts related to these elements, the reader should not conclude that the law is always clear; the law, in reality, is often ambiguous.[1]

Domestic Overland Common Carriage

The current laws governing the liability of domestic overland common carriage are rooted in our Anglo-American common law heritage. Common law refers to nonstatutory law. It comprises the principles that continually evolve from customs, traditions, and practices, and from court decisions.

Our common law tradition recognizes the liability for carrier loss and damage as part of the more general law of bailments. A *bailment* is a legal relationship created by the transfer of possession of goods without the transfer of ownership or title. A carrier, for example, may take possession of the shipment, but the shipper or receiver retains ownership. Common law traditionally has viewed the carrier as an extraordinary bailee that must use the highest degree of care in handling a shipment. This special responsibility was established by the common law because, after taking possession of the cargo, the carrier is the only party having knowledge and control of the property. In other words, the shipper must rely primarily on the carrier for a factual explanation if loss or damage occurs.

The law governing carrier liability for freight loss and damage, including the common law tradition of imposing strict standards of conduct on liability of common carriers, was codified in federal statute by the *Carmack Amendment of 1906*. The amendment originally became Section 20(11) of the Interstate Commerce Act, but it is Section 11707 of the Revised Interstate Commerce Act (RICA). Section 11707 applies to interstate and export/import common carriage provided by domestic rail, rail-water, express, and motor companies.

In the absence of federal statutory provisions, the *Uniform Commercial Code* (UCC) governs carrier liability. The UCC is a compendium of the best commercial laws and practices. Every state except Louisiana has codified the UCC in statutes and, through this process, has adopted the same basic principles of common law liability as the federal government.

The requirements of statutory and common law shape the contract of carriage. In addition, the rules in the tariffs filed with the ICC are binding elements of the shipping contract. Thus, the terms and conditions of the bill of lading, as well as the tariff rules and regulations, are important features of carrier liability.

Basis of liability. Historic liability standards for overland common carriers are based on strict standards of conduct. The basic duty of common carriers requires delivery of the goods (1) to the correct party, (2) in the same condition as received, and (3) with reasonable dispatch. After taking possession and control of the freight, and until delivery, the carrier becomes an extraordinary bailee. In this role, the carrier assumes liability for loss, damage, or delay for all causes except the five common law defenses printed on the back of the standard bill of lading contract. These defenses address events beyond the control of the carrier and may be briefly described as follows:

1. Act of God. An Act of God is a natural disaster that can neither be caused by human intervention nor prevented by human prudence. It refers to disasters that the carrier could not reasonably anticipate and avert, such as floods, tornadoes, earthquakes, and similar events.

2. Act of public enemy. This exception refers to the belligerent actions of another nation's military forces. It does not include acts of terrorists, hijackers, or rioters.

3. Act of public authority. When legal processes against owners are used to seize their freight from the carrier, the carrier is not liable for the loss. In addition, states may use their police powers to seize or destroy contraband, impound cargo as evidence, or quarantine contaminated goods. Carriers obeying such authority are not liable for any loss.

4. Act of shipper. Acts of shippers most frequently involve improper loading or packaging. The general principle, however, is that the fault of the shipper must not be apparent through ordinary observation; if it is, the carrier has the duty to reject the shipment.

5. Inherent vice. Carriers are not liable for loss and damage from the "inherent vice" of a commodity. Inherent vice refers to such factors as natural defects, decay, or disease that cause deterioration over time. Examples include rusting, spoilage, shrinkage, and fermentation.

Burden of proof. In freight claims for loss or damage, the claimant must prove the carrier's liability. This is the initial burden of proof, and it rests with the shipper to establish (1) the goods were in apparent good order when tendered to the carrier at origin, (2) nondelivery or damage at destination, and (3) the actual value of the loss. Generally, a bill of lading free of notations about visible damage or defects (as illustrated in **Exhibit 12.1**) raises the presumption that the cargo is in good order.

After meeting these three conditions, the shipper creates a prima facie case that the carrier is responsible for the loss or damage. The burden of proof shifts to the carrier to disprove the shipper's assertion. Against such claims, the carrier has three lines of defense.

1. Nonreceipt of cargo at origin. The first line of defense is to assess the facts to see if, in the event of nondelivery, the goods were actually received from the shipper at origin. Even though the carrier has signed the bill of lading — which functions as the shipper's receipt for the number of cartons, packages, or other customary units shipped — the facts may prove fewer units were actually tendered to the carrier. The law of evidence applies to the proof offered at trial, and the contractual elements of the agreement cannot be changed without the written assent of both parties. Nevertheless, the receipt portion of the bill of lading (see **Exhibit 12.2**) is noncontractual and may be modified by evidence presented at trial representing the true facts.[2]

2. Bill of lading exception. When loss, damage, or delay occurs, the carrier is presumed to have breached the shipping contract. Escape from liability for the actual value of the loss is possible by proving that one of the bill of lading exceptions lawfully applies.

To be applicable, the exception must be the *direct* cause of the loss or damage. For example, suppose an unforeseen flash flood (Act of God) caused a trucker to take a detour through remote countryside where some opportunistic thieves hijack the truck. The Act of God exception would not apply because the direct cause of the loss was the theft, not the flood.

Furthermore, an exception must be the *sole* cause.[3] In other words, the carrier must be free of negligence. For example, suppose a shipper improperly loads the cargo in a freight car. If a train accident causes the car to derail, the carrier will have difficulty proving that improper loading was the sole cause of the damage. Thus, even when the shipper is partially at fault, the sole-cause test may preclude the carrier's escaping liability for loss and damage.

To recapitulate, in the second line of defense, the carrier must prove that one of the bill of lading exceptions is both the *direct* and the *sole* cause of the loss or damage. Because this avenue of escape is narrow, the transportation community and the courts recognize surface common carriers as "virtual insurers" of the goods. Like the insurance company that sells protection against injury to goods in transit, the carrier must pay freight claims when loss or damage occurs — unless, that is, an exception is the direct and sole cause of the injury.

Exhibit 12.1 Bill of lading free of notations

RULES

(To be Printed on White Paper)

UNIFORM STRAIGHT BILL OF LADING

ORIGINAL—NOT NEGOTIABLE—Domestic

Shipper's No. ....
Agent's No. ....

Carrier. ....

(SCAC) ....

RECEIVED, subject to the classifications and tariffs in effect on the date of the issue of this Bill of Lading.

From ...., Date ...., 19 ......

Street, ... City, ... County, ... State .... Zip ....

the property described below, in apparent good order, except as noted (contents and condition of contents of packages unknown) marked, consigned, and destined as shown below, which said company (the word company being understood throughout this contract as meaning any person or corporation in possession of the property under the contract) agrees to carry to its usual place of delivery at said destination, if on its own railroad, water line, highway route or routes, or within the territory of its highway operations, otherwise to deliver to another carrier on the route to said destination. It is mutually agreed, as to each carrier of all or any of said property over all or any portion of said route to destination, and as to each party at any time interested in all or any of said property, that every service to be performed hereunder shall be subject to all the conditions not prohibited by law, whether printed or written, herein contained, including the conditions on the back hereof, which are hereby agreed to by the shipper and accepted for himself and his assigns.

Consigned to ....

On Collect on Delivery Shipments, the letters "COD" must appear before consignee's name or as otherwise provided in Item 430, Sec. 1

Street, ...

City, ... County, ... State .... Zip ....

Routing ....

Delivering Carrier .... Vehicle or Car Initial .... No. ....

Collect On Delivery $ .... and remit to: ....

C. O. D. charge } Shipper ☐
to be paid by } Consignee ☐

Street .... City .... State ....

No. Pack- ages	O HM	Kind of Package, Description of Articles, Special Marks, and (Exceptions)	• Weight (Subject to Correction)	Class or Rate	Check Column

Subject to Section 7 of conditions, if this shipment is to be delivered to the consignee without recourse on the consignor, the consignor shall sign the following statement:

The carrier shall not make delivery of this shipment without payment of freight and all other lawful charges.

.... (Signature of consignor)

If charges are to be prepaid write or stamp here "To be Prepaid."

Received $ .... to apply in prepayment of the charges on the property described hereon.

Agent or Cashier

Per ....
(The signature here acknowledges only the amount prepaid)

* If the shipment moves between two ports by a carrier by water, the law requires that the bill of lading shall state whether it is "carrier's or shipper's weight."

Note—Where the rate is dependent on value, shippers are required to state specifically in writing the agreed or declared value of the property.

The agreed or declared value of the property is hereby specifically stated by the shipper to be not exceeding .... per ....

Charges advanced:

$ ....

Shipper .... Agent. ....

Per .... Per ....

Permanent address of Shipper: Street, ... City, ... State ....

** Recommended C. O. D. Section to be Printed in Red.

O Mark with "X" to designate Hazardous Materials as defined in the Department of Transportation Regulations governing the transportation of hazardous materials. The use of this column is an optional method for identifying hazardous materials on bills of lading per Section 172.201(a)(1)(iii) of Title 49, Code of Federal Regulations. Also, when shipping hazardous materials, the shipper's certification statement prescribed in Section 172.204(a) of the Federal Regulations must be indicated on the bill of lading, unless a specific exception from this requirement is provided in the Regulations for a particular material.

Source: American Trucking Association, *National Motor Freight Classification* NMF 100-L, May 18, 1985. Reprinted by permission.

Exhibit 12.2 Receipt segment of bill of lading

RULES

(To be Printed on White Paper)

UNIFORM STRAIGHT BILL OF LADING

Shipper's No.

ORIGINAL—NOT NEGOTIABLE—Domestic

Agent's No.

Carrier.

(SCAC)

RECEIVED, subject to the classifications and tariffs in effect on the date of the issue of this Bill of Lading.

From .... , Date .... , 19 .....

Street, ... City, ... County, ... State .... Zip ....

the property described below, in apparent good order, except as noted (contents and condition of contents of packages unknown) marked, consigned, and destined as shown below, which said company (the word company being understood throughout this contract as meaning any person or corporation in possession of the property under the contract) agrees to carry to its usual place of delivery at said destination, if on its own railroad, water line, highway route or routes, or within the territory of its highway operations, otherwise to deliver to another carrier on the route to said destination. It is mutually agreed, as to each carrier of all or any of said property over all or any portion of said route to destination, and as to each party at any time interested in all or any of said property, that every service to be performed hereunder shall be subject to all the conditions not prohibited by law, whether printed or written, herein contained, including the conditions on the back hereof, which are hereby agreed to by the shipper and accepted for himself and his assigns.

Consigned to ....

On Collect on Delivery Shipments, the letters "COD" must appear before consignee's name or as otherwise provided in Item 430, Sec. 1

Street, ...

City, ... County, ... State .... Zip ....

Routing ....

Delivering Carrier .... Vehicle or Car Initial .... No. ....

Collect On Delivery $ .... and remit to: | C. O. D. charge } Shipper ☐
to be paid by } Consignee ☐

.... Street .... City .... State

No. Pack-ages	O HM	Kind of Package, Description of Articles, Special Marks, and Exceptions	*Weight (Subject to Correction)	Class or Rate	Check Column	
						Subject to Section 7 of conditions, if this shipment is to be delivered to the consignee without recourse on the consignor, the consignor shall sign the following statement:
						The carrier shall not make delivery of this shipment without payment of freight and all other lawful charges.
						(Signature of consignor)
						If charges are to be prepaid write or stamp here "To be Prepaid."
						Received $ .... to apply in prepayment of the charges on the property described hereon.
						Agent or Cashier
						Per .... (The signature here acknowledges only the amount prepaid)

* If the shipment moves between two ports by a carrier by water, the law requires that the bill of lading shall state whether it is "carrier's or shipper's weight."

Note—Where the rate is dependent on value, shippers are required to state specifically in writing the agreed or declared value of the property.

The agreed or declared value of the property is hereby specifically stated by the shipper to be not exceeding .... per ....

Charges advanced:

$ ....

Shipper .... Agent ....

Per .... Per ....

Permanent address of Shipper: Street, ... City, ... State ....

** Recommended C. O. D. Section to be Printed in Red.

Ⓞ Mark with "X" to designate Hazardous Materials as defined in the Department of Transportation Regulations governing the transportation of hazardous materials. The use of this column is an optional method for identifying hazardous materials on bills of lading per Section 172.201(a)(1)(iii) of Title 49, Code of Federal Regulations. Also, when shipping hazardous materials, the shipper's certification statement prescribed in Section 172.204(a) of the Federal Regulations must be indicated on the bill of lading, unless a specific exception from this requirement is provided in the Regulations for a particular material.

Source: American Trucking Association, *National Motor Freight Classification* NMF 100-L, May 18, 1985. Reprinted by permission.

Nonetheless, the experience of many practitioners today is that truck and railway companies generally ignore the "direct and sole cause" standard.[4] Further, some railway companies do not consider themselves subject to this standard when they provide exempt carriage. It remains to be seen what deregulated surface freight forwarders will do in this area.

3. *Ordinary bailment.* The third line of defense is to show loss or damage occurred while the carrier was acting as an ordinary bailee, that is, as a warehouse operator. As a warehouse operator, the carrier is responsible only for the ordinary care of the property. As viewed by the courts, the level of care provided by the "ordinary prudent person" sets the standard. In meeting this standard, the carrier is essentially liable for the loss and damage caused only by its own negligence.

Thus, with the transition from common carrier to warehouse operator comes a significant reduction in the carrier's responsibility for the cargo. The carrier has only to explain the circumstances surrounding the loss or damage. In most instances, the shipper or receiver must then prove that the carrier failed to take ordinary care of the shipment or that the carrier was negligent.[5]

It is important, therefore, to ascertain exactly when the warehouse phase is activated during the course of a shipment. This is a challenging task because the transition to an ordinary bailment depends on a number of different factors: the mode of transportation, the type of carriage, and the provisions of the shipping contract and tariffs.

For overland common carriers, the extraordinary bailment begins with possession and control of the cargo (including shipping instructions) and ends when the transportation is completed. Three situations generally activate the ordinary bailment or warehouse phase. The first occurs when the carrier has possession of the cargo at origin, but something remains to be done by the shipper, such as provide shipping instructions. The second occurs when the shipper stops or stores the goods in transit. Strict liability begins again following a reasonable period after the receipt of new shipping instructions and resumption of the move.

The third situation that activates a warehouse responsibility occurs when carriers store goods at destination for the convenience of the shipper—for instance, when a motor carrier tenders the shipment for delivery, but the receiver refuses it, perhaps because dock space is unavailable. This situation also occurs when shipments are left in railcars on public team tracks or public delivery sidings after the notice of arrival and the expiration of free time for unloading.

Regarding delay claims, the basic duty of the carrier is to

- Deliver the goods with reasonable dispatch.
- Inform the shipper of known delays.
- Take prudent measures to protect goods against increased damage during periods of delay.

Reasonable dispatch generally refers to the normal or customary transit time for similar shipments. What is normal or customary, however, may be tied to special schedules offering expedited service by carriers to attract traffic. In addition, the standard for reasonable dispatch is higher for perishable than for nonperishable traffic.

The carrier must both explain the delay and show it is without fault. The burden is on the shipper to show (1) the delay was unreasonable and (2) an injury was sustained as a consequence. Aside from the issue of reasonable dispatch, a delay claim may still be filed when there is evidence of carrier negligence. For example, a carrier failing to inform the shipper of a known delay is negligent.

Limitations on liability. The Carmack regime makes overland common carriers liable for the actual value of the property. It requires the carriers to issue bills of lading, and it prohibits clauses, receipts, or tariff rules that limit liability or the amount of recovery, except as provided by *released rate* authority in Section 10703 of the RICA. As stated in Chapter 5, released rates are reduced rates that limit the amount of recovery for loss and damage. In addition, the RICA assigns liability to connecting carriers and establishes time limits and venue for claims actions. Whether the shipper chooses to pay full-value (nonreleased) or released-value rates for a shipment determines the damages the shipper can recover. In other words, full-value rates allow the shipper to recover the actual value of the loss or damage, whereas released-value rates place a limit on the amount of recovery.

The measure of loss and damage for actual-value and released-value rates entails a number of guidelines:

1. Actual value. The lack of unanimity among numerous court decisions, as well as practical considerations, makes it difficult to define a single, applicable measure of the actual value of loss or damage. Nonetheless, the rules of contract law establish guidelines for assigning actual value.

Overall, the principle of damage computation allows the court to restore the shipper to the same position that is expected to be occupied when the carrier delivers the goods without injury or delay. This restoration is accomplished through claims for general or special damages.

General damages represent the actual value of the loss or damage that a shipper in the same situation would experience. Thus, the potential injury is reasonably foreseeable at the time of the contract, which is important because the carrier is liable only for reasonably foreseeable damages.[6]

The measure of damages also depends on the facts and circumstances of each case. When it is practical to secure the market value of the cargo at its destination, this measure usually determines the general damages. In other instances, the amount shown on the invoice may indicate the appropriate measure. Further, other circumstances may warrant the inclusion of miscellaneous expenses, such as taxes, paid freight charges, inspection costs, and earned profit.

In certain situations, mostly involving late deliveries, shippers may file claims for *special damages*. These damages are not ordinarily foreseeable. Claims arise when a carrier has been notified that action is necessary but fails to take appropriate steps. For example, suppose the shipper gives notice to the carrier that special circumstances necessitate urgent delivery. If the shipment arrives late, the carrier may be held liable for special damages, including lost profits, penalty payments, reshipment costs, or lost wages. Such notice makes the potential injury arising from special circumstances foreseeable, and therefore, the carrier becomes liable for the special damages.

2. *Released value.* Unlike the measure of actual value, the measure of released value is relatively simple to define. Carriers publish released-value rates in a way that clearly indicates limits on the amount of recovery. For example, the National Motor Freight Classification (NMFC) shows the released-value and full-value ratings for cathode ray tubes as follows:

Description	Rating
Sub. 1: Released < $3/lb	110
Sub. 2: Not released	150

To use the 110 rating, the released value of the cathode ray tube must be declared in writing on the bill of lading not to exceed $3 per pound. In the event of total loss, the ceiling for recovery is the gross weight of the shipment multiplied by the value per pound declared on the bill of lading (provided the actual value is not less than the declared value). Should partial loss or damage occur, the amount of recovery depends on the wording in the tariff. In some instances, the shipper may recover any actual losses that do not exceed the ceiling. In others, recovery is confined to the specific units (articles, pieces, or packages) actually lost or damaged. For each unit, the weight times the released value per pound sets the maximum amount recoverable.

In addition, the Staggers Rail Act of 1980 permits rail carriers to include a "deductible" clause in released-rate agreements. This clause allows the carrier to deduct a certain amount from the claim payment. Although the exclusive authority for deductibles is found only in the Staggers Rail Act, the ICC has determined that trucking firms and household goods carriers may also publish these clauses in released-rate provisions.[7]

The RICA also gives carriers greater flexibility to publish released rates independently. Some carriers have used this flexibility to publish "autorelease" clauses such as follows:

> *Unless the shipper declares a higher value in writing at the time of the shipment, shipment shall be considered to have a declared value not exceeding \$_____per pound, per article, per shipment.*[8]

If the shipper exercises the option to declare a higher declared value, additional charges must be paid. Since the common law principle is for carriers to give the shipper a reasonable choice between released-value and actual-value rates, the publication of autorelease provisions in tariffs raises several questions: (1) Is the shipper able to make an informed choice? In other words, did the shipper select the released rate? Or was the shipper simply unaware of the autorelease provisions? (2) Is this choice reasonable if premium charges are assessed as a condition for the carrier's assumption of Carmack liability? Case law presents mixed guidelines on these issues, prompting one authority to recommend that shippers ask carriers to indicate in writing whether any liability limitations exist.[9]

Water Common Carriage

The law governing cargo liability for water transportation has its roots in ancient Mediterranean Sea law. Although liability was relatively strict in these early times, a trend began in the seventeenth century to relax the standards of conduct on liability for cargo loss and damage. This trend in Admiralty Law accelerated during the 1700s and reached its zenith during the 1800s, when water carriers were permitted to contract out of liability even for their own negligence. The relaxation of standards partially reflected a recognition of the perilous nature of water transportation, especially for seaborne trade. Competing maritime nations found it in their best interest to relax standards to promote their merchant marine industries and, thereby, to encourage trade and investment and gain competitive advantage.

The Harter Act was passed by Congress in 1893 to put an end to the practice of contracting out of liability. It still governs *domestic* water carriage. The Carriage of Goods by Sea Act of 1936 (C.O.G.S.A.) establishes the liability regime for seagoing transportation to or from U.S. ports in *foreign* trade. C.O.G.S.A. embodies the Hague Rules, which were forged by major seafaring nations in the Netherlands in 1924.

Both of these U.S. statutes, however, contain similar provisions. In fact, domestic water carriers may incorporate a reference in the shipping contract that allows them to apply the C.O.G.S.A. regime. Since most carriers actually invoke this option, C.O.G.S.A. has supplanted the Harter Act in practice.

As in other regimes, the contract of carriage also controls cargo liability. The conditions expressed in the water carrier's shipping contract, moreover, apply to truck service incidental to waterway terminal operations. Of course, the common law applicable to maritime practices also serves as a major source of liability for cargo loss, damage, and delay.

Unlike the strict liability of the Carmack regime, the liability of water common carriers is based on permissive standards of conduct and fault. Water carriers are not considered "virtual insurers" of the goods. Instead, the transportation community judges marine insurance as a virtual necessity.

As stated in C.O.G.S.A., the duty of the carrier is to exercise due diligence to

- Make the ship seaworthy
- Properly man, equip, and supply the ship
- Make the holds safe and suitable for cargo
- Properly and carefully load, stow, and discharge the goods

The carrier is *not* liable for an unseaworthy vessel, unless the problems are caused by lack of due diligence. The carrier is also *not* liable for uncontrollable losses, including the errors of navigation and management of the ship and the sixteen other exceptions shown in **Exhibit 12.3.** Actual illustrations of several of these exceptions are shown in **Exhibit 12.4.**

By proving delivery of the goods to the carrier at origin, and by proving nondelivery or damage at destination, the shipper establishes a presumption

■■■■■■ *Exhibit 12.3* **Exceptions to water carrier liability — Uncontrollable causes of loss in the Carriage of Goods by Sea Act**

1. Act, neglect, or default of the master, mariner, pilot, or the servants in the navigation or in the management of the ship
2. Fire, unless caused by the actual fault or privity of the carrier
3. Perils, dangers, and accidents of the sea or other navigable waters
4. Act of God
5. Act of war
6. Act of public enemies
7. Arrest or restraint of princes, rulers, or people, or seizure under legal process
8. Quarantine restrictions
9. Act or omission of the shipper or owner of the goods, agent, or representative
10. Strikes or lockouts or stoppage or restraint of labor from whatever cause, whether partial or general: provided that nothing herein contained shall be construed to relieve a carrier from responsibility for the carrier's own acts
11. Riots and civil commotions
12. Saving or attempting to save life or property at sea
13. Wastage in bulk or weight or any other loss or damage arising from inherent defect, quality, or vice of the goods
14. Insufficiency of packing
15. Insufficiency or inadequacy of marks
16. Latent defects not discoverable by due diligence
17. Any other cause arising without the actual fault and privity of the carrier and without the fault or neglect of the agents or servants of the carrier

Source: 46 U.S.C. 1304.

■■■■■ *Exhibit 12.4* **C.O.G.S.A. exemptions from carrier liability**

1. Stranding due to proceeding across a river bar in a storm without a pilot — *Wilbut-Ellis Co. v. M/V Captayannis "S"*
2. Tipping a ship to examine the propeller — *The Indiani*
3. Leaving port without a pilot — *The Oritani*
4. Failure to use, or improper use of, charts and light lists, resulting in stranding — *Daisy Phillipine Underwear Co. v. U.S. Steel Products Co.*
5. Failure to close a tank while taking on sea water as ballast during typhoon season, causing water damage to cargo — *Leon Bernstein Co. v. Wilhelmsen*
6. Failure to alter course to avoid severe weather — *Hershey Chocolate Corp. v. The Mars*
7. Failure to heed the warning of a government light, indicating the location of a reef — *Mangold v. E. A. Shores, Jr.*
8. Leaving port in the face of a storm warning — *Hanson v. Haywood Bros.*
9. Causing steam and water to be injected into a cargo hold without first inspecting it, under the mistaken belief that a fire had started — *Ravenscroft v. U.S. (The West Imboden)*
10. Negligence of the master in undocking a steamship without tug assistance — *The Harry Lukenbach*
11. Failure to properly secure the manhole covers of a cargo hold after cleaning the after peak tank for the stowage of cargo — *The Steel Navigator*
12. Failure to inspect the vessel's holds after discovering leakage in drums of sulphuric acid — *The Milwaukee Bridge*

Source: John Betz, ed., "Visby/Hamburg: Still a Split Decision," *Distribution* (October 1985), p. 77. Reprinted with permission from *Distribution Magazine*, Chilton Co., Radnor, Pa.

that the carrier is responsible for the loss or damage. The carrier then must present the facts to support one of the numerous exceptions to liability. To collect, the shipper must later prove the carrier at fault.

Like overland carriers, water carriers cannot use contract clauses to lessen the level of liability provided by statute. Unlike the Carmack provisions for overland transportation, however, the C.O.G.S.A. regime limits the amount of recovery to the market value of the lost or damaged freight up to a ceiling of $500 per package or customary unit. By agreement with carriers, shippers may elect to pay higher rates for full-value coverage.

The $500 ceiling and the ambiguity of a customary unit are two principal reasons that shippers are dissatisfied with the C.O.G.S.A. regime. The ceiling, established in the 1930s, is outdated. In addition, C.O.G.S.A. does not define a customary unit. The widespread use, since the 1960s, of $8' \times 8' \times 40'$ containers for intermodal transportation has stirred additional controversy about the proper definition of customary units. The issue is whether the packages, cartons, and so forth inside the containers or the containers themselves constitute customary units. These deficiencies, among others, have led to the development of several international protocols, which amend the C.O.G.S.A. regime.[10] Thus far, however, the United States has not ratified these protocols.

Air Common Carriage

Decontrol of the domestic air cargo industry in 1977 and 1978 effectively removed federal statutory law as a source of liability for loss, damage, or delay.[11] The nature of domestic air carrier and air freight forwarder liability is determined both by the conditions stipulated in individual air waybills and by the tariff rules that technically form part of the contract of carriage. The courts must decide the lawfulness of contested contractual provisions.

The liability regime for the air cargo industry also covers the truck shipments of air freight that are incidental to air terminal operations. When truck pickup, delivery, or transfer service makes up part of a continuous through movement by a domestic air carrier, that service is subject to the liability established by the air waybill.

International agreements wield authority over the shipping contracts for international air freight service. The Warsaw Convention establishes liability standards for air freight carriers participating in foreign commerce and serving U.S. points. Most major countries, including the United States, ratified this treaty during the period from 1929 to 1934. The Warsaw Convention regime also governs the shipping contracts of air freight forwarders and generally applies to truck service that is part of a through movement by air.

Like water carrier liability, the cargo liability of domestic or international air freight carriers is based on permissive standards of conduct and fault. Generally speaking, air freight companies may contractually limit liability to (1) classical common law exceptions, (2) other exceptions such as ''perils of the air'' and errors of piloting and navigation, and (3) their own negligence.

When loss or damage occurs, the carrier is most often presumed at fault. To escape liability, the carrier must show the cause of injury falls within one of the authorized exceptions. The shipper then must prove the carrier's negligence.

The Warsaw Convention prohibits contract provisions that attempt to reduce liability below the treaty's standards. Although no specific statutory guidelines prevail on the domestic side, common law does not allow domestic air carriers to contract out all liability for loss and damage. Nonetheless, these carriers have incorporated contract clauses and published tariff rules that greatly restrict liability.

A ceiling on the amount of recovery, moreover, is a standard feature of the air waybill. No uniformity exists domestically, although many carriers have adopted a ceiling of $0.50 per lb times the total weight of the shipment. The international air carrier limit is $9.07 per lb, per article lost or damaged.

Exempt and Contract Carriage

Shipping contracts for contract and nonrailway exempt carriage are generally not subject to the strict standards of the Carmack regime. Carriers may publish the terms of liability in "exempt carriage circulars," or they may negotiate agreements for contract carriage that include the conditions of liability (see Chapter 8). Without these arrangements, ordinary care is the required standard of conduct.[12]

The Staggers Rail Act gave the ICC broad powers to exempt rail traffic and services. Section 11505(e) of the RICA (referred to as the "Matsui Amendment"), however, prohibits the commission from relieving "any rail carrier from an obligation to provide contractual terms for liability and claims which are consistent with provision of section 11707" (Carmack Amendment liability). The Matsui Amendment qualifies this prohibition as follows:

> *Nothing in this subsection or section 11707 of this title shall prevent rail carriers from offering alternative terms nor give the Commission authority to require any specific level of rates or services based on the provisions of this title.*

Some railway companies have been exploiting the imprecise standards established in the Matsui Amendment. For example, exempt rail carriage circulars have attempted to limit traditional Carmack liability in the following ways:

- Autorelease provisions
- Ceilings on the amount that can be recovered per car
- Minimum amounts required to process a claim
- Provisions, such as "the carrier is liable only if carrier negligence is shown," that shift the burden of proof to the shipper[13]

Because of practices like these, there is a good deal of confusion and controversy about exempt railway carriage liability for freight loss and damage. While the courts settle contested issues, shippers need to give special attention to the conditions of liability when selecting among transportation alternatives.

Summary

The sources, scope, basis, and limitations of liability define cargo liability regimes. These regimes are neither uniform nor simple. Common law, statutory law, international treaties, government regulation, shipping contracts, and tariffs create a complex array of principles that govern cargo liability of carriers.

The sources of law that influence domestic overland common carriage liability include the common law, the law of bailments, the relevant statutes, the

UCC, and the contract of carriage. The Carmack Amendment codified cargo liability law in the RICA and applies to interstate and export/import common carriage by domestic rail, rail–water, express, and motor companies. For these transportation companies, liability is based on strict standards of conduct and breach of contract; they become extraordinary bailees and assume liability for loss, damage, or delay for all causes except for five common law defenses. A bill of lading free of notations about visible damage generally establishes a prima facie case that the carrier is responsible for loss or damage. A carrier may escape liability by proving that (1) it did not receive cargo at origin, (2) a bill of lading exception was the direct and sole cause, or (3) it was acting as an ordinary bailee when loss or damage occurred. Liability for delays involves the concept of reasonable dispatch, which refers to the customary transit time of similar shipments. The carrier must both explain the delay and show it is without fault. The shipper must prove (1) the delay was unreasonable and (2) an injury was sustained as a consequence. The Carmack regime makes carriers liable for the actual value of the loss or damage and prohibits limitations except as provided by released rate authority.

The law of admiralty, the Harter Act, the Carriage of Goods by Sea Act (C.O.G.S.A.), and the contract of carriage are principal sources of law governing cargo liability for water transportation. Unlike the strict liability of the Carmack regime, the cargo liability for water common carriage is based on permissive standards of conduct and fault. The shipper establishes a presumption that the carrier is responsible for loss and damage by proving (1) delivery of the goods to the carrier at origin and (2) nondelivery or damage at destination. The carrier escapes that responsibility by presenting the facts to support one of many exceptions to liability. To collect, the shipper then has to prove the carrier at fault. In addition, C.O.G.S.A. limits the amount of recovery to the market value of the lost or damaged freight up to a ceiling of $500 per package or customary unit.

Domestic air carrier and air freight forwarder liability is determined by waybill conditions and tariff rules. The Warsaw Convention establishes liability standards for air cargo carriers participating in foreign commerce and serving U.S. points. The liability regime for both domestic and international air freight carriers is based on permissive standards of conduct and fault. The burden of proof follows the pattern established for water carriers. The air cargo regime also permits a ceiling on the amount of recovery.

Shipping contracts for exempt and nonrailway contract carriage are generally not subject to Carmack standards. Without specific agreements about liability, ordinary care is the required standard of conduct. Although the RICA requires rail carriers to provide contractual terms of liability that are consistent with Carmack standards, the imprecise language in the RICA has encouraged some rail carriers to limit their cargo liability.

Selected Topical Questions from Past AST&L Examinations

1. One exception to the absolute liability assigned common carriers found in uniform bills of lading is when the loss or damage resulted from acts or negligence of the shipper. However, despite this exemption, a carrier may still be held liable. Under the following situations, indicate the conditions under which a carrier would be liable for loss or damage:
   (a) Shipment was improperly loaded by the shipper causing damage in transit
   (b) Shipper purposely misdescribes the goods to obtain a lower rate
   (c) Shipper fails to comply with tariff packaging requirements, which results in damage in transit (Fall 1979)

2. Describe the extent of common carrier liability under each of the following:
   (a) Common law
   (b) Uniform bill of lading
   (c) Warehouse operator
   (d) Released value (Fall 1980)

3. Section 2 of the uniform bill of lading states, in part, "No carrier is bound to transport said property by any particular train or vessel, or in time for any particular market or otherwise than with reasonable dispatch." Carefully explain the concept of reasonable dispatch by discussing how it is measured, who has the burden of proof in delay cases, and the types and measures of damages that may occasion an "unreasonable" delay. (Fall 1980)

4. Under the Staggers Rail Act of 1980, railroads are permitted to enter into agreements with shippers providing for reduced liability in the form of "deductibles" or minimums for claims. What are the major arguments for and against the use of deductibles in the transportation industry? (Spring 1982)

5. The definition of a "package," as it affects marine transportation loss and damage liability, continues to be a concern of container shippers engaged in international commerce. Discuss the effect "package" has on container transportation for both carriers and shippers, and what may be done to resolve the issue. (Spring 1985)

Notes

1. Excellent sources for further study include William J. Augello, *Freight Claims in Plain English*, rev. ed. (New York: Shippers National Freight Claim Council, 1982); Richard R. Sigmon, *Miller's Law of Freight Loss and Damage Claims*, 4th ed. (Dubuque, Iowa: Wm. C. Brown Co., 1974); Saul Sorkin, *How to Recover Loss or Damage to Goods in Transit* (New York: Matthew Bender and Co., 1981); Marvin L. Fair and John Guandolo, *Transportation Regulation*, 9th ed. (Dubuque, Iowa: Wm. C. Brown Co., 1983). Besides writing a comprehensive and readable treatment of freight claims, Augello includes a section on how to read and evaluate legal decisions. Sigmon provides an in-depth study of the terms and conditions of the bill of lading. Sorkin offers the most extensive coverage of ocean and air carrier liability. Fair and Guandolo address reparation and misrouting issues.

2. See Sigmon, *Miller's Law of Freight Loss and Damage*, pp. 21–22.
3. See John Betz, "Filing claims with 'Big Brown'," *Distribution* (May 1986), pp. 59–60.
4. The landmark case to see is *Missouri Pacific Railroad Company* v. *Elmore & Stahl*, 337 U.S. 948; see also the discussion of the "Comparative negligence rule" in Augello, *Freight Claims*, pp. 113–114.
5. For the five specific instances where the carrier must prove it is free of negligence, see Augello, *Freight Claims*, p. 214.
6. For more details, see Augello, *Freight Claims*, pp. 153–166.
7. Interstate Commerce Commission, *Annual Report* (Washington, D.C.: U.S. Government Printing Office, 1983), p. 57; see also *Shippers National Freight Claim Council, Inc.* v. *ICC*, Civil No. 80–4243 (2d Cir., June 13, 1983).
8. Colin Barrett, "More thoughts on declared value," *Distribution* (April 1986), p. 58.
9. Barrett, "More thoughts on declared value," p. 59.
10. See Visby Amendment, Hamburg Rules, Multimodal Transport Convention, and Montreal Protocol No. 4 in Sorkin, *How to Recover Loss or Damage* and Augello, *Freight Claims*, appendix H; see also Fred Loftin, "Proposed ocean rules boost liability coverage," *Canadian Transportation and Distribution Management* (October 1983), pp. 46–48; John Betz, ed., "Visby/Hamburg: Still a split decision," *Distribution* (October 1985), p. 77.
11. Air Transportation Regulatory Reform Act of 1977 (Public Law 95–163).
12. See Francis J. Mulcahy, "Motor carrier cargo liability—An overview," *ICC Practitioners' Journal* 49 no. 3 (March–April 1982), pp. 263–264; Augello, *Freight Claims*, pp. 213–214.
13. For a description of some of the problems that shippers have experienced in this area, see Augello, *Freight Claims*, pp. 64, 463; see also, Joseph Michael Roberts, "Freight claims liability—Where are we today and where are we heading," *Traffic World* (August 4, 1986), pp. 77–81.

chapter | 13

Claims Management

chapter objectives

After reading this chapter, you will understand:
How to make freight loss and damage claims.
Two key tasks for controlling freight loss and damage claims.
How to make overcharge, reparation, and misrouting claims.

chapter outline

Introduction

*Making the Claim for Loss,
Damage, or Delay*
Filing the claim
Time limits
Preparing the claim
Concealed loss and damage
Settling disputes

*Controlling Freight Loss
and Damage Claims*
Selection and acquisition of
carrier services
Claims prevention programs

Other Claims Actions
Overcharges
Unlawful rates or practices

Summary

Questions

Introduction

Claims generally grow either out of cargo loss, damage, or delay or out of the application of erroneous rates in the assessment of freight charges. Besides freight claims, shippers file complaints before courts or regulatory agencies concerning the lawfulness of rates and practices. Such actions involve reparation and misrouting claims. Although technically not "freight claims," these actions are discussed here as part of claims management.

The chapter first investigates the administration of loss and damage claims. Next, it focuses on how to control freight claims. The chapter closes with a discussion of the claims actions that arise from the erroneous application of rates and charges or from the imposition of unlawful rates or practices.

Making the Claim for Loss, Damage, or Delay
Filing the Claim

The injured party is entitled to recover damages. The owner, usually the shipper or the receiver, normally sustains the injury and files the claim. However, someone other than the owner with an interest in the shipment may also give notice to the carrier by filing the claim. For example, large companies with sophisticated traffic departments may file freight claims for the customers that have already taken title to the goods (see Chapter 7). Interested parties may include the shipper, receiver, insurer, and owner. The important thing to remember is that the carrier must be given proper notice in writing.[1]

The Carmack regime allows claims to be filed against the originating or delivering carriers, regardless of where the injury occurred. Claims may be filed with an intermediate carrier when that carrier causes the loss. The railway or motor carrier that issues the bill of lading originates the shipment. The delivering carrier is the one that performs the linehaul service nearest to the destination; this definition excludes railway companies that perform only destination switch service. On the other hand, the domestic freight forwarder is considered both an originating and a delivering carrier. By contrast, air or water carriers essentially confine liability for loss, damage, or delay to service on their own lines.

Time Limits

Shippers lose millions of dollars each year because they file claims after contractual deadlines expire. Section 11707 requires carriers to provide at least nine months to file claims, and two years to file suit. Virtually all ICC-regulated surface common carriers have incorporated terms in the bill of lading that limit claims actions to these statutory minimums. These carriers *contractually* limit (1) the claims-filing period to nine months after delivery or, in the event of nondelivery, after a reasonable time for delivery has elapsed; and (2) the period for civil actions to two years after disallowance of the claim.

Among air and water carriers, time limits for claims and civil actions lack uniformity. For domestic transportation, the individual tariff provisions and the contract of carriage establish filing deadlines. It is left to the courts to determine the reasonableness of these time limits. The Carriage of Goods by Sea Act and the Warsaw Convention set minimum periods for international water and air transportation, respectively.[2] However, it is still important to check the shipping contract and related tariffs because some carriers permit time limits to exceed the statutory minimums.

Preparing the Claim

Rail and motor carriers have created organizations that develop and promulgate rules and guidelines for claims administration. The railroad industry established the Association of American Railroad's Freight Claim Division in the 1890s. During the 1930s, the American Trucking Association organized the National Freight Claims Council. More recently, shippers also formed claims organizations. In 1974, the Shippers National Freight Claims Council (SNFCC) was incorporated. Although these organizations assume various responsibilities, they all have one purpose in common: to formulate rules that improve procedures and support the prompt, equitable, and lawful settlement of claims. It is also noteworthy that SNFCC strives to increase professionalism in claims management. Like the American Society of Transportation and Logistics, SNFCC sponsors educational programs and has a professional certification program for practioners.

Significantly, the ICC's regulations and tariff rules that guide claims preparation and administration are products of organizations such as these. The ICC's regulations serve as a useful benchmark for all modes of transportation and may be outlined as follows:[3]

1. Basic requirements. The claim must be prepared in writing. Essential information includes:

- The facts sufficient to identify the shipment
- The reason for filing—that is, an assertion that the carrier is liable and a demand for payment
- The amount of damages
- The documentation to support an investigation, including

    Original bill of lading or certified copy
    Freight bill
    Bond of indemnity instead of the original copies of the bill of lading or the freight bill if either is lost or misplaced
    Original invoice or certified copy
    Copy of freight bill showing exceptions if loss or damage was noted at time of delivery
    Inspection report of concealed loss and damage, if applicable

2. Duties of the carrier. The carrier must acknowledge the claim was received within 30 days and promptly investigate it. In addition, the claim must be paid, declined, or settled within 120 days; if it is not, the carrier must tell the shipper why and report the status of the claim every 60 days thereafter.

3. Duties of the receiver. The receiver should inspect delivered shipments and note loss or damage exceptions on the delivery receipt. Generally speaking, the receiver has the legal obligation to accept damaged freight and mitigate losses, unless the damage is so extensive that the goods are ''practically worthless.''[4]

Standard claim forms are available to aid preparation (see **Exhibit 13.1**). Often carriers or shippers customize these forms to fit their particular operations and to set up efficient methods of processing paperwork (see **Exhibit 13.2**). Significantly, increasing numbers of shippers and carriers are filing and processing claims by electronic means.

Exhibit 13.1 Standard claim form

STANDARD FORM FOR PRESENTATION OF LOSS AND DAMAGE CLAIM

To: _____   _____
         (Name of Carrier)   (Date)

_____   _____
         (Street Address)   (Claimant's Number)

_____   _____
         (City, State)   ☐ Damage   (Carrier's Number)

This claim for $ _____ is made against your company for ☐ Loss in connection with the following described shipment

_____ (Shipper's Name)	_____ (Consignee's Name)
_____ (Point Shipped From)	_____ (Final Destination)
_____ (Name of Carrier issuing Bill of Lading)	_____ (Name of Delivering Carrier)
_____ (Date of Bill of Lading)	_____ (Date of Delivery)
_____ (Routing of Shipment)	_____ (Delivering Carrier's Freight Bill No.)

DETAILED STATEMENT SHOWING HOW AMOUNT CLAIMED IS DETERMINED
(Number and description of articles, nature and extent of loss or damage, invoice price of articles, amount of claim, etc.
ALL DISCOUNT AND ALLOWANCES MUST BE SHOWN.)

TOTAL AMOUNT CLAIMED	

The following documents are submitted in support of this claim:

☐ Original Bill of Lading
☐ Original paid freight bill or other carrier document bearing notation of loss or damage if not shown on freight bill
☐ Carrier's Inspection Report Form (Concealed loss or damage)

☐ Consignee concealed loss or damage form
☐ Original invoice or certified copy
☐ Shipper's concealed loss or damage form
☐ Other particulars obtainable in proof of loss or damage claimed:

_____
_____

(Note: The absence of any document called for in connection with this claim must be explained. When impossible for claimant to produce original bill of lading, or paid freight bill, a bond of indemnity must be given to protect carrier against duplicate claim supported by original documents.)

Remarks: _____
_____
_____

The foregoing statement of facts is hereby certified as correct.

_____
(Claimant's Name)

_____
(Address)

_____

FORM CS-1 (Rev. 1/81)

AB 101901

Source: Courtesy of the American Trucking Association.

<i>Exhibit 13.2</i> Customized claim form

INSPECTION REPORT OF LOSS OR DAMAGE DISCOVERED AFTER DELIVERY

Terminal.. Date.........................195........ Report Number....................

Shipper ....Origin ....

Consignee ....Destination ....

F/B No. ....Prepaid ( )......Collect ( ) ....Date Consignee requested inspection ....

Date of Mailing...................19......	Date Delivered............19.....	Loss or Damage...........noticed at time of delivery?............ Could loss or damage have been
Date Un-Packed...................19......	Date of Call..............19.....	Were goods unpacked before the inspection was made?...... Were containers and packing available?......

What evidence was there of Pilferage before Delivery?...

Was there sufficient space in Package to contain Missing Goods?........................... | What material Occupied the Remaining Space?.....................

Did Comparison of Check with Invoice or Weighing Package, Verify loss............................. | If Released Valuation, Show Weight of Articles Damaged or Short.....................

Kind of Container............................ New or Old................... Wired ( ) Corded ( ) Strapped ( ) Nailed ( ) Sealed ( )
          (Carton, Box, Crate, Etc.)

Box Maker's Gross Weight Limit...............Loaded Carton....... | Gross Weight of | If Carton Were Flaps Glued........................... | Were Seams or Edges Split?..................

How Were Goods Packed?...

Do you consider Adequately Packed or Protected?......................... | What condition of container or contents indicated loss or damage occurred with carrier?...........................

To prevent comparable damage in the future, how in your judgment should they have been packed or prepared for shipment..
....
....

Did Shipment Have Prior Transportation?........................... | If so, Is Merchandise Still Packed in Original Container?................... | Original Point of Shipment..............

No. of Articles	Describe fully nature & extent of loss or damage	Invoice Price
	(If necessary use other side of this form)	

Will there Be Salvage?................... | What Disposition willbe made of the Salvage?...................

Consignee ....Carrier ....

By ....By ....
                                                                    Inspector

This Report is Merely a Statement of Facts and Not an Acknowledgement of Carrier's Liability.
When presenting claims for loss and damage, attach the following documents:

1. This Inspection Report
2. Original Paid Freight Bill
3. Original Bill of Lading

4. Original Invoice or an exact certified copy showing all discounts
5. Your Bill showing nature and amount of claim
6. Shipper's and Consignee's Concealed Loss and Damage Forms

Claim blanks and other necessary forms to properly present your claim may be obtained from carriers agent.

(If additional space needed, please use reverse side.)

Concealed Loss and Damage

Sometimes carriers may deliver goods that have been pilfered or damaged in transit. Yet the containers give no indication of the problem, and the consignee, because there is no visible loss or damage, signs a clear delivery receipt. A clear delivery receipt establishes only a presumption that the goods were delivered in apparent good order, but the facts may indicate a case of concealed loss or damage.

When concealed loss or damage is discovered, the consignee must promptly report it to the carrier and request an inspection. Tariff rules require shippers to confirm the request in writing. Furthermore, motor carrier (and air carrier) tariffs contain "15-Day" rules that make it essential to notify the carrier within fifteen days of delivery. Otherwise, these rules place the additional burden on the consignee to produce reasonable evidence that the loss or damage did not occur after delivery.[5]

The inspection report serves as documentary support of the claim. The claimant additionally may introduce other documentary evidence such as photographs or affidavits by dock personnel about handling procedures, shipment condition, packaging, and other pertinent factors.

Settling Disputes

Carrier will disallow some claims. The ICC does not arbitrate such disputes; its principal role in the claims arena is to promulgate rules and regulations. Claims actions involve breach of contract, which is a matter for the courts to address.

Both federal and state courts have concurrent jurisdiction over freight claims arising from interstate transportation. Claimants may file suit in either jurisdiction, although the federal court imposes a $10,000 threshold for the cases that it will hear. Federal law, however, prevails in either court.

As previously discussed, the claimant must file suit within the lawful time limits set in the bill of lading. The liability regimes also establish venue that determines the court district in which the civil actions for damages may be brought. Claimants generally can sue the carrier anywhere they conduct their businesses. In 1980, however, the Staggers Rail Act placed restrictions on venue for actions brought against rail carriers.[6] Since then, various shipper organizations have vigorously attacked this change. After studying the issue in 1981, the ICC recommended repeal of the restrictions imposed by the Staggers Act,[7] but at this writing, these restrictions still apply.

As an alternative to civil action, shippers and carriers may voluntarily participate in two formal arbitration plans. The first involves the Transportation Arbitration Board (TAB). The SNFCC and the NFCC joined forces in 1975 to create this neutral alternative to the high cost of litigation. The board consists of two claims experts who are selected from SNFCC and NFCC rosters, and who represent shippers and carriers. Each party pays a nominal charge (about $50) for the service. Awards require a unanimous decision, and once made, are binding on the parties. TAB serves all modes of transportation except water, but almost all claims involve motor carriers.

The Association of American Railroads sponsors the second plan, which involves the American Arbitration Association (AAA) and handles railroad-related claims. With several exceptions, the AAA operates like TAB. Unlike TAB, though, the AAA board consists of three experts selected from a national panel of freight claim arbitrators. In addition, the parties indicate their preference in the selection of arbitrators. Furthermore, decisions need not be unanimous.

Both plans offer the following advantages over litigation:

- Less cost
- No wait for trial date
- Hearings held anywhere
- No public record required
- No restrictive rules of civil procedure or evidence

Controlling Freight Loss and Damage Claims

Freight loss and damage retard the productivity of shippers, receivers, and carriers. The millions of claims filed each year require the transportation industry to pay out billions of dollars. These direct losses are only the tip of the iceberg. Indirect costs amount to six or seven times the actual damages. These indirect costs arise from factors such as

- Claims administration
- Capital tied up in claims proceedings
- Replacement goods movement
- Related production and inventory costs
- Injury to customer service
- Insurance premiums

However, such costs are within the grasp of the carriers and shippers to control. Nearly 70 percent of the total costs of loss and damage could be avoided.[8] From the perspective of the traffic organization, two managerial tasks address this area of claims: (1) evaluating the relationships between direct and indirect costs and transportation alternatives for use in planning the selection and acquisition of carrier services and (2) developing claims prevention programs.

Selection and Acquisition of Carrier Services

With respect to the first task, **Exhibit 13.3** illustrates how cargo loss and damage cost relate to the various elements of transportation. Analyzing these relation-

█████████ *Exhibit 13.3* **Transportation elements in relation to freight loss, damage, and delay costs**

Element of Transportation	Loss- and Damage-Related Cost
Freight rate	Carrier claims payout reflected in rate structure
Transit time	Delayed shipments may cause special damages, especially to perishable goods. Delays have an adverse impact on production and inventory costs, and on customer service.
Equipment condition	Affects dunnage, bracing, patching, cleaning, and possibly disinfecting the equipment.
Bracing and dunnage	An inverse relationship: The greater the amount, the less frequent the occurrence of loss or damage.
Packaging	Inverse relationship as above
Cooperage	Necessary for salvaging damaged goods
Replacement goods	Freight charges may include part of claims settlement; if not, the extra cost of LTL (when the original shipment may have been TL) is born by the shipper.
Freight claims	Costs absorbed by settling for amounts less than actual value of loss. Claim administration, including clerical, training, inspection, photography costs. Settlement time represents a lost opportunity cost on capital invested in lost or damaged goods. Many firms increase purchases and inventories to compensate for expected loss or damage. Loss of customer good will; lost sales.
Insurance	An added cost to the shipper or receiver that does not necessarily reduce the freight rate or the incidence of damage, but does reduce settlement time and processing.

Source: Adapted from Joseph L. Cavinato, ''Loss and Damage from a Shipper's Standpoint: A Provocative Assessment of Key Factors Requiring Analysis in the Deregulation Arena,'' 13 *Transp. L. J.* 347 (1980).

ships has always been difficult because the costs and risks of concern depend on what liability regime governs and on the incidence of loss or damage. Moreover, as noted previously, a particular liability regime's application depends on (1) the mode of transportation, (2) the legal form of carriage, and (3) the type of commerce (foreign, interstate, or intrastate). Likewise, the risk or incidence of loss and damage depends on the mode of transportation. The different modes, as well as the different carriers, experience loss and damage in varying degrees for standard categories such as shortages, wrecks and catastrophes, delays, thefts, and damage (concealed or visible).[9]

Deregulation has made the analysis of these relationships increasingly complex. Since 1980, the transportation community has witnessed the growth of a nonuniform system of liability. Concomitantly, the range of liability options available to shippers has expanded significantly.[10] Released-value rates have assumed increased importance. Exempt and contract carriage, like deregulated air transportation, presents a growing array of contractual choices for carrier liability. Today's traffic manager faces more options than in the past and must judiciously apply tradeoff analysis to the choice of contract terms.

With the growth of released rates, the tradeoff between lower rates and reduced liability has become more important.[11] The analysis must consider four key factors: (1) the value per pound of the product, (2) the rates, (3) the incidence of loss and damage, and (4) the insurance premium. Assume, for example, the following situation for a particular product:

Full-value rate	$3.00 cwt
Released-value rate (at $0.25/lb)	$2.00 cwt
Actual-value per pound	$0.55 lb.

On the surface, the released-value rate appears to offer substantial savings. When using this rate, however, the shipper must assume more of the risk and cost of loss or damage. In a maximum loss situation, the tradeoff analysis for a 1000-lb shipment is as follows:

Regular rate − 10 cwt × $3.00 =	$30.00
Shipment value	$550.00
Amount of recovery	$550.00
Direct shipper loss	$ 0.00
Released rate − 10 cwt × $2.00 =	$20.00
Shipment value	$550.00
Amount of recovery ($0.25/lb)	$250.00
Direct shipper loss	$300.00

The shipper, of course, does not expect the maximum loss. Only a small percentage of shipments will experience loss or damage. Assume, for example, that full-claim payments are made over the course of a year. Further assume that the shipper does not settle for amounts less than claimed. The shipper would find the relative costs for one million pounds, as shown in **Exhibit 13.4.** The released rates enable the shipper to reduce the annual freight bill by $10,000. Direct losses on claims, however, must be absorbed. The tradeoff produces both positive and negative net benefits depending on the incidence of loss or damage.

■■■■■■ *Exhibit 13.4* **Spreadsheet analysis of tradeoff between released-value and actual-value rates**

Annual volume (lb)	1,000,000
Product value	$5.50
Product (unit) weight (lb)	10
Settlement % — Actual value	100%
Settlement % — Released value	100%
Actual-value rate	$3.00
Released-value rate	$2.00
Released value/lb	$0.25

	Incidence of Damage				
	1%	**2%**	**3%**	**4%**	**5%**
Actual-value option					
Units damaged	1,000	2,000	3,000	4,000	5,000
Loss and damage value	$5,500	$11,000	$16,500	$22,000	$27,500
Claims recovery	$5,500	$11,000	$16,500	$22,000	$27,500
Direct shipper loss	$0	$0	$0	$0	$0
Freight charges	$30,000	$30,000	$30,000	$30,000	$30,000
Total	$30,000	$30,000	$30,000	$30,000	$30,000
Released-value option					
Units damaged	1,000	2,000	3,000	4,000	5,000
Loss and damage value	$5,500	$11,000	$16,500	$22,000	$27,500
Claims recovery	$2,500	5,000	$7,500	10,000	$12,500
Direct shipper loss	$3,000	$6,000	$9,000	$12,000	$15,000
Freight charges	$20,000	$20,000	$20,000	$20,000	$20,000
Total	$23,000	$26,000	$29,000	$32,000	$35,000
Released-rate savings	$7,000	$4,000	$1,000	($2,000)	($5,000)

This analysis leads to several observations:

1. The shipper can save total transportation-related costs if the spread between full-value and released-value rates is large and if the incidence of loss and damage is low. Estimates of the incidence, though, are more difficult to obtain now that the ICC and DOT no longer require carriers to file quarterly loss and damage statistics.
2. This simple example assumes full-claims payment from carriers; in reality, only a portion of all claim payments reflect the actual value of cargo loss and damage. Often shippers will settle for lesser amounts rather than

initiate costly civil actions. In addition, small amounts, even if recovered in full, may not offset the cost of processing the claim.

3. The opportunity cost of the funds tied up in the claims process should also be considered.

4. The shipper should assess the opportunity to purchase commercial transit insurance with a portion of the savings made by the use of released rates. In fact, this tradeoff formed the congressional rationale for giving carriers additional released-rate flexibility. Congress wanted to give shippers the freedom to pay released rates and to shop for insurance (including self-insurance) rather than pay full-value rates and have the carrier act as a virtual insurer. SNFCC, however, has argued that the likely effect of reduced carrier liability will be rising commercial insurance premiums. This change will take place because commercial insurance firms cannot pay shippers and then make back-up claims against a carrier having no liability.[12]

Claims Prevention Programs

Another important task for the traffic function is developing policies and procedures to (1) research cargo loss and damage and (2) educate, train, and supervise staff in ways to prevent such losses. Specific attention needs to be given to shipment preparation and receiving, as well as to security measures. Key areas for the shipper include packaging, marking, labeling, and loading. Receivers must watch handling and storage. Firms may be helped in these matters by working with organizations such as the Shippers National Freight Claim Council, the National Industrial Traffic League, the Association of American Railroads, and the National Freight Claim Council.

Cargo security, of course, is essential to any prevention program. Public and private organizations have drawn up useful checklists for this task, and a useful compilation of these measures was published by *Traffic World*.[13] This comprehensive checklist still offers an excellent starting point for cargo security planning.

Other Claims Actions

Overcharges

Carriers overcharge shippers when the amount collected exceeds the legally correct charge. The RICA specifically makes the carriers subject to its authority liable for overcharges.[14] Similar overcharge liability is attached to water carriers for the common carriage of goods by sea. For contract, exempt, or deregulated freight transportation, the contract of carriage (including any applicable tariff provisions) and the common law establish rules for overcharges. The following discussion primarily relates to ICC-regulated carriers.

Only one of several rates or charges, published in tariffs on file with the ICC, is legally applicable to an individual shipment. An overcharge occurs when the shipper pays an amount in excess of the legally applicable rate or charge.

Overcharges may constitute about 2 percent of a firm's annual freight bill.[15] The shipper might pay an incorrect amount for any of the following reasons:[16]

1. Incorrect extension of charges. This may occur if, for example, the shipper uses pounds instead of hundredweight or tons when multiplying the weight times the rate. Such simple arithmetic errors, of course, cause incorrect totals.

2. Improper descriptions. If the wrong description, such as a brand or trade name instead of the correct article description in the classification tariff, is used in the rate search an overcharge may result. Similarly, when incomplete descriptions are given— such as when terms like ''KD'' (knocked down), ''SU'' (set up) or ''NESTED'' are not included on the bill of lading— erroneous rates often result. Another common error is failing to annotate the density of a shipment when required to do so.

3. Wrong shipping weights. Weight lists, which contain information about unit weights (pounds per box, barrel, or other containers) of standard products, may require updating. Obsolete measurements, when multiplied by the number of units shipped, produce incorrect totals. Incorrect weighing procedures also produce erroneous shipping weights and charges.

4. Incorrect rates. Tariffs are difficult to interpret. Even experienced shipper and carrier personnel may have trouble locating the proper tariff or set of tariffs to use, or applying the correct base points, rules, and rate tables to determine the charges.

The ICC does not define duplicate billings as overcharges.[17] This point is important because the statutory time limits for filing complaints or taking civil actions to recover overcharges do not apply to duplicate billings.

Filing the claim. Whoever pays the freight bill should file the overcharge claim with the carrier that collected the freight charges. For a prepaid shipment, the shipper must file. If the shipment is sent collect, the consignee must file. Third parties might play a role when freight bill auditors are used or when terms of sale such as ''prepaid and charged back'' or ''collect and allowed'' are used.

Time limits. Claimants must start a civil court action against ICC-regulated common carriers within three years of the date of delivery to recover overcharges. A complaint may also be brought before the commission against rail and water carriers.[18] The complainant, however, may begin a civil action within six months of the date of disallowance of any part of a written claim, if the claim was presented to the carrier within the three-year limitation. Thus, shippers may sue the carrier even after the three-year period has expired—for example, when the date of disallowance falls between the thirtieth and the thirty-sixth month of the three-year statute-of-limitation period.[19]

Following deregulation, no statutory time limits apply to civil suits against air freight companies, including air freight forwarders. Shippers must carefully check the tariffs and shipping contracts for time limits.

How to file. Overcharge claims must be presented in writing to the carrier. Carriers encourage shippers to use either standard forms created by industry organizations or reasonable facsimiles. Since standard forms duplicate much of the information already contained in the freight bill, many firms use abbreviated forms to cut the cost of paperwork.

1. Basic requirements. The burden of proof is on the claimant to show that the original charges are incorrect. Generally speaking, the claim should contain enough information and documentation to allow the carrier to conduct an investigation and to pay or decline the claim. For simple errors, say in calculating freight charges, the original freight bill, together with an explanation of the error, usually is sufficient. Other errors require further documentation and supporting information. Specific elements include the following:[20]

- Name and address
- The reason for filing the claim and the facts, including

  Tariff authority, article description and weight
  Original rate and payment
  Correct rate and charges
  Amount of refund, if known
- Documents

  Original freight bill, or in its absence, a bond of indemnity (in which the claimant agrees to protect the carrier against another claim filed on the same shipment and based on the original freight bill)
  Anything else that will provide documentary evidence to show the original freight charges are incorrect (invoices, work orders, catalog pages, or letters from the weighing and inspection bureaus or classification committees that attest to weights or descriptions)

Although not required by statute, the convention is to include interest charges in the amount of the refund when the period under consideration exceeds thirty days.

2. *Duties of the carrier.* Regulations governing motor carriers and freight forwarders require carriers to acknowledge the claim within thirty days of its receipt, unless it is paid or declined within that period. After the acknowledgement, carriers must establish a file, initiate an investigation, and dispose of the claim within the next sixty days, except when an extension is agreed to by both parties. Written notification of disallowance is required.

Although the ICC has not established specific regulations for rail and water carriers, these carriers have similar duties. Individual tariffs and shipping contracts govern the duties of air freight carriers.

Unlawful Rates or Practices

The RICA makes carriers liable for damages sustained from unlawful rates or practices.[21] Such violations of the act give shippers cause to bring complaints before the ICC or the courts for reparations or for misrouting damages. The RICA requires complaints to be initiated within two years of delivery or tender of delivery.[22] In this type of action, unlike for overcharges, the date of any written disallowance has no effect on the two-year limit.

Reparations. Reparations refer to an award of damages made because a legal rate — that is, one properly filed and published — is unlawful. Although this action is proper from a procedural standpoint, the rate might violate the RICA because it is unreasonable (excessively high) or creates unreasonable discrimination. (The majority of rates go into effect without ICC scrutiny.)

The commission has authority to hear disputes and to award damages in cases involving rail and water carriers. In contested cases, the shipper has the option of either filing the complaint with the ICC or of initiating a court suit, but not both.[23] Before making a decision, the courts typically will ask the ICC, as the expert agency on rate matters, to rule on the reasonableness of rates.[24] The shipper must prove that the rate is unlawful. Since many of the statutory and common law barriers to excessive rates have been removed for rail carriers (see Chapter 3), this is especially difficult.[25] Furthermore, even if the commission should find new or existing rates unlawful, it does not necessarily follow that it will award damages; this outcome is especially likely when disputes involve regional or national rate structures.

In uncontested cases, carriers must petition the commission for approval to make refunds voluntarily. Such petitions are processed through special docket procedures that permit rail and water carriers to seek authority for the refund of, or for the waiver of collection of, charges determined to be unreasonable. In 1982, the commission began to streamline its informal special docket process for cases that show both parties in agreement and that involve insignificant amounts.[26]

The ICC has no authority to settle such disputes for motor carriers. Thus, the shipper must sue in state or federal courts. In these actions, the commission assumes only an ancillary role; if requested, it will advise the courts on the lawfulness of rates.

Misrouting. Misrouting refers to the carrier's unlawful use of a higher-rated route. Misrouting claims usually involve rail shipments and generally occur in one of two situations:

1. In the absence of routing, the carrier fails to use the cheapest practical route.
2. With routing given in the shipping contract, the carrier fails to comply and moves the shipment over a higher-rated route.

Since the rate over the actual route is legally applicable, the shipper must both pay this rate and seek remedy through a misrouting claim. The difference between (1) the charges for the actual route used and (2) the charges for the proper route over which the goods should have moved defines the damages. Moreover, when the carrier breaches the shipper's routing instructions, damages may extend to injuries caused by delay in delivery.

A number of issues, encompassing the following questions, complicate misrouting actions:

- When does the shipper have the legal right to select the route?
- What constitutes the "cheapest practical route?"
- What are the carrier's duties when the shipper's routing on the bill of lading is incomplete or inconsistent?

Not surprisingly, a considerable body of legal literature addresses questions like these.[27] In the past, traffic managers spent a good deal of time mastering the technical and legal principles that govern misrouting issues to protect the company's interests. Today, regulatory reforms and deregulation have greatly reduced the significance of this responsibility, allowing traffic managers to devote more time to managerial tasks.

Summary

Loss and damage claims administration involves five fundamental issues, including: (1) filing requirements, (2) time limits, (3) claim preparation, (4) concealed loss and damage, and (5) disputes. The owner of lost or damaged freight normally has responsibility for filing claims. Someone other than the owner with an interest in the shipment may also give notice to the carrier by filing the claim. The important thing is that the carrier is given proper notice in writing.

The Carmack regime allows claims to be filed against the originating or delivering carriers. Claims may be filed with an intermediate carrier when that carrier causes the loss. Air and water carriers confine cargo liability to service on their own lines.

ICC-regulated surface common carriers must provide at least nine months to file claims and two years after disallowance of the claim to file suit. Among air and water carriers, time limits for claims and suits are not uniform. It is important to check the shipping contract and tariffs for filing deadlines.

ICC rules and regulations for claims administration serve as useful guidelines for all modes of transportation. These guidelines set forth (1) the basic requirements for preparing a claim, (2) the duties of the carrier, and (3) the duties of the receiver. When concealed loss and damage is discovered, the consignee must promptly report it to the carrier and request an inspection. Otherwise, tariff rules require the consignee to produce reasonable evidence that the loss or damage did not occur after delivery.

The ICC does not arbitrate claims disputes, which are matters for the courts to address. Both federal and state courts have concurrent jurisdiction over freight claims arising from interstate transportation. Claimants may file in either jurisdiction, although the federal court has a $10,000 threshold for the cases that it will hear. As an alternative to civil action, shippers and carriers may voluntarily participate in two arbitration plans: (1) the SNFCC and NFCC plan administered by the Transportation Arbitration Board, and (2) the Association of American Railroads plan administered by the American Arbitration Association.

Actual freight losses and damages run into the billions of dollars for shippers and carriers. Indirect costs, moreover, may exceed direct costs by a factor of seven. Yet about 70 percent of those costs are avoidable. Cost-control planning must focus on two prime tasks: (1) the evaluation of tradeoffs between cargo loss and damage cost factors and transportation alternatives and (2) the development of a claims prevention program. The first task always has been difficult because of the lack of uniformity in cargo liability. Since 1980, legislative and regulatory changes have nurtured even less uniformity and have made this task more complex.

Shippers make overcharge claims when rates or charges, higher than the legally applicable amount, are collected. Claims for reparations or misrouting damages arise when carriers charge unreasonable rates or commit unreasonable practices.

▆▆▆▆▆ *Selected Topical Questions from Past AST&L Examinations*

1. One of the many important functions of a shipper's traffic department is the processing of claims. Regarding common carriage by railroad, truck, and air, what are the time requirements for filing and documentation necessary to file claims for loss or damage and for overcharge claims? (Fall 1979)

2. Section 12 of the Motor Carrier Act of 1980 encourages the use of released (or released-value) rates. What is a released rate? What effect will released rates have on the firm's logistics costs? (Fall 1980)
3. Identify and briefly define each of the existing types of freight claims. Which type refers specifically to the case where a published rate is thought to be unlawful? (Spring 1981)
4. Discuss the provisions of the Carmack amendment and the impact deregulation may have on the application of this law. (Spring 1985)

Notes

1. For further reading on what constitutes "proper notice," see William J. Augello, *Freight Claims in Plain English*, rev. ed. (New York: Shippers National Freight Claim Conference, 1982), p. 277.
2. See Section 1303, C.O.G.S.A.; Articles 26 and 29 Warsaw Convention.
3. See 49 CFR 1005, ICC Regulations on Freight Claims.
4. For a discussion of what "practically worthless" means, see Augello, *Freight Claims*, pp. 283–287.
5. See *National Motor Freight Classification* Item 500135; *Official North American Local Cargo Rules and Rates Tariff*, Rule G60(B)(2)(b).
6. 49 U.S.C. 11707(d)(2)(A).
7. Interstate Commerce Commission, *Ex Parte No. 403 Rail Carrier Cargo Liability Study*, decided September 28, 1981.
8. See "Minimizing the transportation loss factor," *Inbound Traffic Guide* (July 1983), pp. 44–45.
9. See Lawrence B. Wilson, Paul O. Roberts, and James T. Kneasfsey, "Models of freight loss and damage," *Traffic World* (April 9, 1979), pp. 59–65.
10. For the shippers' reactions to some of these changes, see Frederick J. Stephenson and John W. Vann, "Air cargo liability deregulation: Shippers' perspective," *Transportation Journal* 20 no. 3 (Spring 1981), pp. 45–58.
11. See Joan M. Feldman, "Liability: Are shippers paying too high a price for low rates?", *Handling and Shipping Management* (June 1984), pp. 45–48.
12. Laurie A. Safer, "Looser liability looms for freight shippers," *Handling and Shipping Management* (May 1981), p. 64.
13. "Cargo security checklist," *Traffic World* (April 4, 1977), pp. 57–59.
14. 49 U.S.C. 11705(b)(1).
15. Lester A. Probst. *The "Freight Payment Problem": An Analysis of Industry Applied Solutions in the Nineteen Eighties*, 1st rev., (Metuchen, N.J.: Transportation Concepts and Services, Inc., 1980).
16. For a more extensive discussion of causes, see "Overcharges: How to prevent them, how to collect them," *Transportation and Distribution Magazine*, 1967 reprint (Washington, D.C.: Traffic Service Corp.); see also G. Lloyd Wilson, *Industrial Traffic Management Part I* (Chicago, Illinois: Traffic Service Corp., 1941 ed.), pp. 129–130.
17. 49 CFR 1008. These regulations apply to motor carriers and freight forwarders.
18. 49 U.S.C. 11706(b).
19. 49 U.S.C. 11706(d). It appears that the ICC now interprets this section to mean six months from the last disallowance received by the shipper. See "Limitations of actions — Actions seeking overcharges," *Traffic World* (November 7, 1983), p. 14; see also John Guandolo, *Transportation Law*, 4th ed. (Dubuque, Iowa: Wm. C. Brown), p. 50.
20. 49 CFR 1008 contains specific regulations governing overcharge claims against motor carriers and freight forwarders.
21. 49 U.S.C. 11705(b)(2) and 11705(b)(3).
22. 49 U.S.C. 11706.

23. 49 U.S.C. 11705(c)(1).
24. The procedure follows from the doctrine of primary jurisdiction. For further discussion, see Guandolo, *Transportation Law*, pp. 637–662.
25. See also Jurgen Basedow, ''Common carriers continuity and disintegration in U.S. transportation law — Part I,'' *Transportation Law Journal* 13 no. 1 (1984), pp. 1–42; ''Common carriers continuity and disintegration in U.S. transportation law — Part II,'' 13 no. 2 (1984), pp. 159–188.
26. 49 CFR 1130.
27. For further reading, see Marvin L. Fair and John Guandolo, *Transportation Regulation*, 9th ed. (Dubuque, Iowa: Wm. C. Brown Co., 1983), pp. 192–201.

chapter 14

Fleet Management

chapter objectives

After reading this chapter, you will understand:
Why private carriage is used.
The costs of private carriage.
The underlying economics of leasing versus buying.
The opportunities and operating problems in private carriage.

chapter outline ▰

Introduction

Private Motor Carriage
Costs of private truck service
Fleet purchase versus lease decision
Transfer pricing
Opportunities and
operating considerations

Private Rail Operations

Summary

Questions

Introduction

Fleet management, as used here, is a general term applying to the planning, operations, and management by traffic managers of transportation vehicles or equipment. This management ranges from the operation of company trucks and automobiles to the use of shipper-owned-or-leased railcars, barges, and even airplanes. Fleet management is often referred to as "private carriage."

The Revised Interstate Commerce Act (RICA) defines private motor carriage as transportation service when

(B) the person is the owner, leasee, or bailee of the property being transported; and (C) the property is being transported for sale, lease, rent, or bailment, or to further a commercial enterprise.[1]

As defined further in Section 10524 of this act, private carriage means engaging in transportation when

(1) the property is transported by a person engaged in a business other than transportation; and (2) the transportation is within the scope of, and furthers a primary business (other than transportation) of the person.[2]

━━━━━ *Exhibit 14.1* **Reasons for private carriage use**

- Cost reduction over currently available services (outbound and for purchased goods)
- Service improvement over currently available services
- Special handling and shipping
- Special routing needs
- Tight control over interplant work-in-process goods movement
- Loss and damage reduction or prevention
- Need for product control during movement
- Competition against for-hire carriers
- Emergency transportation needs
- Assurance of equipment availability
- Corporate advertising on the highways
- Contingency against deregulation shakeout effects

The evolution of transportation law, especially since air freight deregulation in 1977, makes clear definitions of private carriage difficult. For shippers, many transportation options are available in this area today. Shipper options pertain to the provisions of equipment, labor, and other aspects of the transportation operation.

As shown in **Exhibit 14.1**, shippers use private carriage for a variety of reasons. Often, there is no single reason that a firm wants to engage in private transportation but rather a blend of reasons. Even when a shipper has used private carriage for some time, ascertaining the reasons for having started it is often difficult. Many firms continue to use private carriage without periodically reviewing its ability to fulfill a needed role. Deregulation in the area of for-hire transportation today enables shippers and receivers to obtain any reasonable rate or service movement from for-hire carriers. It is important to evaluate private operations today because of the opportunities available elsewhere. Private operations might have been necessary for one or more reasons in the past, but because outside services can be obtained with less capital commitment or risk, private operations may not be so easily justified today.

This chapter explores the key decisions facing traffic managers when considering private carriage by examining the management of private truck and railcar fleets.

━━━━━ ## Private Motor Carriage

Private truck fleets involve risk for the firm, because they entail capital commitments and, often, labor requirements that are less flexible than regular for-hire forms of carriage. Though private carriage can benefit a firm in ways outlined in Exhibit 14.1, other considerations exist that must be taken into account. These elements (shown in **Exhibit 14.2**) are not presented as negative aspects of

■■■■■■■   *Exhibit 14.2* **Management elements of private motor carriage**

- Driver hiring, qualification testing, etc.
- Driving time requirements and payroll administration
- Unionization matters
- Fleet specification determination
- Lease versus buy analysis
- Compliance with federal safety regulations
- Maintenance activity (hiring personnel and purchasing parts and equipment) or contracting for outside maintenance services
- Tire management
- Dispatching and communications
- Highway tax administration
- Budget activity
- Movement of goods for other firms
- Rate negotiations with company plants and product divisions
- Private fleet budgeting and reporting
- Capital budgeting for the fleet

private carriage but simply as an indication of the many other management and clerical activities that must be considered when using fleets. Often, these considerations are overlooked or not considered when justifying a fleet for the first time, and they become part of the unexpected escalation of overhead cost during the first years of fleet operations.

A five-step approach to evaluating the benefits of private motor carriage can be described as follows:

1. Define specific transportation needs. Firms should define their specific transportation needs before starting a private fleet operation. Private carriage is generally started because of a problem with existing transportation services, and this problem should be carefully defined. High transportation cost, poor transit time, inflexible transportation needs, or lack of control over the product movement are typical problems. The reason for defining the specific need is that private carriage designed to solve one problem might not be effective in attacking another. A fleet designed to reduce costs over a high-density route may not provide the same service over a low-density route in the company shipping pattern; for instance, if the original fleet was justified using special equipment, and it no longer is a part of the operation, or vice versa.

2. Analyze traffic flows. Traffic managers should analyze the firm's traffic flows by product and by origin-destination patterns. Private truck fleets need to operate at capacity for a large percentage of their total miles. Further, the fleets must be utilized fully over the course of the year; therefore, the period analyzed should be at least a year to show where a fleet might be used fully.

Exhibit 14.3 illustrates two situations where tonnage analyses are important. In situation 1, the firm with a large fleet but seasonal traffic will find some of the fleet not used during most of the year, resulting in overinvestment, poor equipment utilization, low manpower use, and low return on investment. On the other hand, situation 2 illustrates the condition of a fleet size matching the base traffic load for the year. The traffic that exceeds the fleet capacity is handled by other carriers.

A flow analysis of all the firm's traffic patterns— that is, plant to plant, plant to warehouse, vendor to plant, and warehouse to customer—should be performed. To balance the flow as much as possible will require gathering data from purchasing, production, and distribution information systems. For example, a good operation of matched flows might consist of a triangular movement from a plant in Chicago to a warehouse in Kansas City. From there, a move might be made to a customer in Nashville, Tennessee, followed by an empty, or "deadhead," move to Murfreesboro, Tennessee, for pickup at a company supplier for movement on to the Chicago plant.

Exhibit 14.3 **Analysis of fleet utilization**

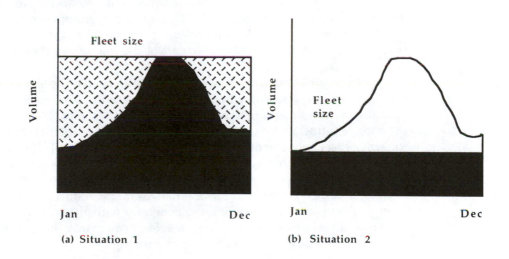

(a) Situation 1

(b) Situation 2

■ Portion of fleet being used

▨ Portion of fleet not being used

The flow analysis must be examined realistically. For example, a decision must be made about movement distance. Often, firms do not consider that outbound moves of beyond one thousand miles may need two drivers (in either a sleeper cab or relay). The ten-hour driving limit forces a cost increase at the relay point or the use of two alternating drivers in the same truck. Such additional cost might cause the fleet to be uneconomical for distances beyond one thousand miles.

If the fleet is being justified as a cost-reduction investment, a high ratio of loaded to total miles must be possible. Some firms find that unless the fleet is 90 percent loaded, few economies exist in comparison to for-hire forms of carriage. If the firm seeks service improvement, it must do simulations of vehicle movement to ensure movements can physically be made within the required time that are better than current for-hire means. Fleet managers should avoid holding up shipments to fully load the private fleet lest customer service deteriorates, inventory holdings increase, customer cash payments slow up, and negative performance reviews result.

Analyzing the firm's traffic patterns should determine if outside carrier sources might correct the company's problem more efficiently than a private fleet. Quite often, presenting a contract carrier with a three-way move, as previously described in the Kansas-to-Tennessee-to-Chicago example, results in a favorable contract-rate arrangement.

Further, as another example of the benefits of a flow analysis, trailers might be the only equipment needed. Outside owner-operators can then be hired to haul the firm's trailers. With no need for tractor investment and driver hiring and use, the firm can avoid a large part of the risk of private carriage.

3. Examine alternative transportation services. Traffic managers should investigate and experiment with outside transportation services to determine if alternative sources might help correct the transportation problem. The firm might look into contract motor carriage, rail contracts, pool-truck distribution at the end of a piggyback movement, and other intermodal movements. Further alternative services include different packaging (if loss and damage is a problem with current carriers), shippers' associations, freight forwarders, and use of supplier or customer vehicles. One of these alternatives might reduce or eliminate the current problem without involving the company with the overhead commitment and risk of private carriage. A firm, therefore, can justify the private fleet option only after outside alternatives have been investigated.

4. Determine the least costly form of private carriage. A traffic manager should determine the least costly method of initiating private carriage, if the firm decides such an operation is the best method for solving its transportation problem. Besides equipment, private carriage requires supervision and management, personnel, and a control system. Equipment is generally leased or bought. Supervision and management are often overlooked aspects of private carriage. For instance, a simple routing system might be managed by existing traffic department personnel; however, a complicated route structure, with many rigs and trailers, often will require staff personnel assigned just to manage the fleet. Some of this is simplified today with computer routing systems. However, a fleet staff is especially needed when the firm decides to buy the equipment and perform the maintenance in house. Hiring and training drivers, who receive pensions and other benefits, may also be necessary. Most firms operating private truck fleets, however, make lease arrangements only for drivers or drivers with equipment. Finally, a control system for private carriage includes budgeting, cost collection, pricing and charging, and tax administration. All these aspects are part of the implementation details and overhead costs of the private fleet.

5. Expand fleet incrementally. A firm initiating private carriage should start on a small scale. By starting small, many of the management mistakes can be handled, and adjustments are easy to make without severely disrupting the firm. A key here is being able to withdraw from private carriage if it is difficult to implement or if the original problem appears to have been solved through other means.

Several common pitfalls to avoid are:

1. Failing to investigate alternatives. Not considering alternatives to private carriage can cause the firm a good deal of expense and difficulty. Private carriage is only one possible alternative to for-hire common carriage. By not considering others, the firm stands to switch from one form of carriage, and a set of specific problems, to another form with the same or additional problems. Again, private carriage is perhaps the riskiest of all forms of carriage, since it requires a long-term commitment of capital (even when leasing equipment) and a dedication to a specific form of transportation. Existing carriers might lose interest in serving the firm for its leftover freight, if the firm embarks on private carriage; therefore, switching back to all for-hire carriage might take much time and effort.

2. Sacrificing purchasing flexibility for balanced loads. Justifying the fleet on balanced loads, using specific vendors for backhauls, can limit a firm's purchasing flexibility. An outbound movement pattern might be clearly established because of sales and warehouse distribution patterns. But vendors at specific locations have fewer inbound options. Therefore, a fleet pattern justified on a few of these vendors might cause the firm to use these sources to degrees that weaken the firm's purchasing power. Further, switching to other vendors in other locations might cause the fleet to lose its inbound move.

3. Placing fleet administration in the traffic organization. Not separating the fleet administrator from the traffic organization can diminish the effectiveness of the private fleet. The traffic manager directly or indirectly responsible for the performance of the fleet may make that responsibility an overriding concern without considering the firm's or department's other missions. Holding shipments to fully utilize the fleet, or delaying inbound pickups to meet the convenience of the truck fleet, might be to the firm's detriment. The use of the fleet by the traffic manager should, therefore, be an independent decision to minimize cost. The private fleet needs to present a unique, valuable service to the firm, or else it should be abolished or reduced. The corporate fleet of PPG Industries, for example, must always compete for traffic against for-hire carriers in both price and service.

4. Underpricing private transportation service. Underpricing private carriage service can also be detrimental to the firm. When a firm does not charge enough to intracorporate users of the service, the demands for service might become uneconomically high. That is, production, marketing, and other departments might use the fleet at less charge than the firm is incurring in total costs. In this setting, the fleet can become large and, in total cost, not be as economical as other choices available to the firm.

Costs of Private Truck Service

Private motor carriage requires looking at a host of different costs, for it is vastly different from for-hire carriage where a single-trip freight bill is submitted and paid. **Exhibit 14.4** presents many of the costs of private carriage. These costs behave differently, depending on the service. For example, as the truck is driven more miles each month, the total depreciation or lease cost for that month will remain the same, but it will decrease on a per-unit basis by being spread out over the miles or over the tonnage of freight being hauled. The same phenomenon occurs with licenses, insurance, and administrative costs. Driver's wages and fuel, on the other hand, are variable in that they increase in total costs according to each mile hauled.

Private truck costs can also be viewed in another context. As shown in **Exhibit 14.5**, the major cost categories of private carriage will change over the life or total miles of a truck. The operations curve represents fuel, taxes, wages, insurance, tires, and so on. Downtime increases as the vehicle ages, as does maintenance. Parts inventory costs exist in the form of the investment in parts and the lost-opportunity cost of the cash tied up in these parts. Since depreciation decreases over the life of the asset, the depreciation curve illustrates the accelerated forms of depreciation. If a straight line for depreciation were used, it would suddenly end in the last of year of depreciable life.

Obsolescence, as shown in Exhibit 14.5, is a variable concept of private car-

███████ *Exhibit 14.4* **Private carriage cost analysis**

Cost Area	Comments
Driver's wages	Often per-mile on over-the-road operations; per hour for local delivery operations
Fringes and bonus	Fringes often are fixed in nature; bonus depends on plan
Tractor-trailer cost	Usually fixed in nature as depreciation; in lease can be a fixed-cost component with a per-mile charge as well
Fuel	Variable over distance; can be different according to driver's driving habits and terrain
Tolls	Assessed according to route
Meals	Variable over distance on over-the-road operations
Motels	Same
Highway taxes	Some are fixed and some are variable according to distance or ton-miles hauled
Supplies and tools	Fixed
Maintenance	Vary according to age of vehicle, driving habits, and type of operation, as well as degree of preventive versus breakdown maintenance
Communications	Depends on extent of movements and distance away from base
Insurance	Fixed per period
Tires	Variable according to distance; can wear more through improper inflation and speeds above 60 mph
Licenses	Fixed per period
Vacation replacement	Fixed
Administration and clerical	Fixed

Source: Adapted from Thomas R. Henke, *Managing Your Private Trucking Operation* (Washington, D.C.: Traffic Service Corporation, 1976), p. 6. Reprinted by permission.

riage that compares newer, better models of trucks and trailers to older models. For instance, during the 1970s, obsolescence actually favored older vehicles, because of the increasing requirements for antipollution devices in trucks, making newer trucks more costly to purchase and operate than older ones. This is a factor today with the use of 48-foot trailers and 28-foot twins, both of which were restricted in most states; today, they are common on interstate highways.

The total cost of the vehicle over its useful life is shown in Exhibit 14.5 by the top curve. As initial training of drivers and maintenance personnel is carried out, and as the high initial depreciations are written off, total costs decrease.

Exhibit 14.5 **Generalized private fleet cost behavior**

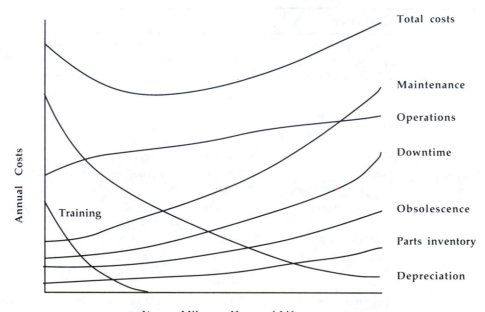

Source: Based on J. L. Cavinato, *Finance for Transportation and Logistics Managers* (Washington, D.C.: Traffic Service Corp., 1977), Chapters 9–11.

When a low point is reached, total costs will begin to increase as maintenance and other costs begin to rise. This same behavior of total cost is noticed by managers who also have responsibility over company sales-staff car fleets; automobile cost structures behave in the same relative manner.

Since the Surface Transportation Act of 1982, highway-related taxes have become more significant to fleet managers. This law imposed changes in the highway taxes for trucks. A summary of these taxes is shown in **Exhibit 14.6.**

The Surface Transportation Act is designed to provide funds to help maintain and improve the nation's highway system, which is deteriorating faster than anticipated. Because many states do not have sufficient tax revenues to maintain the highways, this law also provides assistance to the states.

The costs of operating trucks over the highways will fall on both for-hire and private fleets. One estimate, made by the American Trucking Association, is that total taxes will increase by 126 percent for five-axle-tractor semitrailers and 136 percent for five-axle-twin trailers. The total taxes on the latter unit over average use will be about $4400 per year.[3]

Financial responsibility is another major cost area for private fleets. In 1981, the amount of insurance or other financial responsibility that had to be carried

Exhibit 14.6 **Highway taxes affecting all trucks**

User Type Fee	Rate/Tax	Effective
Gas or diesel fuel	$0.09 per gallon	April 1, 1983
Tires	0–40 lb: no tax 40–70 lb: $0.15 for each pound over 40	January 1, 1984
	70–90 lb: $4.50 + $0.30 for each pound over 70	
	Over 90 lb: $10.50 + $0.50 for each pound over 90	
Truck parts	0, unless purchased within six months of purchase of new tractor or trailer, then 12%	
Truck sales	12% for tractors over 33,000 lb gross-vehicle weight (GVW) (at retail price level)	
	12% for trailers over 26,000 lb GVW (at retail level)	
Heavy-vehicle use fee	Trucks at 33,000 to 55,000 lb of GVW: $50 + $25 for each 1000 lb over 33,000 lb	
	Trucks at 55,000 to 80,000 lb of GVW: $600 + $40 each 1000 over 55,000 lb	July 1, 1984
	$600 + $44 each 1000 over 55,000 lb	July 1, 1986
	$600 + $48 each 1000 over 55,000 lb	July 1, 1987
	$600 + $52 each 1000 over 55,000 lb	July 1, 1988

Source: Surface Transportation Assistance Act of 1982 (Public Law 97–424).

on trucks was increased, and then, on July 1, 1983, it was again increased. For-hire and private trucks moving oil must carry at least $1 million public liability insurance, or other proof of financial responsibility. When hazardous goods are carried, the minimum responsibility is $5 million.[4] This factor is important in considering private carriage, as well as in selecting for-hire carriers.

Another major management concern for fleet operators is safety. All trucks, whether for-hire or private, fall under the Department of Transportation's safety regulations, found in Title 49 of the Code of Federal Regulations. This agency regulates highway safety of trucks and drivers by imposing maximum driver time in the truck, as well as log requirements. Vehicle safety is also included in the regulations. **Exhibit 14.7** shows a breakdown of the major violations of safety regulations, found by DOT field inspectors.

■■■■■ *Exhibit 14.7* **In-field safety violations — 1979**

Violation Area	Total Number of Violations	Percent of Total Violations
Lighting and electrical systems	2961	40.41
Brake systems	1827	24.93
Tires	419	5.72
Emergency equipment	392	5.35
Wheels	386	5.27
Hazardous materials	205	2.80
Exhaust systems	182	2.48
Suspensions	171	2.34
Other	785	10.71
Total	7328	100.00

Source: John Dolce, *Fleet Management* (New York: McGraw-Hill, 1984), p. 139. Reprinted by permission of McGraw-Hill.

These costs—including those for taxes, insurance, and safety—are all part of the administrative overhead needed to operate a fleet of private trucks. They are shown here not only because they are major costs but also because they are often overlooked or not considered when planning the start-up of a fleet. The burden of handling many of these separate costs, as well as vehicle safety concerns, can be minimized through vehicle leasing.

Fleet Purchase versus Lease Decision

The question of financing the lease is a major part of the preliminary analysis and implementation of private carriage. Leasing has long been a financing alternative in for-hire carriage firms and is common in private carriage operations as well.

The decision to lease or purchase involves different cash flows and different risks for the acquiring firm. **Exhibit 14.8** presents some of the major considerations of each form of financing. Many of these considerations will be enough to make a firm select one form of financing over another. For instance, maintenance can be a high fixed cost that is economical only when using many trucks. Another maintenance consideration is when the fleet operates thousands of miles from its base. Maintenance is then best handled through arrangements with other firms.

There are many types of truck leases, and each presents a different financial responsibility and cash commitment for the firm. Some of the most common types of leases are shown in **Exhibit 14.9**.

One major leasing consideration involves distinguishing between an oper-

━━━ *Exhibit 14.8* **Lease versus purchase analysis**

Factors favoring purchase

1. Funds are available for both working capital and the equipment procurement.
2. Money can be borrowed at reasonable rates.
3. The trucking investment yields a satisfactory return.
4. Financial and operating ratios are favorable.
5. Maintenance services are available.
6. Overhead costs of the fleet are not excessive.

Factors favoring leasing

1. Lessor can provide nearly all the financing.
2. The company is short of cash or does not care to put cash into fleet venture.
3. Equipment is needed only temporarily.
4. Flexibility is provided in the operation, which is important in seasonal periods.
5. A hedge against risks of obsolescence is needed.
6. Accounting, maintenance, and administrative personnel are limited.
7. Bank financing for a purchase requires compensating balances or other restrictions.
8. Minimal ''bail-out'' risk to the firm is provided when first entering into private carriage.
9. Buying power of leasing firm permits more economical acquisition than the firm could attain on its own.
10. Tax benefits cannot be used through the purchase option.

ating lease and a capital lease. An *operating lease* appears only on the firm's financial statements each year as a lease expense. A *capital lease* appears as an annual expense on the income statement as well as a form of asset and a form of liability on the balance sheet. By showing it as an asset, the capital lease reduces the firm's return on asset computation. By showing it as a liability, the capital lease reduces the firm's financial position with regard to debt, versus equity and other ratios used in credit analysis. One of the reasons firms often lease is to use the assets, but they do not have to show them on the records.

The distinction between a capital lease and an operating lease is made by the accounting profession through the Financial Accounting Standards Board statement 13. Known as FASB13, this document states that a lease will have to be ''capitalized'' on the firm's financial statements if any one of the following features exists in the lease:

- Transfer of ownership at the end of the lease term
- Bargain purchase option is in effect
- Lease term is 75 percent or more of the economic life of the asset
- Present value of minimum-lease payments is 90 percent or more of the fair value of the vehicle, less investment tax credit

Exhibit 14.9 Types of leases available

1. *True lease*
   Recognized by both law and tax authorities as providing lessor with benefits and risks of ownership. Lessee acquires use of asset without building any equity in the asset.
2. *Conditional sale*
   User is treated as owner from outset; an installment purchase.
3. *Lease/purchase agreement*
   Similar to conditional sale, but lessee has option to purchase asset at bargain price at end of term.
4. *Full-payment lease*
   Lessor receives full cost plus financing, overhead, and a profit return on investment.
5. *Nonfull payment lease*
   Contract in which lessor depends on an unguaranteed portion of the residual value of asset to recover costs plus a return on investment.
6. *Operating lease*
   Lessor provides maintenance, or some other service, in addition to the use of the asset at a price that fills the requirements of FASB13. Similar to nonfull payment and true lease.
7. *Finance (capital) lease*
   Lessee makes payments over the useful life of the asset and meets the requirements of FASB13. Similar to full-payment lease.
8. *Net lease*
   Similar to finance (capital) lease except that all other payments, such as taxes, maintenance, and insurance, are paid by lessee.
9. *Master lease*
   An agreement for the leasing of certain equipment with options for lessee to lease additional equipment as necessary at a predetermined rate without a new contract.
10. *Rental contract*
    Short-term contract for the use of assets, including maintenance and insurance.
11. *Direct lease*
    Could be almost any type of lease, but lessor holds title to the asset.
12. *Leverage lease*
    Group of investors may provide portion of purchase price for the assets, and the balance is borrowed from banks or institutional investors.
13. *Sale/leaseback contract*
    Company sells some assets to a firm and leases them back under a direct lease. Usually done by a company needing cash.
14. *Full-service lease*
    Lessor can provide everything, such as maintenance, insurance, taxes, and other incidentals. Can be tailored to provide or eliminate any type of service or expense.

Though the FASB13 is the strict accounting guideline used in dealing with leases, and technically reduces several of the reasons that firms lease rather than purchase, many long-term leasing arrangements are made that still enable the firm to treat the lease as an operating lease.

Several methods for specifically evaluating the lease versus purchase option are available. These methods generally entail a cash flow analysis of all payments, tax benefits, and costs associated with both purchasing and leasing. The evaluation of the two options requires discounting the flows according to the present value factor of the firm's cost-of-capital rate. If the sum of discounted costs associated with the lease is greater than the sum of discounted costs associated with the purchase, the favored option from the standpoint of cash flow is purchase. Otherwise, lease is the favored option. Still, many firms will ignore this difference, unless it is great, and lease, because of the convenient maintenance and administrative duties offered through leasing.

Several aspects of leases warrant critical evaluation by the prospective lessee.[5] One major element of leasing to evaluate is the cost, since leasing arrangements often use a fixed fee plus a cost per mile. The any-escalation clause of leases also must be examined carefully to determine how additional costs will be added, what index is used to trigger escalation, and on what basis the lessee can verify the extent of the need for escalation. Finally, either the lessor or the lessee can benefit from the investment tax credit.

Another major element of leasing agreements is the available service features, including routine and emergency maintenance. Some leasing firms have strong, nationwide maintenance networks, whereas others offer only sparse or concentrated maintenance operations. One important feature of maintenance is comparing that performed for for-hire firms with that performed for private fleets. Labor problems, as well as priority scheduling within the leasing firm, make important differences between maintenance for large for-hire firms and for small private firms. Some leasing firms even arrange to provide maintenance by common carriers at their own terminals.

Labor constitutes yet another important element in leasing agreements. The difference between union and nonunion personnel can make a difference in the carrying out of the lease agreement.

The fuel supply for the leasing firm is another feature that a lessor should investigate. Contracts that the leasing firm has with its fuel supply firms should be studied since a nationwide fuel supply can offer a competitive price.

A final aspect of leasing involves drivers. In the past, the Interstate Commerce Commission prohibited leasing drivers from the same source as vehicles. In 1982, though, the ICC reduced some of the restrictions on single-source leasing; doubtlessly, this practice will continue as further deregulation of the industry evolves. To date, the requirement states that the private firm control both the vehicle and the driver, even though they can be leased from the same source.

This restriction on this form of leasing has long been in place because an owner-operator or any other firm would, in effect, be providing for-hire service through single-source leasing.

Driver leasing can be beneficial for the shipper for many reasons. Such benefits range from the leasing company handling parts of the payroll function—such as pension administration, insurance coverage, separation of labor contracts from the firm, driver safety programs, credit and background investigations, and licensing—to handling all the payroll function.

Transfer Pricing

One of the major decisions in operating a private fleet is on what basis the firm will charge for its services. Transfer pricing is the accounting term that applies to charging for any intracorporate activity, such as private carriage service or private warehousing. Although a for-hire firm charges to obtain a profit for the firm, this is not always desirable or feasible in intracorporate settings where one division charges another for services. Some of the several forms of transfer pricing that firms use in the private carriage area are as follows:[6]

1. Market price. This is a price based on prevailing tariff rates charged by for-hire carriers in the area. Market price is useful to the firm if the private truck fleet is an investment center responsible for its own costs, profit, and return of profit to the corporate parent. For market price to work, the private fleet must be free to carry goods for outside firms; this price is valid for charging these outside customers. Moreover, market price is valid only when the firm is able to freely select outside carriers and to use the fleet when hauling its own goods.

2. Market price adjusted for marketing costs. This is a tariff-based price used by for-hire firms, but the costs of marketing and profit are adjusted out of this price so the firm is not charged for these cost elements. Many private carriers use a pricing method similar to this, charging whatever the prevailing commodity or other for-hire rate is, less, say, 15 or 18 percent.

3. Full-cost pricing. This method of pricing attempts to recoup the fleet's total (variable and fixed) costs. Judgments will have to be made about how to allocate fixed costs to specific runs (see Chapter 9). In addition, practical difficulties may arise because full-cost-per-unit estimates are sensitive to the volume of traffic (units) that will be transported. If the actual volume is less than projected, the full cost per unit will be underestimated and will not enable the firm to recoup the fleet's total cost. If the volume is larger than projected, the users of the fleet will be overcharged. Therefore, this method of pricing may be difficult to implement without interdepartmental dissatisfaction.

   4. *Variable-cost pricing.* The variable-cost method of pricing includes charges for only the variable costs of transportation, such as fuel, direct labor, tolls, and some equipment maintenance. It does not include recouping depreciation or lease expense, licenses, insurance, taxes, or general and administrative expenses. By subtracting the variable costs of a run from projected revenues, fleet managers can determine the contribution to the fleet's fixed costs. Determining which costs are fixed and which are variable can be difficult, because that determination depends on the nature and scale of fleet operations. In addition, if this pricing method results in a low charge (approaching variable costs) to the customer, it can create heavy demands from other departments seeking to reduce their transportation costs. This pressure can lead to the firm's investing in a larger fleet than is economically justified.

   5. *Two-part pricing mechanism.* This approach assesses separate charges for recovery of fixed costs and variable costs. General Electric designed this pricing scheme into what it called TRUXBUX, which charges for tractor and trailer mileage (mileage-related maintenance and lease fees), labor, fuel, and tolls according to the miles. Period costs, or those that arise over time rather than by mileage, are assessed according to the time the truck is involved in the movement. These include daily depreciation or lease expenses (whichever is assessed), time-related maintenance and repair, licenses, taxes, insurance, administration, and supplies expenses.

   TRUXBUX computes a fleet's profit by including the amount of for-hire charges avoided by using the fleet as revenue. Total costs are subtracted from this amount to show a fleet profit for the firm using it.

   6. *Negotiated price.* This price can be any sum settled on by the fleet managers and a user in the firm. A negotiated rate should at least be a standard rate that the fleet will use as a charge during the coming period as opposed to an actual rate in which the fleet could pass on any cost variances to the user during the period. This latter system leads to costs not being controlled to the degree they should be by a "seller" of the service. Standard rates, on the other hand, represent a commitment for a period by the fleet manager that the rates will be so much per unit. Therefore, any cost overruns caused by the lack of fleet management control are not passed on to the using departments.

   The general cost behavior of private fleet runs is presented in **Exhibit 14.10**, which shows that total costs rise at a decreasing rate over increasing miles. There is a fixed-cost element present at the start that represents dispatching, position, and loading costs. The cost of unloading is also represented in that fixed cost, because it will have to be incurred, regardless of the miles involved in the run. The variable costs over each mile will tend to decrease slightly as additional miles are incurred. A slight jump in these costs will occur when the driver must layover, or when the run is continued with a second driver — either of which will happen every four hundred to five hundred miles. Some owner-operators will not charge for this increment, but it is still a cost incurred by the motor carrier operator.

On a cost-per-mile basis, the cost is fairly high for a short run, but it decreases as the run lengthens. This cost can be made lower in some interplant shuttle runs, when the truck is loaded and unloaded quickly and it is fully utilized throughout the day. As with any operation, the more intensely utilized the asset, the less per-unit fixed costs will become. That is, insurance, licenses, administrative overhead, truck depreciation, and so forth will be spread out over more units of freight or miles of operation.

Exhibit 14.10 **Profiles of private truck costs**

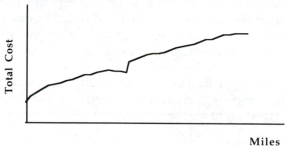

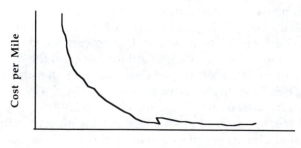

Opportunities and Operating Considerations

Several new opportunities are available for private fleets, most of which have come about from the loosening of regulations in private and for-hire carriage. Some of the opportunities can be represented as follows:

1. For-hire operating authority for private fleet operations. In 1978, the Toto Purchasing and Supply Company applied to the ICC for operating authority on the empty-haul portion of its private fleet. To the surprise of many in the transportation community, the ICC approved the application with the following conditions:[7] (1) the operations had to maintain separate records; (2) the for-hire carriage could only be incidental, or filled in for the empty-haul portion of a private-haul trip run; and (3) ICC and DOT regulations had to be adhered to for the total operation. One of the restrictions, however, that prevented many firms from using this new ability is the need to seek ICC review of securities when financing from outside sources is used. This meant that a firm had to seek Securities Exchange Commission (SEC) and ICC approval of securities, but this was removed in 1982 when the bus industry regulatory reform law was enacted.

2. Compensated intercorporate hauling. Compensated intercorporate hauling (CIH) is the movement of freight via private carriage for two or more parts of one company. In the past, private carriage could legally take place within only one corporate sphere. That is, compensatory movements could be handled for several divisions but not for other corporations owned by the corporate parent firm.

The Motor Carrier Act of 1980 amended Section 10524 of the RICA with a provision that exempted from regulation intercorporate hauling for compensation between firms that are 100 percent owned by a parent firm.

3. Separate trucking firm ownership. Many firms, finding the Toto provision of ICC security review and the 100 percent intercorporate ownership requirements too restrictive, sought another avenue of obtaining in-house carriage, namely, the establishment of a trucking firm as a subsidiary of the entire corporation. This subsidiary then applied for ICC rights to handle freight for its parent firm and for others.

The separate trucking subsidiary concept has several benefits. One, the subsidiary can use unionized drivers on this system that would shield the rest of the firm, often non-union, from the trucking operation of the corporation. Two, the subsidiary can report its separate accounting practices more easily than in typical private carriage operations. This accounting benefit reduces one of the prime criticisms of private carriage in the form of complete-cost separation. That is, much of a regular private carriage operation incurs overhead costs and other costs in the firm that are not usually fully captured as part of the private carriage reporting system. Three, the subsidiary can deal with the ICC's security regulations without involving the corporate parent. Four, the separate subsidiary can handle freight for corporations that are only partly owned by the parent firm. Five, the operation can sell its services easily to outside shippers and be managed as a stand-alone trucking operation with profit responsibilities to the corporate parent.

One example of a separate subsidiary trucking operation is Techfleet, which is owned by the former Burroughs Corporation and is now part of the BDC Corporation. This operation was initiated to handle sensitive computer equipment with quality carriage under tight control of Burroughs and was prompted by problems with the many trucking firms and household-goods carrier agents that were the mainstay of computer shipping. Techfleet was incorporated; it applied for ICC operating authority and complied with the other regulations required by trucking firms. It started operating early in 1984 with high-quality tractor-trailer rigs and moves freight not only for Burroughs but also for other firms requiring special handling of sensitive goods.

4. Backhaul allowances. Section 10732 of the Revised Interstate Commerce Act was added in 1980 by the Motor Carrier Act. This new provision allows private carriers of grocery products to pick up its purchased product from a vendor that normally uses zone pricing and to receive compensation for performing carriage that the vendor would have had to provide via for-hire carriers. The law requires that the compensation be available to any customer of the grocery product firm and that the compensation not be greater than the actual costs to the seller of delivery to such customer. This provision also allows greater utilization of private fleets, because they can obtain backhaul shipments in this manner.

5. Single-source leasing. This proposal by the ICC permits private carriers to lease drivers from the same source as vehicles. The opportunity here is in the use of owner-operators. The proposal required that the driver be paid via payroll separately from compensation for vehicle costs. Further, the firm must direct and control the operations of the owner-operator in alike manner as an employee as a separate carrier firm. This new freedom opens many new sources of carriage besides the regular for-hire carriers available to shippers, and it is a low-risk alternative to regular private carriage.

6. Captive for-hire firm using owner-operator. Many shippers have struck arrangements with owner-operators by supporting them for operating authority before the ICC, including the support of loan applications for equipment. In this manner, the firm has available a for-hire carrier firm, usually through a contract carriage arrangement. These arrangements are helpful when the shipper has stopoff and other special routing carriage needs.

7. Trip leasing of private fleet equipment to for-hire carriers. This ICC proposal will permit private carriers to trip-lease equipment to for-hire carriers on backhaul movements. Thus, a private carrier can avoid an empty haul and earn revenue from a for-hire carrier that used the equipment for its shipments. This practice switches the relationship between the firm and the carrier, whereby the carrier is a form of customer to the firm. A problem might occur in this case in reciprocal purchasing contexts if the joint relationships become coercive.

Private carriage will no doubt become less restrictive as more deregulation occurs. There are proposals to eliminate the ICC in the near future. If this occurs, private carriage will have the freedom to act without operating rights, tariffs, and many other current requirements in for-hire situations. However, safety regulations will always continue, as will hazardous material requirements.

Private Rail Operations

Traditionally, private rail operations have encompassed situations where shippers obtained and supplied railcars for their own shipments. This practice has long been required in tank-car movement, since rail carriers did not want to be responsible for the empty backhauls and the cleaning and contamination problems of this type of service. Shippers have also provided cars of other types, like covered hoppers and boxcars, to ensure a certain carrying capacity when these types of railcars are in short supply. Further, many utilities have provided their own hopper cars for unit train movements.

Traditionally, the shipper initiates the use of private railcars by either purchasing or leasing a fleet of cars from a railcar builder. These cars are then assigned to the shipper by specific car initials and numbers. This practice is common with chemical companies' tank cars, grain firms' covered hopper cars, and even the military's specially equipped boxcars. If these cars are leased from a car builder, the shipper pays a monthly rent to the builder. On using the car, the shipper pays the rail carrier the regular rate for movement. The rail carrier keeps track of the miles the car traverses on its line and then submits a mileage allowance (depending on the car's age and distance on the road) back to the car lessor. Some of these allowances cover only loaded miles, whereas others cover both loaded and empty miles. The car lessor then adjusts the next monthly rental by the allowance received. Through high-loaded mileage utilization of the railcar, the rental can approach zero dollars a month.[8]

This system provides the shipper with a supply of cars when needed, but there is a risk that the cars might sit idle for some periods, and a car rental will still accrue. Some large shippers have sought to sublease excess cars to other shippers or even coerced rail carriers into subleasing the cars for these slack periods. The rail carrier at origin, however, has to sign an ''OT-5'' agreement before a shipper can use private freight cars. During slack periods, these agreements (which may last only six months) cannot be renewed. The cancellation can place shippers in an awkward position if they have long-term lease agreements of a year or more.[9]

Rail contracts and deregulation, however, have opened up opportunities in this area. For example, some railroads will sell cars to shippers for use in contract arrangements.

Another aspect of private carriage becoming more important today is actual rail operations. The abandonment freedoms established for rail lines permit rail carriers to drop unprofitable lines more easily than in the past. However, many of these lines still have some traffic, and some shippers still need the carrying capacity and capabilities of railcars on these lines. Occasionally, shippers, or groups of shippers, have taken over responsibility for operating these lines. Firms in the United States even operate abandoned rail lines under contract arrangement with the shipper to and from junction of the railroad that abandoned the lines. Hammermill Paper and Goodrich are just two firms that have engaged in these types of operations in the past few years.

The future will see further changes in the traditional relationship between the railroad and the shipper. The laws are less restrictive today, and opportunities for creative arrangements are present now that were unheard of in the 1970s. Rail contracts, deregulation of various forms of rail service (like boxcars, export coal, piggyback), and easier abandonments are just some of the reasons for these changes.

Summary

Private carriage is just one alternative form of transportation available to the firm. With the advent of deregulation and the new carriage forms and rate freedoms, some firms are reducing or eliminating the use of private carriage. This activity has often been perpetuated in firms without an objective analysis, deregulation has prompted in-depth studies of the situation. In other firms, private carriage is seen as a way of maintaining quality service and a contingency of service if one eventual shakeout of carriers eliminated needed services for reasonable costs. And still other firms have sought to enter the carrier business by starting trucking operations that serve both the firm and others on for-hire bases.

The key to sound management of private carriage is that the firm be aware of its transportation needs and that it uses the forms of transportation best suited to those needs. Private carriage is one form of transportation that often has been brought into existence and then not evaluated periodically on whether it still fulfills the original and additional needs.

Selected Topical Questions from Past AST&L Examinations

1. Discuss the advantages and disadvantages of the use of private carriage. (Spring 1980)
2. What are some of the facts, information, and data which should be reviewed

before implementing a private trucking operation? (Assume a manufacturing operations with dispersed, multiplant facilities). (Fall 1981)

3. Does the Interstate Commerce Commission permit private carriage by a company for its own subsidiary? Discuss this situation, including procedural requirements, under the current statute. (Fall 1981)

4. You are the general traffic manager of a multidivisional corporation with several operating subsidiaries engaged in the manufacture of diverse product lines. One area of your responsibility is the management of a trucking division. At a recent policy meeting of your top management, you are asked to explain what the change in recent regulatory attitude would have on your trucking operation. What would you tell this group? (Spring 1982)

5. Your immediate superior has asked you to submit a memo to her outlining the advantages and disadvantages of converting the present private fleet of twenty owned tractor-trailers to a contract motor carrier subsidiary. Please comply with her request. (Fall 1982)

6. Why have some shipping firms established separate carriers as corporate subsidiaries in place of their former private carriage operations? (Fall 1984)

7. In 1978, the Toto decision by the ICC permitted private carriers to apply for for-hire common or contract operating rights that were incidental to the prime move. Since that time, many new freedoms were granted private trucking. List and describe at least three new practices that private trucking operations may now conduct. (Fall 1984)

8. What is "single-source leasing"? What are the ICC-mandated minimum requirements to be contained in the lease agreement? What are the perceived advantages and disadvantages of this concept? Identify and explain. (Spring 1985)

9. Golden Gate Corp. (GGC) currently leases a tractor and trailer at an annual fixed charge of $10,000, plus $0.50 per mile. GGC has determined that it can purchase a comparable vehicle and incur an annual fixed cost of $16,000, plus a per mile operating cost of $0.35. (All costs are after-tax.)

   (a) How many miles per year must GGC generate to warrant the purchase of a vehicle from a cost perspective?

   (b) Assume the GGC is not required to capitalize the lease and that GGC must borrow the capital to finance the purchase. Compare the impacts of leasing and buying on GGC's debt ratio. (Spring 1985)

Notes

1. 49 U.S.C. 10102 (14).
2. 49 U.S.C. 10524.
3. Public Law 97-424 *Surface Transportation Assistance Act 1982.*
4. 49 CFR 387.
5. For a good analysis of lease evaluations, see Warren R. Ross, "Equipment leasing," *The Private Carrier* (August 1981), p. 45.

6. Joseph L. Cavinato, "Price strategies for private trucking," *Journal of Business Logistics* 3 no. 2 (November 1982), pp. 72–84.

7. *Toto Purchasing and Supply Co., Inc. Common Carrier Application No. MC-141414 (Sub.-1)*, 128 MCC 873 (1978).

8. See John E. Tyworth, "The analysis of shipper operating policies and private freight car utilization," *Transportation Journal* 17 no. 1 (Fall 1977), pp. 51–63.

9. See John E. Tyworth, "Pricing private freight car service: An analysis of transportation use and the problem of standby car capacity," *ICC Practitioners' Journal* 46 (September–October, 1979), pp. 789–800; "On the transportation use controversy: An analysis of car service and boxcar utilization in the movement of food products," *Transportation Journal* 19 no. 4 (Summer 1980), pp. 1–14.

chapter 15

International Traffic Management

chapter objectives

After reading this chapter, you will understand:

The forms of international carriage available to the firm.

The need for the facilitating firms in the movement of goods.

The methods of pricing carrier services and the terms of sale and the methods of payments for international transactions.

The role of key documents and marine insurance.

chapter outline

Introduction

Export-Import
Transportation Services
Canada-Mexico movements
Overseas movements
Intermodal movements
NVOCC
Tramps

Export-Import Facilitators

Pricing and Cost Factors
Carrier rates
Terminal charges
Insurance customs duties

Method of Payment

Documentation

Marine Insurance

Export-Import Traffic Issues

Summary

Questions

Introduction

Export-import traffic management involves the same basic concepts of product flow and control activities as domestic processes but with many refinements for special required elements. Exporters and importers evaluate mode, carrier, rate, and service features, as do domestic shippers. But the port, forwarder, customs broker, means of financing title terms, and type of cargo insurance are additional factors of export-import traffic management. The subject is increasingly important today as more companies either elect to import goods from overseas suppliers, establish their own plants and import the goods, or engage in global sourcing that includes coordinating overseas sources and manufacturing services for importing into the United States or for shipping to third nations for distribution and sale. Further, international movements require additional documentation and government-related activities. This chapter presents the available forms of carriage, processes, and factors important to traffic managers who have international distribution responsibilities. Many of the cost components to consider in an export-import move are outlined in **Exhibit 15.1.**

■■■■ *Exhibit 15.1* **Cost components of an export-import move**

Component	Factor to Consider
Land transportation	Mode, carrier, and routing from plant to port
Port	Differences among ports arising from land routing, ocean carrier services, wharfage, handling, and other terminal costs; cost of inventory-in-transit to port, including port delays
Packing/packaging	Higher costs for ocean than for air transportation
Terminal	Cost of terminal storage and handling at transloading areas
Method of shipping	Breakbulk versus container
Documentation	Number of documents and processing costs
Transit insurance	Type and scope of coverage; purchase location (export or import nation) can affect coverage.
Overseas carrier	Freight charges often depend on ocean-conference agreements or other contractual arrangements with carriers.
Import duties	Duties depend on designed use of product and product configuration.
Terms of sale	Assign responsibilities for method of payment, insurance, claims, and freight charges.
Method of financing	Costs and risks of alternative methods
Delivery	Means of delivery from port or terminal to final destination

■■■■ *Export-Import Transportation Services*

From the perspective of the U.S.-based firm, international goods movements fall into two distinct groups: Canada–Mexico movements and overseas movements.

Canada–Mexico Movements

Shipments to and from Canada and Mexico can be moved by motor, rail, air, and water. Since ample capacity for overland movement exists by railway and highway, through shipments by these modes is commonplace. In addition, a good deal of water transportation takes place for coal and other bulk commodities on the Great Lakes.

Movements to and from Canada can often be made by through shipments via single motor carriers or by through railway hauls. Shipment clearance, documentation, and terms-of-sale processes are relatively simple. Mexican movements mostly require another domestic carrier in that country, as well as a broker or forwarder facilitator for the actual cross-border movement. Through railcar shipments can be made to and from Mexico or Canada, unless car supply problems force carriers to embargo cross-border moves. Many motor shipments require change of equipment at the border, since few U.S. carriers have interline or through route arrangements with their foreign counterparts.

Overseas Movements

Shipments to and from nations that lie beyond North America require, of course, either water or air transportation. **Exhibit 15.2** indicates the types of transport available for overseas linehaul movements.

Air. Like domestic traffic, overseas air traffic often consists of relatively small shipments that have high value and are time sensitive. Many of these shipments move via passenger aircraft as ''belly'' freight in loose or containerized form. Regular airlines, such as Pan American and British Air, allow this type of freight. Other airlines, such as Flying Tiger, offer all-freight service for cargo that can be loose, packaged, containerized, or palletized. Aircraft also can be chartered for an entire plane-load movement.

Air transportation provides the benefit of speed and the ability to serve inland points rather than simply ports that ships serve. Service can be booked directly with a carrier or through a freight forwarder that uses airline service, and charters may be arranged directly or through intermediaries like forwarders or brokers. All other factors, like sales terms, documentation, financing, and clearances, are similar to those for seaborne trade.

Water. Liner-ship firms offer scheduled services via fixed routes for moves that do not require an entire ship. These shipments may range from a few pounds to several thousand tons. Shipments using liners can be made in *breakbulk* form, that is, packaged, crated, and possibly palletized. These shipments are loaded by crane into a ship and stored on one of the ship's decks along with other cargo. Breakbulk shipments must often be crated to conform to carrier packaging specifications.

Some liner ships also carry *containerized cargoes* that must be put together at the plant, port, or packaging house and then lifted aboard the ship by a crane. Containerized freight avoids much handling en route and accounts for some of the crating expense of breakbulk moves.

Some liner firms offer roll-on-roll-off (RORO) services that allow large shipments, like construction equipment, to be driven directly on board without the use of a crane. RORO ships can also handle containers and crated breakbulk

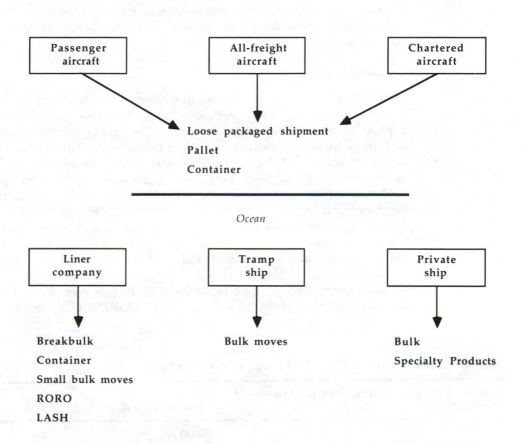

████████ *Exhibit 15.2* **Overseas linehaul services**

goods. These ships are particularly useful for movements to ports where few loading facilities exist.

Another type of specialty liner is called a LASH, for "lighter-aboard-ship." Freight is stowed in lighters or barges on board the mother ship. Near ports, the mother ship transfers the barges by elevator or crane to (or from) the water where they may be towed to (or from) distant inland-water locations for unloading and loading. For example, the LASH ships that serve routes between New Orleans and Mediterranean seaports move lighters from as far inland as Pittsburgh, Pennsylvania, or St. Paul, Minnesota, to Italy. The goods are loaded onto a barge at the inland river port, towed to New Orleans, loaded onto a

LASH ship intact, and then eventually off-loaded at a mooring post in a foreign harbor. A key advantage of this system is that the ship is not affected by delays when no berth is available at a landside site; this improves ship turnaround and shipment transit time. A major drawback is the high capital cost of the barges, one of which is generally needed at each end of the linehaul for every one aboard the mother ship.

Intermodal Movements

In recent years, overseas container movements have been made faster by combining overland (rail) and water movements of containers. *Land-bridge* is a term applied to these water-land-water moves. One application involves the Japan-to–Europe itinerary. The all-water route through the Panama Canal can take a month or more, but the land-bridge alternative— Japan–United States water movement, a rail haul to the East Coast, and a water move to Europe— reduces total transit time. Containerization allows the shipment to avoid much of the handling and delay of breakbulk moves. In addition to the Asia–North America–Europe connection, a land-bridge route is also being developed across Mexico, between the Pacific and Atlantic Oceans, to avoid the delays and tolls of the Panama Canal and, thereby, reduce the transit time from Japan to the East Coast of the United States.

Mini-bridge plays an important role for traffic managers in the United States; some popular mini-bridge routes are shown in **Exhibit 15.3**. Mini-bridges reduce

---

Exhibit 15.3 **Examples of mini-bridge**

Shipment Origin/ Destination	Previous All-Water Route	Mini-Bridge Route	
		Ship	**Rail**
Japan–Korea/New York	Kobe, Puson/New York	Kobe, Puson/Oakland	Oakland/New York
Japan–Korea/ Baltimore– Philadelphia	Kobe, Puson/Baltimore– Philadelphia	Kobe, Puson/Long Beach	Long Beach/Baltimore– Philadelphia
Europe/California	Europe/Long Beach– Oakland	Europe/New York	New York/Long Beach–Oakland
Europe/New Orleans	Europe/New Orleans	Europe/Charleston, South Carolina	Charleston, South Carolina/New Orleans
East Coast South America/Los Angeles	East Coast South America/ Long Beach	East Coast South America/New Orleans	New Orleans/Los Angeles

transit time in comparison to all-water routes and can open up opportunities for service from liner firms other than those serving the nearby port. It also avoids an organized labor requirement that goods be loaded or unloaded from containers by longshoremen if the origin or destination is within fifty miles of the port of ship loading or unloading. This factor can cause delays in shipping and increase costs. As shown in Exhibit 15.3, Baltimore shippers would have to hire longshoremen to stuff a container shipment to Korea, but they could use less experienced plant personnel to stuff a mini-bridge-container move by rail from Baltimore to Oakland, California, and then by ship to Korea. The fifty-mile rule is avoided because the Oakland pier is more than fifty miles from the Baltimore plant.

Mini-bridge reduces transit time over all-water routes, and it allows the water carrier to haul more total freight in a given year because of the reduced water distances. A shipper in Los Angeles, for example, might have only one water carrier available to Brazil. Mini-bridge opens the opportunity to use another carrier through New Orleans.

Micro-bridge, a fairly recent innovation, allows a single carrier to offer through container moves to and from inland cities. Shippers need only to arrange service from one carrier, rather than deal with a railway or motor carrier and the container line. Examples of micro-bridge services are through moves from Chicago to Europe through Montreal, St. Louis to Europe through New York, and St. Louis to South America through New Orleans. Micro-bridge services are often offered at less through cost than separately arranged services, and they have the added advantage of easier shipment tracing with one carrier.

NVOCC

Nonvessel operating common carriers, or NVOCCs, perform roles similar to those of domestic freight forwarders. The NVOCC holds itself out to the general public to move the small less-than-container load lots that are relatively expensive to ship. The NVOCC consolidates many such shipments from many shippers into one container, obtains a full container rate, and passes some of the savings along to the shipper in the form of a lower rate.

The NVOCC generally uses scheduled liner services. Tramps and private ships are used for larger loads, and their scheduling is different from liners' scheduling.

Tramps

Tramps are ships that are generally available for entire shipload use. These ships are owned and operated by firms that seek loads wherever they may be found. Tramp ships are chartered through ship brokers that arrange the agreement between a shipper and a ship operator. The document binding the parties is called a ''charter party.''

Tramp charters may be arranged through several means. A *voyage* charter governs movement from one port to another. The ship is hired for a fixed fee or for a daily rental complete with crew, stores, and so on. Another form of arrangement for tramp ships is a *time* charter, in which the shipper may use the ship for several moves. Still another type of charter is a *bareboat* charter, in which the person hiring the ship must provide crew, provisions, fuel, and so on, usually for a long-time assignment. A final way in which tramp service might be obtained is through a *contract of affreightment* whereby more than one shipper shares the use of a tramp for a voyage charter or for the fronthaul or backhaul of a time charter. A 10,000-deadweight-ton vessel might be shared by two 5,000-ton shippers. Or, one shipper might use the vessel between the United States and Korea with another between Korea and the United States.

A tramp ship seeks the most favorable business wherever it may be found. The tramp-ship trade typically consists of the following commodities and destinations:

Grain	Great Lakes to the Mediterranean
Empty	To Africa
Ore	Africa to Baltimore, Maryland
Chemicals	Baltimore to Venezuela
Empty	To Caribbean island
Fertilizer	Caribbean island to Jacksonville, Florida
Empty	To Norfolk, Virginia
Coal	Norfolk to Japan

Markets for tramp ships are fluid and subject to the laws of supply and demand. Rates might be high along one coast during a month in which few ships are available, yet depressed a month later when many ships terminate their sailings with no outbound shipments forthcoming.

Private carriage by ship is undertaken for cost and service reasons, like that found in domestic land carriage. Firms moving specialty goods, such as automobiles, chemicals, and ores, often use ships designed to operate efficiently in these trades. Shipper firms do not often directly operate tramps as a direct part of domestic manufacturing operations. Instead, tramps are often time or bareboat chartered from other firms or are owned and operated by subsidiary firms. This practice insulates the parent firm from the complicated legal, labor, and liability problems of overseas ship operations.

Export-Import Facilitators

Many firms are used in the export-import process in addition to land, ocean, or air carriers. **Exhibit 15.4** highlights the roles of these firms.

The flow processes in an overseas movement is shown in **Exhibit 15.5.** Be-

Exhibit 15.4 Facilitating firms in export-import shipments

Firm	Who Uses	Activities
Foreign freight forwarder	Importer and exporter	Arranges ocean booking, land transportation, terminal storage, crating, documentation. Provides knowledge of foreign nation requirements and translation services.
Nonvessel operating common carrier (NVOCC)	Exporter and some importers	Consolidates small shipments for lower cost pooled container load lots
Customs authorities	Importers	Government operation in each nation that assesses and collects duties on imported goods
Customhouse broker	Importer	Actually handles shipment clearance and duty payments for importers
Ship broker	Importer or exporter	A firm that arranges the charter of a ship between a tramp-ship owner and a shipper of goods
Ship agent	Either tramp-ship owner or importer or exporter	A firm local to a port that arranges many of the details required while a ship is in port (e.g., a berth, clearances, documentation)
Port authority	All	The public organization that usually builds and supplies piers and terminals for rent or ownership by ship lines and storage firms. It also sets uniform rules and fees for ships, freight, and labor within the port district.
Stevedoring firms	Steamship firm	The labor used to move freight between the ship and the terminal
Terminal/wharf/pier	Steamship firm and importer or exporter	The place where goods are stored between the inland movement and the ocean move
Insurance firm	Importer or exporter	Buyer or seller has protection against "general average" loss and perils of sea-movement-related loss and damage
Consulates	Exporter	Foreign representative in various cities of the United States through whom many document clearances must be obtained before a shipment can be sent to particular countries
Export trading company	Exporter	A firm that helps market, finance, and ship goods for firms seeking to export
Foreign-trade zone	Importer	A site designated by the customs department to which imported goods may be brought without import duties being applied; storage and manufacturing take place at these sites; duties are not assessed unless taken off property; can reexport without duty payment.

Exhibit 15.5 **Export-import flowchart**

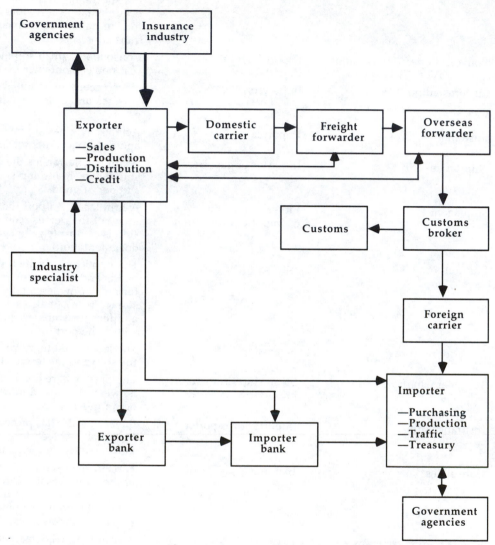

Source: Adapted from ''The Documentation Dilemma,'' *Inbound Traffic Guide*, (July, 1983), p. 50.
Reprinted by permission from Thomas Publishing Co., One Penn Plaza, New York, NY 10019.

sides this general pattern of movement, the crating firms, the terminals and ports, and the letters of credit handled by the banks all affect the export-import operation. Crating firms prepare goods for movement according to national and ocean carrier requirements. Terminals are important because they store goods before ship arrival and loading, which generally involves a storage and handling fee. Port selection can also be an important consideration in export-import movements, since it is affected by both ocean- and land-carrier rates, as well as by the frequency of ship calls and terminal storage costs.

Pricing and Cost Factors

The key pricing and cost factors that concern traffic managers are carrier rates, terminal charges, customs duties, and insurance.

Carrier Rates

Air cargo rates are generally assessed by weight, by per container load, or by a combination of both. Some rates are FAK rates, or freight all kinds, which offer a single rate per unit weight regardless of commodity; other rates may be based on the commodity and the weight. Some container rates are flat charges of a single amount, regardless of the weight or commodity. Because carriers need to recoup a certain cost per cubic dimension of the container, they will charge a rate at or near that amount.

Air cargo rates are subject to fluctuation in markets with over-capacity and competition among the lines. Passenger airlines can carry cargo in plane bellies and containers as a joint cost by-product to passenger movement; when excess capacity exists, carriers often reduce freight prices to attract revenues and minimize plane operation costs. But these attractive rates can disappear with sudden increases in passenger traffic.

Some air cargo markets are controlled by bilateral agreements between nations. Often, limited air service means tight lift capacity and relatively high rates.

In developing rates, liner firms consider potential traffic volumes, the nature of commodities, competition, regularity of traffic flow, density, susceptibility to loss and damage, loading and unloading characteristics, and stowability. Liner breakbulk rates are generally published according to weight or measure. For example, a rate of $50 W/M means the shipment will be charged $50 per weight ton or measure ton, whichever produces the greater revenue for the carrier. The possible measures used by carriers are as follows:

Weight

Short ton	2000 lb
Long ton	2240 lb
Metric ton	2204.62 lb

Measure

U.S. cubic ton	40 cu ft
Metric cubic ton	35.314 cu ft

Assume the $50 W/M rate applied to a shipment that weighs 10,000 lb and measures 4 ft × 6 ft × 10 ft, and that the tariff states short tons and U.S. cubic tons apply. Therefore, there are 5 weight tons (10,000 lb/2000 lb) and 6 measurement tons (240 cf/40 cf), and the shipment charges are $300 (6 tons at $50 per ton). If the tariff entry had read $50W, the charges would be based on the 5 weight tons.

Container charges are assessed on the basis of weight of the loading or by a flat fee per container. On the flat basis, there is an incentive to fully load the container to reduce the unit charges. **Exhibit 15.6** presents other liner charges with brief descriptions.

Liner rates are published in tariffs of either an individual carrier or a rate

■■■ *Exhibit 15.6* **Liner charges**

Rate Charges	Purpose
Fuel surcharge	To cover fluctuations in fuel costs when basic tariff charges cannot be charged as quickly
Currency surcharge	To account for frequent changes in currency levels
Congestion surcharge	An added fee over the regular rate to ports in an area to cover the extra time required in berthing, loading, or unloading due to congestion
Out-port differential	An added fee to pick up or drop off cargoes to a port not normally served by a liner firm
Time-volume	A lower rate is offered in exchange for a guaranteed high volume of traffic over a specified period
Transshipment charges	A fee assessed by a liner firm for dropping off cargoes at an intermediate port and arranging for subsequent movement by another liner firm
Minimum charge	A minimum charge assessed regardless of the rate and weight per cubic total charges—for small shipments
Oversize/weight charges	Added charges for large shipments
Refrigeration charges	Added fee for this protective service
Dangerous cargo	Added fee for special handling needed on these cargoes

conference. A conference is a group of carriers serving the same ports and general trade routes, and they collectively set rates. There are approximately 360 conferences in the world. Though they have provided stability in the rate structures, they have often been criticized for being inflexible and restrictive with certain rate innovations and in meeting some specific customer needs.

Rate conferences often provide both contract and noncontract rates. A noncontract rate is applicable to any shipper, whereas a contract rate charges up to 15 percent less than the noncontract rate and applies to moves by shippers that have signed exclusive patronage agreements to use only conference carriers. Contract rates are loyalty inducements that apply only to nonbulk movements. Rates of all liners and conferences must be filed at the Federal Maritime Commission (FMC) in Washington, D.C., and are subject to provisions of the Shipping Act of 1916.

The Shipping Act of 1984 permitted the use of time-volume rates, which are similar to domestic railway or motor contract rates. Time-volume rates, which are lower than normal rates, are provided in return for a shipper moving a guaranteed amount of freight via the contracting carrier or conference over a period. This form of pricing has been popular in the container, LASH, RORO, and breakbulk trades since mid-1984.

Tramp ship charges are determined through the market forces of supply and demand, the location of available ships in relation to shipments, crew costs, taxes, port fees, and fuel costs. The world market for tramp ships can increase because of ship shortages in relation to high demands or it can be depressed from ship oversupply, low oil demand in the Mideast, or the lack of large grain movements. In depressed or rising tramp demand markets, shippers can lock in low rates through long-term "time charters." In a decreasing market, however, it is wise to use voyage charters and wait for more favorable cost conditions.

Terminal Charges

Export cargoes are consigned to ocean terminals for storage until they are loaded onto the ocean carrier. Terminals also must be able to store goods inbound before load movement. Therefore, goods are handled out of trucks and railcars to the terminal storage site, or vice versa. Tramp-ship movements require the arranging of a port terminal to transfer the goods from the domestic carrier to the ship. Traffic managers often select the carrier and port based on terminal operations available at a port. Some packaging, palletizing, and container loading is preferred as well. Other factors include security, efficiency, and charges for storage. Some free-time days are allowed, but most terminals charge for longer storage. The New Orleans, Houston, Philadelphia areas have "tank farms" for chemicals and elevators for grain that serve this purpose.

Customs Duties

Import shipments are subject to customs clearance and duty assessment, arranged through a customhouse broker. The exact amount of the duty is determined according to the commodity and country of origin. Duty is assessed according to the value of each item or on a per-unit basis, or a combination of both. Duties and customs processes are found in Part 19 of the Code of Federal Regulations.

Drawback and foreign-trade zones offer key advantages for importers. The long-standing importing process included paying for duties on arrival at the port. An importer could also move them away from the terminal, where storage charges were high, and have them held in a bonded warehouse, a facility where goods could be stored without having to pay duties on them. This tactic enabled the importer to "clear" the shipment at a later date or reexport the goods to another country. With *drawback*, goods may be imported, stored, manufactured, and exported, and nearly all the original customs duty may be retrieved. This national policy feature serves to reduce the cost of goods that can be reexported. Goods that will be used for drawback must be registered as such; records of their disposition must be kept, and a filing must be made to claim duty drawback after export.

Foreign-trade zones also offer important opportunities. These zones are sites, designated as such by the U.S. Treasury Department, where factories and warehouses may be built. Goods not subject to duty until removed for domestic use may be imported and moved to the foreign-trade zone without duty payment. Therefore, goods may be reexported without any original duty payment. Further, goods may be stored here, used in manufacturing, and sold with duties payable rather than earlier upon import. This advantage can greatly reduce the cost of holding inventories, since no duty payment is made until close to when customer payment would be received.

Insurance

Insurance is another cost factor for exporters and importers of goods. The Carriage of Goods at Sea Act provides few protections for shippers of goods, unlike the traditional carrier obligations found for domestic common carriage. Further, goods are handled by many parties in an international move, so it would be difficult or impossible to attach responsibility for damage to any one of them. For example, there is the land carrier moving the goods to the port, the warehouse or terminal at the port while awaiting ocean movement, the actual pier or terminal involved in the loading, the ocean carrier, and a similar set of like parties involved in the off-loading and ultimate delivery to the consignee. Shippers and receivers take responsibility for seeking out and purchasing insurance for shipments so that compensation may be obtained for losses and damage regardless of which party might have been responsible for it.

Method of Payment

Two major aspects of the exporting and importing processes are the terms of sale and the method of payment. The terms of sale define at what point payment shall be made, and where the title to the goods transfers from one owner to another. The method of payment addresses how the seller will be paid. In domestic settings, where credit can be established with open accounts, firms can seek legal recourse against a buyer that does not pay. In international settings, assurances of proper goods transfer and payment are generally handled by banks through letters of credit.

Terms of sale are gradually evolving worldwide into a set of uniform options. The International Chamber of Commerce has established internationally accepted terms, referred to as INCOTERMS. **Exhibit 15.7** shows the responsibilities of both buyer and seller for the activities from export plant to import facility. This table illustrates that the point at which the actual bill transfers can be at one end at the seller's factory (ex works), at the inland carrier pickup (F.O.T.), at the port of export (F.A.S., F.O.B. versus export port, or ex ship), at the destination port (ex quay, C&F), at an intermediate international boundary other than the final import country (delivered frontier), at the final import country boundary (delivered duty paid), or at the customer site (freight paid to destination, or freight and insurance paid to destination).

The terms-of-sale or point-of-title transfer determines whether the buyer or seller is responsible for each detail of the move. Generally speaking, these terms most often apply at the ports or airports. Most of these details are handled by forwarders and customhouse brokers hired by the exporter and importer.

Letters of credit ensure payment for the goods and that the buyer is to receive the goods so ordered. Once the buyer and seller agree to a transaction of specific goods, say an export transaction, the buyer obtains a letter of credit from its bank, backed up by a deposit, an account, or a loan. This document is a statement from a foreign bank that it will transfer funds on completion of certain tasks, such as packaging, moving to port, obtaining documents, or anything else required by the seller under the specified term of sale. Once a dock receipt, bill of lading, loading, or other requirement takes place, the letter of credit is satisfied, and the foreign bank will transfer funds to the domestic seller's bank. In this way, both buyer and seller use the trust of banks and other intermediary party actions to affect title transfer.

Letters of credit (L/C) work in several ways. A revocable L/C is one that can be revoked at any time up to title transfer. An irrevocable L/C cannot be revoked

Exhibit 15.7 **INCOTERMS responsibilities and risks**

Activity	Ex Works		FOT		Free Carrier		FAS		FOB U.S. Port		Ex Ship		Ex Quay		C&F		CIF		Del. Frontier		Del. Duty Paid		Frt. Paid to Dest.		Frt. and Ins. Paid to Dest.	
	S	B	S	B	S	B	S	B	S	B	S	B	S	B	S	B	S	B	S	B	S	B	S	B	S	B
Export license	X		X		X		X		X		X		X		X		X		X		X		X		X	
Import license		X		X		X		X		X		X		X		X		X		X	X			X		X
Supply product	X		X		X		X		X		X		X		X		X		X		X		X		X	
Inspect shipment	X		X		X		X		X		X		X		X		X		X		X		X		X	
Certificate of origin	X		X		X		X		X		X		X		X		X		X		X		X	3	X	3
Export pack	X		X		X		X		X		X		X		X		X		X		X		X		X	
Load on vehicle		X	X		X		X		X		X		X		X		X		X		X		X		X	
Move to port		X		X		X	X		X		X		X		X		X		X		X		X		X	
Ocean documents		X		X		X		X		X	X		X		X	2	X	2	X		X		X	3	X	3
Select vessel		X		X		X		X		X	X		X		X	2	X	2	X		X		X		X	
Load ship		X		X		X		X	X		X		X		X		X		X		X		X		X	
U.S. pier costs		X		X		X		X	X		X		X		X		X		X		X		X		X	
Ocean transport		X		X		X		X		X	X		X		X		X		X		X		X		X	
Ship to quay		X		X		X		X		X		X	X		X		X		X		X		X		X	
Foreign customs		X		X		X		X		X		X	X			X		X		X	X			X		X

Foreign transport

Marine insurance

War risk insurance

Unload foreign vehicle

Risks

On U.S. vehicle

To U.S. pier

On board vessel

Ocean transport

Ship to foreign quay

On foreign vehicle

Foreign transport

Foreign vehicle to buyer

S = Seller

B = Buyer

1 = Buyer must pay costs for B/L

2 = Seller must furnish buyer with clean B/L at his or her own expense

3 = Buyer must pay all costs for documents including certificate of origin

4 = To frontier

5 = From frontier to destination

Source: Joseph V. Barks, "Exporting Without Tears," *Distribution* (October 1980), p. 80. Reprinted with permission from *Distribution Magazine*, Chilton Co., Radnor, Pa.

unless a maximum time has passed. A confirmed L/C provides the added assurance from a domestic bank that the seller will receive payment from the foreign bank. Therefore, the safest form of letter of credit for an exporter is an irrevocable, confirmed letter of credit.

A sight draft is a document like a check that can be presented for goods payment. It is a flexible document that can be used for dollar amounts that are unknown until all activities have taken place and costs have been determined.

Documentation

International movements require more intermediaries and steps than domestic movements. **Exhibit 15.8** lists the majority of the documents, highlighting their general requirements for exporting and importing. Each nation uses some of these forms and steps, all of which require many documents, much process handling, and constant cross-checking for accuracy. The cost of documents and paper processing is high, and this sometimes discourages overseas buying and selling.

Exhibit 15.8 Export-import documentation

Document	Purpose/Use
Booking request	Official request from shipper or forwarder to steamship line to reserve space on forthcoming voyage; request for empty container to load outbound move
Booking acceptance or confirmation	Notice of reserved space on specific vessel at date and port as shown; from carrier to shipper or forwarder
Cargo manifest	Official ship listing of all cargoes on each voyage
Certificate of origin	Documents required by United States and other nations attesting to exact contents and national origin of ingredients, parts, or all goods being imported; used in control of prohibited goods or to prevent goods from certain nations (e.g., North Korea, North Vietnam, Cuba)
Charter party	The document used for the agreement between a shipper and tramp-ship owner or operator; the lease document
Commercial invoice	An invoice of the seller stating the value or price of goods; includes terms of sale, exporter, importer, packages or pallets, price, etc.
Consular invoice	A document obtained from the consulate of an importing nation located in an export nation; one of the documents needed to clear goods into a country; gives carrier assurance that goods shipped can be off-loaded in importing nation
Consumption entry	Prepared by importing customhouse broker for shipments requiring customs clearance into the country

(continued)

Exhibit 15.8 (continued)

Document	Purpose/Use
Customs invoice	Invoice document containing assessment information on shipments requiring duty charges
Dock receipt	Receipt given by ship terminal for goods delivered to it for holding until loading onto ship
Domestic bill of lading	Motor, barge, or rail bill of lading for domestic move; title transfer with foreign trade definition terms
Domestic shipping instructions	Instructions for land moves given to freight forwarder or import agent to arrange this lap of the trip
Exclusive patronage agreement	Document signed by shipper thereby resulting in lower conference rates being applied
Export declaration	Form filed with agency of export country describing goods; used to collect trade statistics
Export permit	A permit necessary to export certain goods; a permission from the government to export
Foreign freight forwarder invoice	A bill for forwarder services
Freight invoice	Carrier bill for freight services
Hazardous material manifest	A distinct listing of hazardous cargoes on a ship
Import broker instructions	Instructions to import broker from importer about delivery of shipment
Inspection certificate	Document of an independent inspector attesting to exact description, count, condition, and sometimes quality of a shipment; seller must arrange for this to be provided
Insurance request	A request for insurance covering goods while en route
Letter of credit	Document obtained by importer from bank used to affect goods payment and title transfer
Letter of instruction	From shipper to freight forwarder about how shipment is to be made
Ocean bill of lading	Bill of lading of ocean carrier; if negotiable, carrier will deliver at destination only to party acquiring receipted copy of some bill of lading from shipper
Packing list	Document describing goods, packages, weight, etc., that accompanies the shipment
Power of attorney	Official document indicating that import or customhouse broker is to act in behalf of importer to clear goods upon importation
Pro forma invoice	Generally required by importers to prepare preliminary import documents; used in some nations to convert local currency into currency needed to pay for goods
Purchase order	Document from buyer to seller indicating offer to buy
Surety bond	Bond obtained by a customhouse or import broker for an amount equal to duties for an import movement

Source: Adapted from Jessie C. Jessen, "Anatomy of an Import," *Distribution* (October 1982) p. 74. Reprinted by permission from *Distribution Magazine*, Chilton Co., Radnor, Pa.

A systems called CARDIS, Cargo Data Interchange System, has been developed to smooth the flow of data and documents through electronic means. Steps are being taken to implement and expand CARDIS on a broader scale, which includes

■ Assisting in document preparation
■ Distributing documents electronically
■ Using trade shipments
■ Speeding up letter of credit collection and payment
■ Producing copies of important codocuments like bills of lading, packaging lists, and so on

Marine Insurance

International carriage obligations by air and ocean carriers are substantially different from those liability and claims processes found in domestic rail and truck movements (see Chapters 12 and 13). The liability regimes are slanted toward the carrier. Many acts of damage or loss that would be recoverable from domestic railway and motor carriers are not recoverable from water transportation companies. Thus, insurance protection for every movement is essential.

Export-import traffic planning must take notice of the following terms that apply to marine insurance:

1. General average. This term stems from the French word, *avarie*, which means damage to cargo and ship. If a ship must jettison cargo or parts of the ship to save the ship, other cargoes, or life or property of others at sea, maritime law requires the remaining cargo owners and ship owners must make up the loss to the person(s) whose goods were sacrificed. Thus, a $100,000 loss in this manner is made up proportionately by the other cargo owners and the ship owners for relative values invested in the voyage. This obligation of the goods owner can arise through no action or fault of this person. General average insurance protects the shipper from being assessed for such losses.

2. Particular average. This type of insurance coverage extends to loss or damage that does involve losses by others. In translation, it means particular loss or damage.

3. Free of particular average. This coverage extends to losses resulting from perils of sea, including collision, burning, sinking, and stranding.

4. With average. Coverage of this sort is broader than the above kinds and extends to shipment loss or damage above 3 percent of its value. Additional coverages can be obtained throughout the "with average" clauses that include theft, nondelivery, and some water damage.

5. All risks. This coverage is quite inclusive in that it covers any external cause except war, strikes, riots, seizure, and detention.

The exporter or importer usually seeks the most extensive coverage possible. The litigation of determining who is responsible for loss and damage (ship company, terminal, domestic carrier), and the limitations on carrier liability that result from specific provisions in the ocean bill of lading, make marine insurance a wise choice.

Marine insurance policies can be written to extend from origin to destination. An open or floating policy covers all shipments in transit and remains in force until canceled, even though each shipment might be limited to a maximum coverage. There is no need to acquire insurance with each individual shipment.[1]

Export-Import Traffic Issues

Several export-import issues of major importance in the 1980s bear specific study. These are national maritime policies, U.S. cargo preference laws, and the growth of export trading companies.

The Shipping Act of 1984 changed maritime policies affecting economic regulation of mostly liner traffic to and from the United States. Its major provision of interest to shippers is the freedom to enter into time-volume rates. Additional changes made by the law include easier independent action by carriers within conferences. In the past, a carrier wishing to depart from a conference rate had to completely withdraw from the conference. The new act allows partial withdrawal, or independent action, in areas where negotiations with a specific shipper or receiver results in a rate or service different from that offered by the conference. This change opened the way for ocean carriers to offer through intermodal services for both land and water moves. The act also permitted shippers to form shippers' associations for seeking favorable rates and services.[2]

An important maritime policy of the United Nations Conference on Trade and Development (UNCTAD), which came into effect in 1982, relates to the flag of ships that can be used to transport goods between nations. This policy states that trade between nations A and B is to move 40 percent on ships under the flag of nation A, 40 percent on ships of nation B, and only 20 percent on any other national flag ship. The purpose is to foster development of shipping in each nation and to protect traffic of home ship investments and labor. The effect of this will be to create spoke-like shipping patterns from each nation.

The net effect of the UNCTAD agreement threatens to decrease shipping efficiency in the world and possibly to create delays in shipping. Assume, for example, that one U.S. steamship firm operates round-the-world service calling on ports on the east and west coasts of Africa and the United States. The ship line might have monthly service between one particular port and the United States. Inadequate shipping investment by the African nation might cause delays in bookings on that flag carrier after the U.S. ship used the 40 percent estimated U.S. flag tonnage allotment earlier in a year. Similarly, the U.S. ship might not be filled with either that nation's freight or U.S.-bound freight. A sailing of this sort involves loads between many Asian and African nations en route. By being prohibited from picking up much of this "third country traffic," the ship line operates at less than full capacity. In delaying traffic, the UNCTAD agreement might lead to waivers from government agencies to ship via alternative carriers.

The U.S. cargo preference law is another issue of concern for exporting and importing traffic managers involved in U.S. or state government cargoes. This law requires that 50 percent of government-owned-or-sponsored freight moves via U.S. flag transportation firms. A traffic manager of a firm that obtains a government aid construction contract in a foreign nation must check into and obtain U.S. flag shipping capacity from the United States to that nation. The American carrier is often more costly than another carrier, and this difference affects the traffic management transportation budget.

Export trading companies are an area of opportunity for U.S. firms. These organizations were allowed to come into existence through changes in government policy to finance goods being exported; they can arrange the movement and handle the marketing of the goods in a foreign nation. Such companies are convenient for firms that do not have experience in exporting. The act of exporting can be a barrier of time, cost, and effort for domestic firms, so expertise can be used to avoid much of the learning process involved in exporting. Some export trading companies perform many of the traditional shipping, forwarding, and documentation functions for an exporter.

Export-import traffic management involves the same basic activities as domestic traffic. There are, however, additional functions and parties that are a part of the entire movement in international settings. The specific nature of an international movement and some of these necessary additional steps can act to prevent a firm from exporting or importing. These barriers are slowly dropping as documentation, containerization, export trading companies, through services, insurance policies, and so forth, become easier to use and are made available to the shipping public in simple terms.

The ocean transportation industry has been experiencing severe overcapacity and imbalances-of-trade problems in the past few years. The strong U.S. dollar caused heavy importation flows in the United States with little outbound movement. Similarly, overbuilding of ships in the 1970s has caused an excess of ships in the world. These events have caused downward pressures on rates and operating losses for many lines. Though this situation brings benefits to the shipping public in the short term, the instability of rates and services and the viability of the carriers require careful transportation planning by traffic managers.

Summary

International movements involve the same basic principles as domestic shipments. The difference lies in the number of steps and parties often necessary in overseas shipments. The nature and size of the commodities being moved dictate a particular type of mode or carrier. This is particularly the case with containerized goods versus large bulk movements of grain or coal.

There are many firms that assist shippers and receivers with the key details of overseas movements. These facilitators act to secure transportation, move goods across international boundaries, arrange payment, and acquire insurance. These processes are found to lesser degrees or in different forms in domestic trades. Firms that are minimally involved in international movements typically seek out and hire such facilitators to perform the necessary steps. An increasing number of U.S. firms have such volumes of overseas movements that they are staffing their own operations to perform these tasks in-house.

International traffic management was traditionally a small, special part of the field. The scale of international movements today, the globalization of the firms involved in world trade, and gradual simplification of overseas movement processes will all lead to this sector being of major importance in many firms.

Selected Topical Questions from Past AST&L Examinations

1. The growing volume of international trade is creating a greater need for traffic managers to be knowledgeable in the mechanics of export and import business. One major function of the traffic manager is to supervise documentation, a very important part of the export shipment. Define the following:
   (a) Export declaration
   (b) Commercial invoice
   (c) Consular invoice
   (d) Certificate of origin
   (e) Export license (Spring 1980)
2. Why are international freight forwarders needed? What functions do they perform? (Fall 1980)
3. In international transportation, LASH is often mentioned.
   (a) What is meant by it?
   (b) Where is it most appropriate?
   (c) What are the relative pros and cons of such a form of transportation? (Fall 1980)

4. You are the export traffic manager for the ABC Manufacturing Company and an exporter of iron and steel manufactured products. You are asked to participate in a seminar about export documentation with sales trainees. Name and briefly discuss the nature and purpose of those shipping documents commonly used in the U.S. export trade. (Fall 1981)

5. Explain the difference between liner and nonliner shipping. (Fall 1981)

6. International logistical activities are extremely complex. For this reason, many firms use the services of an international freight forwarder (IFF).
   (a) What functions are generally performed by an IFF for shippers?
   (b) What are the sources of income for IFFs?
   (c) Under what circumstances should a firm not use an IFF — that is, when would shippers be better off to perform the IFF functions themselves? (Fall 1981)

7. Please define three of the following terms as they pertain to ocean transportation.
   (a) Open rate
   (b) Loyalty agreement
   (c) Weight/measurement rate (Spring 1982)

8. Assume you are a traffic manager for a machinery manufacturer and that your management is considering entering a foreign market. You are asked to explain to your marketing manager the meaning and liabilities attached to the various export terms (such as FOB, FAS, CIF, and C&F) customarily used in the foreign shipping trade. What would you say? (Spring 1982)

9. There are many distinguishing features specific to the services of liner and tramp shipping. Describe each and note differences where appropriate. (Fall 1984)

10. What is a foreign-trade zone? How can firms benefit from the use of these facilities? (Fall 1984)

11. Several financial arrangements may be made in obtaining payment for sale and transportation of export goods. The most common is the letter of credit. Discuss the use and application of a letter of credit. (Spring 1985)

Notes

1. For further study, see Alfred Murr, *Export/Import Traffic Management and Forwarding*, 5th ed. (Cambridge, Maryland: Cornell Maritime Press, Inc., 1977), chaps. 13 and 14.
2. 46 U.S.C. 1701–1720 (1984).

chapter | 16

Social Regulation Affecting Traffic Management

chapter objectives

After reading this chapter, you will understand:

The need for and role of hazardous goods movement control.

The primary requirements for hazardous goods movement.

The role of hazardous waste-product movement control.

Many of the other social regulations affecting traffic management.

Introduction

Hazardous Materials Transportation
Laws and regulations
Definition of hazardous materials
Scope and magnitude
of hazardous materials movement
Hazardous goods
movement requirements
Hazardous waste products movement

Truck and Driver Highway Safety

*Other Sources of
Social Regulation*
The Clean Water Act of 1977
Occupational Safety and Health
Administration (OSHA)

Fire safety
Food and Drug Administration and
Public Health Service
State regulations

Summary

Questions

*Appendix 16.1—Catalogue of Social
Regulations Relating
to Traffic Management*

*Appendix 16.2—Hazardous
Material Definitions*

Introduction

Since the late 1960s, social regulations of freight transportation have increased. Social regulations do not pertain to the pricing, services, operations, or financial arrangements of carriers; rather, they apply to the safety, environment, or processes employed in the field. The major social regulations affecting traffic come from the hazardous materials movement regulations enacted by the Department of Transportation. These, toxic substance control regulations, and waste movement regulations are strict requirements that must be followed by shippers, carriers, and receivers in the movement of such commodities.

This chapter discusses today's key social regulations concerning hazardous goods movement, hazardous substance storage, and highway safety matters in private carriage.

Hazardous Materials Transportation

Laws and Regulations

Regulations of hazardous goods primarily originated with a law to govern interstate movement of explosives and shipment markings. The ICC set rules for carriage of such goods, based on an act of Congress, which included special

404

markings on the loads and particular handling aboard transportation vehicles. With the creation of the federal Department of Transportation in 1966, jurisdiction of hazardous goods management was transferred from the ICC to the DOT.

In 1974, the Transportation Safety Act brought major changes to the field. This law, which was passed in response to several severe transportation accidents involving dangerous goods, empowered the DOT to create regulations over the transportation of any hazardous goods. The Office of Hazardous Materials Operations, within the DOT, administers and enforces the law and empowers the Environmental Protection Agency (EPA) to develop methods for safely disposing of dangerous waste products, regulating hazardous goods, recording the creation of such materials, following their movement and disposal, and keeping records in the event of long-term storage.

Long-standing DOT regulations dealing with the safety of commercial trucks also affect operations of private and for-hire fleets that haul hazardous materials. These regulations deal with vehicle safety and inspection, as well as with driver qualifications. Both the DOT and EPA are involved with vehicles when they handle hazardous goods or waste products.

The above laws and regulations are promulgated by federal agencies. Each state, likewise, has sets of laws that regulate hazardous goods. This chapter will stress only the federal laws and regulations. Further, **Appendix 16.1** catalogues most of the social regulations affecting traffic and transportation. This appendix shows the area, agency, pertinent regulations, and title of the Code of Federal Regulations that applies to each law.

Definition of Hazardous Materials

A hazardous material is a substance in a quantity or form that presents a more than a reasonable risk to the safety or health of persons or property. It includes any material that may have detrimental effects on transportation and emergency personnel.[1]

Legally speaking, hazardous materials are those covered in Section 104 of the Transportation Safety Act of 1974. This section declares that if the Secretary of Transportation finds the transportation of a particular quantity or form of materials in commerce may pose an unreasonable risk to the health and safety of persons or their property, such material, or group or class of such materials, shall be designated a hazardous material. These materials may include, but are not limited to, explosives, radioactive materials, etiologic agents, flammable liquids or solids, combustible liquids or solids, poisons, oxidizing or corrosive materials, and compressed gases.[2]

Scope and Magnitude of Hazardous Materials Movement

Firms in the United States produce billions of pounds of dangerous chemicals and other substances each year. Examples include sulfuric acid, ammonia, ethylene, chlorine gas, nitric acid, phosphoric acid, toluene, styrene, and vinyl chloride. These substances are needed in the primary operations of many industries and in the production of other goods. The movement of these chemicals cannot be prohibited or severely restricted without hindering the economy or reducing employment.

Experts estimate that more than 100,000 shippers handle dangerous goods a year over 50,000 different carrier entities. In the area of highway transportation, an estimated one billion tons of hazardous cargoes are moved yearly. Included in this estimate are about 600,000 tons of blasting materials, 300,000 tons of poisons, and 100,000 shipments of radioactive isotopes.[3]

Because of the large amount of hazardous material being transported, accidents occur all too often. One major incident took place in Bushkill, Pennsylvania, in 1964. A truck was carrying a dangerous tank load when the trailer tires caught fire. The local fire department responded to the driver's plea for help. Unfortunately, though, upon spraying the tires with water, a leak from the tank ignited, and the explosion killed several firemen and destroyed much surrounding property. The type of danger the load posed had not been marked on the truck.

On April 10, 1975, a tire blowout on a truck tractor in Los Angeles caused both twin trailers to overturn and burn. One trailer carried tires, and the other contained twelve tons of methomyl insecticide. Since there were no hazardous markings on the units, the firemen did not wear self-contained breathing gear, and the burning chemicals released toxic vapors that sent ninety-four firemen and civilians to the hospital for treatment.[4]

Hazardous Goods Movement Requirements

The regulations that apply to hazardous goods movement stem from the Transportation Safety Act of 1974. The specific regulations pertaining to this area, which are created by the Department of Transportation, are found in Title 49 Code of Federal Regulations (CFR). These regulations basically consist of the following:

- Classification of hazardous materials
- Table of hazardous materials showing

    Proper shipping means to use
    Class of hazard for each material
    Identification number
    Label, packaging, and placard requirements

- Manifest requirements and shipping papers

Classifications of hazardous materials are found in 49 CFR 171 and 173. These twenty-five classifications consolidate all possible hazard situations into a limited number of similar groupings. Appendix 16.2 lists these hazardous materials. The general groupings include explosives, combustibles, corrosives, flammables, poisons, radioactive materials, and others. These classifications describe the specific type of hazard posed by the materials, as well as indicate the warning labels and other precautions necessary to ship them. The CFR Hazardous Materials Table is the prime source for determining whether goods are subject to regulations. **Exhibit 16.1** shows a sample page from this table (49 CFR 172.101). Column 1 indicates whether items are hazardous and whether the regulations apply to air or water movement. Column 2 lists the proper shipping name to use on the bill of lading, waybills, manifests, delivery receipts, and any other documents used in the movement. The table further provides the correct shipping names for goods that may contain mixtures of hazardous substances or determines that a material may be referred to by brand name.

The class of each item is indicated in column 3 of Exhibit 16.1. Column 3A includes the identification number to be used in international movements as required under United Nations rules. The required warning label to affix to each package is indicated in Column 4. Such markings warn carrier personnel and others how these goods should be handled in movement, as well as in emergencies. Packaging requirements are presented in Column 5, as specified in 49 CFR 173.

The outside of the vehicle or container carrying hazardous materials must be marked by a placard, the type shown in column 7c of the exhibit. This external marking serves to warn fire, accident, and other personnel of the danger associated with the shipment.

A basic shipping requirement of a hazardous good is the notation on the bill of lading and manifest. The shipper must note the hazardous material, although the carrier is not excused if it picks up an unmarked hazardous shipment. Hazardous goods are to be shown in the following manner:

- Proper shipping name as it appears in column 2 of the Hazardous Material Table. The commodity description must conform to this table. No brand names are to be used.
- The hazard class is to be shown, such as flammable liquids.
- The identification number from Column 3A.
- The number of shipping units, such as boxes, cartons, drums.
- The total weight.
- Shipper's certification. This statement is to be included on the bill of lading. This is a legal requirement that acts to prompt all required steps in such shipments. The wording is as follows:

This is to certify that the above-named materials are properly classified, described, packaged, marked, and labeled, and are in proper condition for transportation according to the applicable regulations of the Department of Transportation.

Exhibit 16.1 **Sample page from 44 CFR 172: Hazardous Materials Table**

(1) +/ E/ A/ W	(2) Hazardous materials descriptions and proper shipping names	(3) Hazard class	(3A) Identification number	(4) Label(s) required (if not excepted)	(5) Packaging		(6) Maximum net quantity in one package		(7) Water shipments		(c) Other requirements
					(a) Exceptions	(b) Specific requirements	(a) Passenger carrying aircraft or railcar	(b) Cargo aircraft only	(a) Cargo vessel	(b) Passenger vessel	
	Phosphorus heptasulfide	Flammable solid	UN1339	Flammable solid	None	173.225	Forbidden	10 pounds	1.2	1	Separate from oxidizing materials
	Phosphorus oxybromide	Corrosive	UN1939	Corrosive	None	173.271	Forbidden	1 quart	1	1	Keep dry. Glass carboys not permitted on passenger vessels

On air shipments, the following additional certification must appear:

This shipment is within the limitations prescribed for passenger aircraft cargo-only air-craft [delete whichever is nonapplicable].

- An ''X'' mark is to be placed in the HM column of the bill of lading indicating which goods in the shipment are hazardous.

Markings of all hazardous packaging is required by 49 CFR 173. Package marking notifies carrier personnel which goods are hazardous and how they should be handled in routing, as well as in emergencies. This packaging requirement allows simple cross-checking with documents and other information systems used by carriers. This marking must include the proper shipping name of the commodity, the hazardous identification number, and the name and address of the shipper or consignee. Some exceptions apply if the goods are bulk liquids, transported in a single vehicle with no other goods, or are not being transferred from one carrier to another.

Package labeling is performed by the shipper. As shown in column 4, labels indicate the specific nature and danger of each hazardous commodity as defined in Appendix 16.2. Examples are poison, corrosive, radioactive material, and so on.

The placard required by DOT must be displayed on the outside of the freight-carrying vehicle for most hazardous shipments. These placards are standardized signs that must be placed on each end and sides of the van or freight car. Vehicles without placards may not be moved, and penalties will apply if they are. If a vehicle has placards but no hazardous goods are being transported, penalties will also apply.

Traffic managers must be concerned with how selected carriers, such as private fleets and shippers' associations, handle hazardous goods. The American Trucking Association states that each terminal should have a ''key man'' familiar with hazardous materials and ways of handling emergencies related to them. Such control provides information and support to the carrier, for it is lack of knowledge that has led to many disasters that could otherwise have been minimized.

Additional suggested carrier activities in this realm include

- Meeting with shippers and consignees of hazardous goods to discuss and understand specific procedures and problems
- Training pickup, handling, linehaul, and delivery personnel
- Developing consistent and sound procedures for routing and emergency situations
- Including vehicle inspection and driver qualifications as routine activities within hazardous processes
- Failure to report leakage of a corrosive liquid on a previous airplane movement

■ Noting shippers who improperly labeled and marked a hazardous shipment

■ Noting trucker runs where hazardous goods moved in same vehicle with food

■ Movement of two hazardous materials that are not to be in the same vehicle together

■ Failure of shipper to provide shipper certification

Each hazardous incident must be reported to the Department of Transportation, Hazardous Materials Regulations Board, Washington, DC 20590. The general elements of hazardous materials regulation are shown in **Exhibit 16.2.**

The Chemical Transportation Emergency Center, or CHEMTREC, provides emergency assistance and information for transportation accidents involving hazardous goods. This agency is supported by the Manufacturing Chemists Association. The CHEMTREC contact person, called a communicator, can be reached at 800-483-9300, and will require the name of the caller, a call back number, the shipper, container type, manufacturer of the goods, location, and other information. The communicator will also provide some immediate assistance information, as well as attempt to get manufacturer's expert assistance at the scene of the accident.

Other hazardous response assistance is available through the Pesticide Safety Team Network, the Transportation Emergency Assistance Program (Canada), Oil and Hazardous Materials — Technical Assistance Data Systems of the Environmental Protection Agency, and the Chemical Hazards Response Information System (CHRIS) of the Coast Guard.

Hazardous Waste-Products Movement

Although the shipment of sold or purchased hazardous goods was subject to DOT's hazardous materials regulations, little control over the final disposition, storage, and handling of the waste goods existed until the 1970s. Waste goods often posed severe dangers to the environment. Major chemical dump sites of hazardous materials were discovered in the Love Canal area of Niagara Falls, New York, and the Valley of Drums in Kentucky. The Resource Conservation and Recovery Act of 1976 (Public Law 94-580) was designed to create distinctions between hazardous materials and substances and "waste." This law empowered the EPA to implement specific regulations governing hazardous waste shipments, storage, and disposal. These regulations are found in 49 CFR 260–265. Movements of these wastes have been included in DOT's hazardous goods movement regulations of 49 CFR 171.

In their basic form, hazardous wastes must be moved by carriers certified for this purpose, and a special hazardous waste manifest must be used for the movement. The EPA requires that the goods be moved properly, and that they be stored and disposed of according to its regulations. The manifest must in-

■■■■■ *Exhibit 16.2* **General elements of hazardous materials regulation**

Incident
   Type of operations (air, highway, water)
   Date of incident
   Location of incident

Reporting carrier, company, or individual
   Full name
   Address
   Type of vehicle or facility

Shipment information
   Name and address of shipper
   Name and address of consignee
   Shipping paper identification number
   Whether shipping papers issued by carrier, shipper, or other party

Deaths, injuries, loss, and damage
   Number of persons injured
   Number of persons killed
   Estimated amount of property loss or damage
   Estimated total quantity of hazardous materials released

Hazardous materials involved
   Classification
   Shipping name
   Trade name

Nature of packaging failure (check all that apply)
   Dropped in handling
   Water damage
   External heat
   Bottom failure
   External puncture

Packaging information
   (Detailed checklist of packaging involved, such as markings, capacity per unit, number
   of packages, etc.)

Remarks
   Name of person preparing report
   Telephone number
   Signature
   Date

dicate the firm that generated the waste, its address and phone numbers, along with the storage treatment, or disposal facility. Only sites certified by the EPA may be used for storage and disposal. Often, the cost of moving hazardous wastes can be very high, and unscrupulous firms hire carriers to dump these goods in the woods at night to minimize waste costs. Drums of hazardous materials have also been discovered in abandoned warehouses where the true owners are not possible to track down.

Several attempts have been made to create hazardous waste clearinghouses. The rationale for this is that ''one man's garbage is another man's resource.'' By acting as information centers, some firms could rid themselves of hazardous waste that might be used by others and reprocessed into usable goods.

Truck and Driver Highway Safety

Extensive social regulations exist concerning highway safety. The DOT regulations, promulgated by the Bureau of Motor Safety, as well as driver qualifications, are found in 49 CFR.

Vehicle safety is a prime concern of private fleet managers and any traffic manager who selects for-hire carriers. The following DOT regulations include requirements for both vehicle and driver safety:

- Vehicle parts and accessories
- Vehicle inspection and maintenance
- Accident reporting
- Driver qualifications
- Driving practices
- Hours of driving service

The DOT requires drivers to maintain a log of driving activity that shows the number of hours in and out of the truck, as well as the number of hours of actual driving. This log must be kept up to date, and failure to do so can bring prosecution. The DOT regulations prescribe maximum hours of driving; for example, no driver is supposed to drive more than sixty hours in a seven-day week. In interstate service, a driver may drive up to ten hours following eight consecutive hours off duty.

Traffic managers are concerned with these DOT regulations in for-hire situations, as well as for their in-house fleet. An improperly maintained truck can break down en route, thereby slowing transit time of valuable goods. Improper driver qualifications or safety problems can precipitate an accident and even expose the company's goods to theft and later product liability.

Other Sources of Social Regulation

The Clean Water Act of 1977

This law covers discharges and run-offs of hazardous material, such as chemicals and biological wastes, that disturbs the physical integrity of the U.S. waterways. Traffic managers, in charge of fleets and large parking and dock areas, can affect waterways in several ways; for example, using sodium chloride on parking lots when deicing, using nonphosphate soaps in fleet washing, and causing oil spills on lots and maintenance facilities. Warehouse drainage fields and parking lot run-off might also directly cause harm to nearby waterways. EPA inspectors examine these areas routinely.

Occupational Safety and Health Administration (OSHA)

These regulations are promulgated by the Department of Labor and concern work-area safety. Traffic managers with responsibility for warehouse and dock areas will find OSHA inspectors looking at dock doors and gates, stairs, forklifts, warehouse racks, forklift chargers, ventilation, asbestos exposure, and many other features of the work area.

Fire Safety

Each municipality has the power to inspect for fire safety within the work area. Further, this safety precaution is often reinforced by the firm's own insurance requirements. Fire personnel may visit and inspect warehouses, terminals, and dock areas for sprinklers, fire-fighting equipment at the site, and related features.

Food and Drug Administration and Public Health Service

These agencies oversee any operations that handle or store food or other items of human consumption and any practices that relate to sanitation in vehicles and warehouses. Public health matters are enforced by local, county, and federal agencies.

State Regulations

Most states have laws and regulations that apply to environmental, toxic product, food, and vehicle safety areas. In some instances, these state laws conflict with the federal laws, and resolving these conflicts may take a long time. Of particular concern today is the movement of radioactive materials over federal highways in states that prohibit such movement.

Summary

Social regulations have grown in the past decade, and they will play increasing roles in traffic and transportation in the future. Social regulations often require compliance. Almost all social regulations carry criminal penalties affecting the shipper, the carrier, and sometimes the consignee.

The regulations involved in this area have arisen from different and sometimes hasty social pressures. They have been created and promulgated in the transportation area through a variety of agencies. Some of them have differing definitions and some even have conflicts in how they are to be followed by industry. For the present, a clean, healthy, and safe workplace and living environment is a major social choice made by most industrial nations of the world. There is no way that a modern society can live without some of these goods. Traffic and distribution personnel will always have an involvement in the safe shipment and storage of dangerous and potentially polluting goods.

Selected Topical Questions from Past AST&L Examinations

1. The Resource Conservation and Recovery Act, which went into effect in 1980, covers the transport of waste products. What are the details of this law as they affect waste products? (Spring 1981)
2. What are the five key documentary requirements on the bill of lading that are necessary when a hazardous shipment is being moved by domestic transportation firms?
3. What are the key items found in the Hazardous Materials Table, 49 CFR 172?
4. In late 1986, there was an effort to provide a national drivers license system for commercial truck drivers. This licensing system would be implemented at the federal level. What are the pros and cons of such a national (versus state) drivers licensing system?

Notes

1. Warren E. Isman and Gene P. Carlson, *Hazardous Materials* (Encino, California: Glencoe Publishing, 1980), p. 2.
2. Transportation Safety Act of 1974, Public law 93-833, Section 104.
3. Isman and Carlson, *Hazardous Materials*, p. 6.
4. Isman and Carlson, *Hazardous Materials*, p. 6.

Appendix 16.1 Catalogue of Social Regulations Relating to Traffic Management

Area of Concern	Agency	Pertinent Regulations	CFR Source
Product safety	Consumer Product Safety Commission (CPSC)	Traffic manager involved in recall of product when products are deemed unsafe	Title 16
Establishing foreign trade zones	Department of Commerce	A positive regulation that enables foreign trade zones to be established. Traffic manager's firm might become supportive in the development of such a zone. Will cause a shift in present transportation patterns and some export-import methods.	Title 15
Employee safety	Department of Labor, Occupational Safety and Health Administration (OSHA)	A regulation body that requires specific work-area standards. Factors concerning the traffic manager are in warehouse and dock settings, including: Walking-working surfaces Means of egress Vehicle mounted work platforms Manlifts Environmental control Hazardous materials Personal protective equipment Medical and first aid Fire protection Compressed gas and air equipment Materials handling Storage Machinery and machinery guarding Tools Special industry regulations Toxic substances Though company engineering and design personnel are involved in planning and constructing the facilities to comply with these regulations, the traffic manager is the person responsible for them after they are in place. OSHA inspectors may come on site with little notice.	Title 29

(continued)

Appendix 16.1 (continued)

Area of Concern	Agency	Pertinent Regulations	CFR Source
Hazardous materials	Department of Transportation (DST), Research and Special Programs Administration	A shipper- and carrier-related body of regulations dealing with how products deemed hazardous are to be packaged and shipped. Traffic manager is the person who must see that products are properly packaged, labeled, and shipped with the proper documentation. He or she might also become involved with a rule-making proceeding dealing with the listing of additional products or the changing of present regulations. The basic regulations are in the form of: Rule-making procedures General information Hazardous materials table and communications requirements Shippers' general requirements for shipments and packages Carriage by rail, aircraft, vessel, and public highway Shipping container specifications and specifications for tank cars Pipelines Shipping papers Marking Labeling Placarding Preparation of HMs for transportation	Title 49

Safety of railroad facilities	Department of Transportation (DOT), Federal Railroad Administration (FRA)	Shippers are often involved or concerned with these regulations, since they deal with the speed and safety specification of railroad lines serving their plants or their customers. Railcar safety standards will come into play for both railroad-owned equipment or cars leased and used by the shipper.	Title 49
Motor carrier safety	Department of Transportation (DOT), Federal Highway Administration	Rules and regulations dealing with any motor vehicles used for carrying freight and passengers. Private carriage vehicles are subject to these regulations which deal with: Motor carrier safety Qualification of drivers Driving of vehicles Parts necessary for safe operations Reporting of accidents Hours of service Inspection and maintenance Movement of hazardous materials	Title 49
Fuel allocation	Federal Energy Administration (FEA)	Regulations dealing with mandatory gasoline and diesel fuel rationing. Currently, these are dormant, but they are the rules that will come into use in the event that rationing is used. They have a direct bearing on private and contract carriage serving the firm. Abolished in early 1980s.	Title 10
Animal and plant inspection	Department of Agriculture, Animal and Plant Health Inspection Service	Regulations stating diseases and conditions inspectors are to look for in interstate transportation facilities. Provides powers of inspection even in private vehicles carrying plants and animals.	Title 9
Toxic substance control	Environmental Protection Agency (EPA)	Regulations dealing with the inventorying and accountability of toxic substances. Traffic managers are a key part of the movement and record-keeping requirements of this regulation.	Title 40

(continued)

Appendix 16.1 (continued)

Area of Concern	Agency	Pertinent Regulations	CFR Source
Movement of toxic waste substances	Environmental Protection Agency (EPA)	The Resource Conservation and Recovery Act requires EPA to establish rules dealing with the disposition of toxic waste materials. When fully implemented, the rules will require firms to dispose of waste materials under specific rules for transporting, packaging, labeling, and accounting for waste materials. Transporting might be performed only by EPA-certified carriers, and storage only at EPA-certified sites. Problem: Marking, labeling, and commodity description of waste materials evolving under these regulations appear to conflict with the Department of Transportation's hazardous materials regulations.	Title 40
Customs	Department of Treasury, Customs Service	Provides regulations dealing with the importation of products into the United States. Regulations deal with: Packing, stamping, marking Transportation in bond Customs warehouses Drawback Customhouse brokers Quotas Use of foreign-trade zones Examination, sampling, and testing of merchandise Antidumping	Title 19
Vessel operations	Department of Treasury, Customs Service	Concerns firms that operate private carriage vessels off or along the coasts	Title 19
Food and drug safety	Department of Health, Education and Welfare, Food and Drug Administration (FDA)	Provides regulations for packaging of food and drug substances	Title 21

Appendix 16.2 **Hazardous Material Definitions**

Explosives

Explosives— Any chemical compound, mixture, or device, the primary or common purpose of which is to function by explosion (that is, with substantially instantaneous release of gas and heat), unless such compound, mixture, or device is otherwise specifically classified in Parts 170–189. (Sec. 173.50)

Class A— Detonating or otherwise of maximum hazard. The nine types of Class A explosives are defined in Sec. 173.53.

Class B— In general, function by rapid combustion rather than detonation and include some explosive devices such as special fireworks, flash powders, and so on. Flammable hazard. (Sec. 173.88)

Class C— Certain types of manufactured articles containing Class A or Class B explosives, or both, as components, but in restricted quantities, and certain types of fireworks. Minimum hazard. (Sec. 173.100)

Blasting agents— A material designed for blasting that has been tested in accordance with Sec. 173114a(b) and found to be so insensitive that there is little probability of accidental initiation to explosion or of transition from deflagration to detonation. (Sec. 173.114a(a))

Flammable Materials

Flammable liquid— Any liquid which, under specified test procedures, has a flash point of less than 100 degrees F (37.8 degrees C); for example, crude oil, petroleum, ether, gasoline.

Flammable solid— Solids, other than explosives, likely to cause fire through absorption of moisture, through spontaneous chemical changes, or as a result of retained heat from the manufacturing or processing of the commodity, including spontaneously combustible or water-reactive materials; for example, charcoal briquettes, lithium metal, strike-anywhere matches.

Flammable gas— Gas that is flammable or explosive under prescribed test procedures; for example, acetylene, hydrogen, propane.

Combustible liquid— A liquid not meeting the definition of any other hazardous materials class having a flash point of 100 degrees F (37.8 degrees C) or more and less than 200 degrees F (93.3 degrees C); for example, crude oil, petroleum, formaldehyde, fuel oil.

Spontaneously combustible material— Any liquid having a flash point above 100 degrees F and below 200 degrees F as determined by tests listed in Sec. 173.115(d). Exceptions to this are found in Sec. 173.115(b).

Oxidizing Materials

Oxidizer— A substance that yields oxygen readily to stimulate the combustion of certain other substances; for example, ammonium nitrate mixed fertilizer, lead nitrate, lead peroxide.

Organic peroxide— A derivative of hydrogen peroxide in which part of the hydrogen has been replaced by an organic material; for example, acetyl benzoyl peroxide, benzoyl peroxide, lauroyl peroxide.

Corrosives

Corrosive— A liquid or solid that causes visible destruction or irreversible damage to skin tissue at the point of contact, or that has a severe corrosion rate on steel; for example, phosphorus pentachloride solid, potassium fluoride solution, sulfuric acid.

Poisons

Poison A— Specifically enumerated gases or liquids of such nature that a small quantity of gas or the vapor of the liquid when mixed with air is dangerous to life; for example, cyanogen, phosgene.

Poison B— Liquids or solids (including pastes and semisolids) known to be so toxic to man as to create a health hazard during transportation, or which are presumed to be toxic to man because of the effects on laboratory animals; for example, arsenic, some insecticides, carbolic acid.

Irritating— Liquids or solids that give off dangerous or intensely irritating fumes when exposed to air or on contact with fire; for example, brombenzilcyanide, grenades, tear gas, xylyl bromide.

Compressed Gas

Compressed gas— Material having an absolute pressure exceeding 40 psi at 70 degrees F, an absolute pressure exceeding 104 psi at 130 degrees F, or any liquid flammable material having a vapor pressure exceeding 40 psi absolute at 100 degrees F.

Radioactive Materials

Radioactive materials— Substances that spontaneously emit radiation capable of penetrating and damaging living tissue and undeveloped film. Fissile radioactive materials are classified according to the controls required for nuclear criticality safety; for example, cobalt 60, gold 198, iridium 192.

Other

ORM-A— A substance having an irritating, noxious, anesthetic, or similar property that would cause extreme annoyance or discomfort to passengers and crew in the event of leakage during transportation; for example, bone oil, carbon tetrachloride, dry ice.

ORM-B— A substance that, when set, would cause significant damage to a transport vehicle and is so designated in the Hazardous Materials Table; for example, calcium oxide, chloroplatinic acid solid, lead chloride.

ORM-C— A substance listed in the Hazardous Material Table not classed as ORM-A or ORM-B, but that has characteristics warranting special treatment during transportation; for example, battery parts, bleaching powder, caster beans.

Flash point— Means the minimum temperature at which a substance gives off flammable vapors, which in contact with spark or flame, will ignite.

Forbidden— The hazardous material is one that must not be offered or accepted for transportation.

Limited quantity— Means the maximum amount of a hazardous material, as specified in those sections applicable to the particular hazard class, for which there are specific exceptions.

part *three*

Control

Strategic, managerial, and operational transportation plans guide traffic activities. Control refers to the task of keeping actual performance in line with the goals established in these plans. Information systems, audits, budgets, and statistical techniques support the control process. These tools help determine whether performance standards are met, that is, the "effectiveness" of management. They also measure differences between authorized and actual expenditures, that is, the "efficiency" of operations. Further, managers can use such tools to track productivity and to identify opportunities for improvement. Chapter 17 presents some insights into information system requirements and the tools for control.

The organizational system needs to mesh with the control process for effective management. Internal consistency in organization, staff, and control system is the key to successful management. Chapter 18 addresses organizational issues generally considered important to traffic management. It also investigates how certain organizational structures facilitate decision making in the traffic area.

chapter | 17

System Support and Control

chapter objectives

After reading this chapter, you will understand:

The control process as it relates to traffic management.

The alternatives available for the management and control of freight payments.

The information system requirements for an effective management control system.

Fundamental techniques of control, including transportation accounting, costing methods, budgeting, freight payment and auditing, and performance measurement.

Current computer usage patterns in the traffic area of the firm.

chapter outline

Introduction

System Support and Control Framework

Information System Requirements
Key information needs
System design elements

Key Techniques
Transportation accounting
Costing methods
Budgeting
Freight payment and auditing

Productivity and performance measurement
Statistical methods for control

Computer Usage Patterns
Electronic data interchange

Summary

Questions

Appendix 17.1—Measuring and Reporting Specific Benefits Resulting from Transportation Cost Control Management

Introduction

Many companies consider the traffic function a prime area for computer-supported information and control systems. Top executives in these firms recognize the tremendous potential that traffic management has today for making a contribution to the firm's profitability. This promise, together with traffic's data-rich management environment, makes it an exceptionally attractive area of the firm for such modernization.

This chapter first presents a framework for information system support and management control in the traffic function. The focus then shifts to the two key aspects of a traffic information and control system: (1) the information system requirements and (2) the techniques of control. A discussion of computer usage patterns of shippers completes the chapter.

System Support and Control Framework

Management control includes measuring actual performance and taking actions to ensure it is brought into line, or kept in line, with desired objectives. **Exhibit 17.1** illustrates the traditional view of this task. As shown, key com-

426

Exhibit 17.1 **Traditional view of the control process**

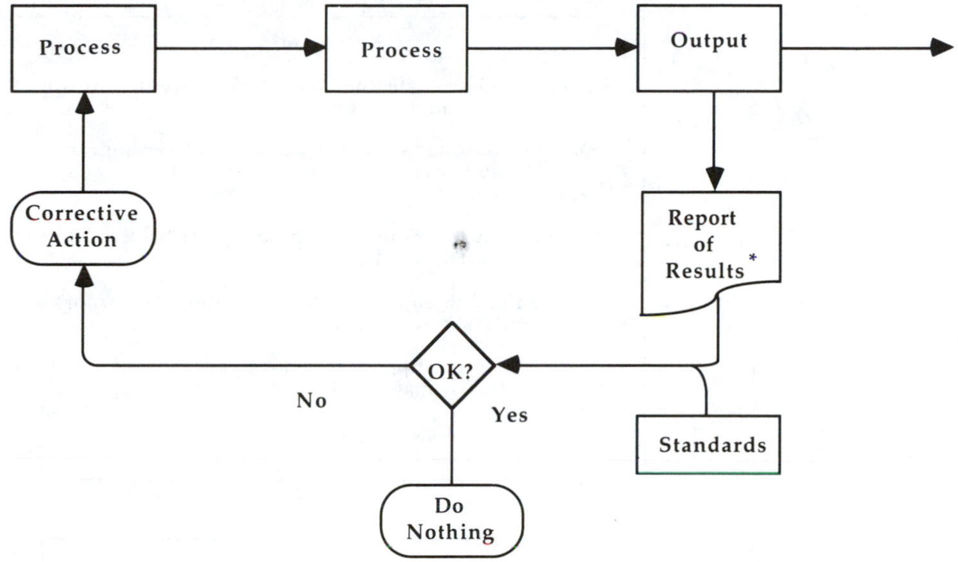

*Summary reports, status reports, or exceptions reports

Source: James H. Foggin, "Improving Motor Carrier Productivity with Statistical Process Control Techniques," *Transportation Journal*, 24, no. 1 (Fall, 1984), pp. 58–74. Reprinted by permission.

ponents of control include (1) input, process, and output; and (2) comparing results with standards and taking corrective action as necessary. The elements needed to make this control process work are as follows:

- A set of standards to measure progress toward organizational goals
- A system of measurements to compare actual performance against standards
- A mechanism for correcting deviations from standards

An information system is an integral part of the control process. **Exhibit 17.2** presents a system support and control framework for traffic management. As shown in the exhibit, this framework has two major components: key requirements and key techniques. The first identifies the principal kinds of information that support management control. The second lists important tools for management control. The next two sections of this chapter will examine these components.

■■■■■■ *Exhibit 17.2* **Framework for system support and control**

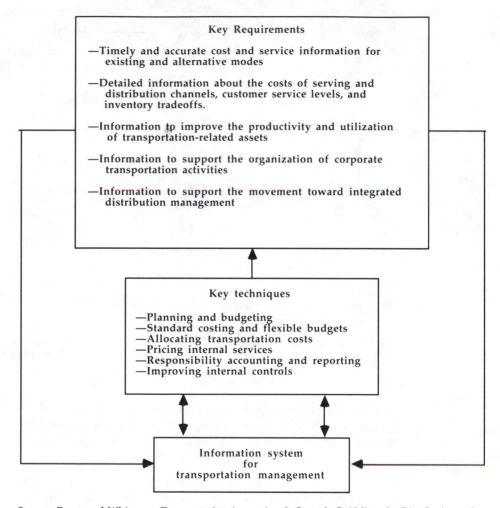

Key Requirements

—Timely and accurate cost and service information for existing and alternative modes

—Detailed information about the costs of serving and distribution channels, customer service levels, and inventory tradeoffs.

—Information to improve the productivity and utilization of transportation-related assets

—Information to support the organization of corporate transportation activities

—Information to support the movement toward integrated distribution management

Key techniques

—Planning and budgeting
—Standard costing and flexible budgets
—Allocating transportation costs
—Pricing internal services
—Responsibility accounting and reporting
—Improving internal controls

Information system
for
transportation management

Source: Ernst and Whinney, *Transportation Accounting & Control: Guidelines for Distribution and Financial Management* (Oak Brook, Illinois: National Council of Physical Distribution Management, 1983), p. xxv. Reprinted by permission.

Information System Requirements

During the design and development of an information system, traffic management must consider the information needs of the traffic function and the elements of good system design.

Key Information Needs

The most important requirement for an effective information system for transportation management is determining key information needs, including[1]

1. Cost and service information for alternative modes. This includes information about traffic lanes, product groups, modes, and carriers.
2. Detailed information about alternative distribution channel costs, customer service levels, and inventory tradeoffs. This information might include the traffic flow patterns for various channels, the types and numbers of channel members coupled with the directly corresponding distribution costs, the location and extent of inventories, and the customer service performance for key traffic lanes.
3. Information about the performance of transportation-related (fleet) assets. Service (transit times) and utilization measurements are important items of information here.
4. Information to support the organization of corporate transportation activities. For example, transportation cost data might be organized by product groups and by regions, divisions, or other organization units.
5. Information to support the movement toward integrated distribution management. This information will link cost and performance data for both inbound and outbound transportation. In addition, it will link transportation costs to other activity centers such as purchasing, order processing, packaging, and material handling.

In effective transportation support systems, such information is (1) comprehensive, (2) consistent across divisions and subsidiaries to facilitate companywide decisions, (3) reliable, and (4) sufficiently detailed to identify corrective actions and control transportation costs.[2]

System Design Elements

Traffic managers need to consider a number of elements in the design and development of information and control systems. These elements are as follows:

1. Paperwork. An effective system will minimize the multiple handling of data and the relatively nonproductive use of personnel for repetitive clerical tasks.

2. Confidentiality. The ability to protect sensitive information is an important consideration that should be addressed early on in the task of designing an effective system. Traffic records such as freight bills and bills of lading provide information regarding sources of raw materials and customer identities and locations. A system must protect the security and integrity of such information.

3. Back-up capability. Whether the present system is manual or computerized, a system needs back-up capability in the event the prime system is unable to function. Even with manual in-house systems, water or fire damage can conceivably incapacitate a facility, thereby requiring at least some degree of support from another department within the firm, or from an outside third party. Firms that use third-party systems should likewise incorporate back-up capabilities. A well-developed contingency plan is a justifiable investment of the firm's resources.

4. Cash flow. Managing cash is a serious matter in any firm. Since cash is an asset that should earn interest or produce a yield either overnight or for longer periods, a company wants to receive cash as early as possible and delay its disbursement as long as possible. In times of high interest rates, many firms have made more profits from the management of cash flow than from marketing and selling the company's primary products or services. Some firms successfully manage to secure supplies from vendors, to process and sell finished products, and to collect on their receivables, all before paying their vendor invoices.

5. Speed and reliability. A system must be reliable so that it can provide results within an acceptable time. The reports and information that arrive late create a time lag that can mask negative trends and delay corrective action. The result could be the loss or dissatisfaction of a customer, or an unnecessarily high cost might be incurred.

6. Financial commitment. Although corporate objectives will place limits on the extent and type of financial outlay for any system, that system must be cost effective. In addition, corporate policies will influence the decision whether to develop system support internally or to rely on outside sources. Financial commitment is an important issue for traffic managers because they frequently seek to minimize the financial investment or long-term commitments to tariff files, computer systems, and office space.

7. Compatibility. These requirements focus on how well the traffic system fits with the rest of the firm's information system and on how much confidence managers can have in the information generated. Ideally, the system that provides information useful to traffic management should be part of the firm's overall information system. Since this arrangement is frequently cumbersome or difficult to accomplish, a growing trend is for the traffic function to develop its own capabilities for capturing and accessing information.

8. Accuracy. Accurate information, of course, is desirable. Nevertheless, the level of required precision may range from fine to crude for different types of decisions. For example, ball park estimates of carrier costs may be sufficient for some rate negotiations. By contrast, a traffic manager wants more precise estimates of, say, duplicate payments and overcharges.

9. Information quality. As used here, information quality indicates how suitable the information is for decision making. An effective information system has to generate accurate and relevant reports. Although a large number of traffic managers are computer literate, few have had formal education in the nature and management of information systems. They will need to work closely with information system specialists to develop information suitable for making effective decisions. Such coordination may be difficult, however, if a traffic manager is uncertain about information needs or has difficulty conceptualizing and verbalizing those needs.

On the other hand, there are information systems managers who could benefit from a better understanding of traffic management. In many firms, it is difficult for individual functional areas of the firm, such as traffic, to develop a meaningful working relationship with the corporate data processing/management information system (DP/MIS) department. Unfortunately, this is more common than the situation where the two departments being able to work together productively and effectively in the interest of seeing that overall traffic, logistics, and corporate goals are achieved. In the final analysis, it will be necessary to see that corporate DP/MIS departments become responsive to the needs of the business at hand, while personnel, in areas such as traffic, learn about the capabilities, benefits, limitations, and so forth, of alternative information system configurations.[3]

Key Techniques

Although each of the key techniques shown in Exhibit 17.1 is relevant, the principal concerns here are with (1) transportation accounting, (2) costing methods, (3) budgeting, (4) freight payment and auditing, (5) productivity and performance measurement, and (6) statistical methods for control.

Transportation Accounting

Ideally, a transportation accounting system should accumulate, organize, and report appropriate financial and operating information to the traffic manager. Key accounting concepts especially applicable to logistics and traffic management include[4]

- Controllable and noncontrollable costs
- Direct and indirect costs
- Fixed and variable costs

- Standards and standard costs
- Actual (or historical) and opportunity costs
- Relevant costs and sunk costs
- Full costs and marginal or incremental costs
- Break-even analysis
- Cost of capital

To find out more about the transportation accounting methods and techniques actually used by shippers, the Council of Logistics Management and the National Association of Accountants launched a landmark study in 1983 that examined the practices of more than sixty corporations.[5] The study reported that

1. A wide range of transportation accounting practices exists, and there is significant inconsistency in the accounting treatment of similar transportation costs.
2. Most transportation accounting systems are oriented toward financial reporting and not toward the needs of distribution or financial management.[6]
3. Most systems are inflexible and are not coordinated.
4. The management of objectives, standards, and controls is generally inadequate.
5. In nearly half of the companies surveyed, effective information about products, customers, shipping terms, and routes was not available.

These findings suggest that transportation accounting systems are not providing traffic managers with information that is sufficiently reliable and accurate to make effective decisions.

Costing Methods

Standard transportation costing and individual shipment costing procedures are two important tools to support the control task in the traffic function.

Standard transportation costing. Standard costing procedures and flexible budgets are techniques that must be understood, if a traffic manager wants to identify, measure, allocate, and control costs properly. Although these two techniques are closely related, they differ in that standard costing is a "unit" concept, whereas flexible budgeting is an "aggregate" concept. For example, a standard cost might represent the unit cost to ship a case of a product from a Chicago distribution center to a sales warehouse in Dallas. On the other hand, flexible budgeting combines the standard cost per unit with the number of units shipped during a particular period to estimate what the total transportation cost should have been.

Four major steps are involved in the development of standard costs:[7]

1. Identify the particular activity that is to take place.
2. Measure the standard physical quantity of the activity that will take place.
3. Determine the standard price, or unit cost, for a single unit of the activity that will take place.
4. Compute the standard cost of the transportation activity by multiplying the physical quantity by the unit cost or price per unit of activity.

If a particular linehaul transportation movement by truck between two points five-hundred miles apart takes an average of nine hours to complete, and if the driver's wages are $12 an hour, the standard cost of the movement is $12 × 9, or $108. Although cost could also be expressed also as a function of distance, in this example time is the more relevant measure of the physical activity that takes place.

Exhibit 17.3 shows a simplified transportation cost accounting system that might be used by a shipper to control transportation costs. This example includes a standard traffic mix of three products, A, B, and C; they move from plant to customer at standard transportation costs per ton of $15, $10, and $12, respectively. The standard cost for this traffic mix is $29,000. Since shipment sizes and product mix can vary from shipment to shipment, it is important to identify instances where actual transportation costs differ from standard costs and to determine why such differences exist.

For the situation depicted in Exhibit 17.3, the total actual cost incurred was $42,000. This amount clearly differs from the $29,000 standard cost. The $13,000 discrepancy consists of a $6000 product mix variance and a $7000 price variance. Although the mix variance is acceptable, the price variance merits further investigation to find out why the cost of transportation was higher than anticipated. For example, overexpensive carriers or modes may have been used, or someone may have failed to capitalize on incentive rates.

Shipment costing systems. It is extremely helpful to have a reasonable estimate of shipping costs before the negotiation of price-service agreements with carriers. Costing software is now widely available from both private and public sources. (See Chapter 9 for a comprehensive discussion of shipment costing concepts and methodologies.)

Budgeting

A budget is one of the most valuable tools for controlling transportation expenditures. Although closely related to standard costing, budgeting is more comprehensive than costing alone.

The purpose of any budget is to establish cost and revenue projections. These

■■■■■ *Exhibit 17.3* **A standard transportation cost accounting system**

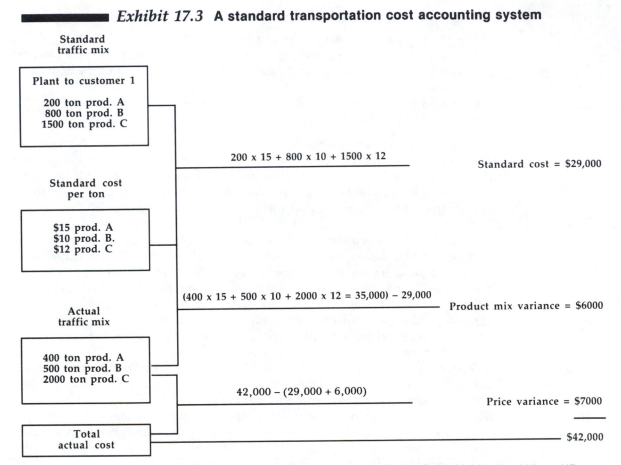

Source: Robert H. Leilich, "Cost Measurement Systems," *Traffic World* (May 24, 1982), p. 117. Reprinted by permission.

projections need to be flexible so that they can be developed over a range of activity levels. Such flexibility will enable traffic managers to answer "what if" questions about alternative levels of activity in the traffic area. An effective budgeting process will also permit a thorough after-the-fact examination of actual versus anticipated results. Such a "variance" analysis will identify the specific areas where a traffic manager needs to search for the underlying causes of the variances.

Developing a transportation budget requires coordination with other corporate departments such as marketing, manufacturing, distribution, and finance.[8] As shown in **Exhibit 17.4**, the key inputs to this process include information about sales and production forecasts, product mixes, geographic dispersion of markets and sources of supply, product sourcing, and mode mixes.

■■■■■■ *Exhibit 17.4* **Key inputs of transportation budget development**

```
                        ┌──────────────┐
                        │    Sales     │
                        │   forecast   │
                        └──────────────┘
                               │
        ┌──────────────────────┼──────────────────────┐
        │                      │                      │
  ┌───────────┐        ┌──────────────┐        ┌──────────────┐
  │  Product  │        │  Production  │        │  Geographic  │
  │    mix    │        │   forecast   │        │  dispersion  │
  └───────────┘        └──────────────┘        └──────────────┘
        │                                            │
   ┌──────────┐                              ┌──────────────┐
   │ Product  │                              │  Shipment    │
   │ sourcing │                              │  routing     │
   └──────────┘                              └──────────────┘
        │                                            │
        └─────────────────┐      ┌──────────────────┘
                     ┌──────────────┐
                     │    Mode      │
                     │    mix       │
                     └──────────────┘
                            │
                            ▼
                     ┌──────────────┐
                     │Transportation│
                     │   budget     │
                     └──────────────┘
```

Source: Gene R. Tyndall, John R. Busher, and James D. Blaser, "Accounting and Control in Physical Distribution — Phase I: Transportation," *Annual Conference Proceedings of the National Council of Physical Distribution Management*, 1982, p. 85. Reprinted by permission.

The transportation budget is a key element of traffic planning (see Chapter 6). A recommended approach to transportation budgeting is shown **Exhibit 17.5.** Because this approach is a comprehensive one, it easily permits an examination of budget variances by factors such as mode, shipment size, traffic lane, product type, and percentage of freight absorbed.

Traffic managers will need to distinguish between fixed and variable costs to develop useful budgets. The total cost of virtually any activity comprises both fixed and variable cost components. Knowledge of the relationship between these components will enable traffic managers to calculate the total cost for various anticipated activity levels.

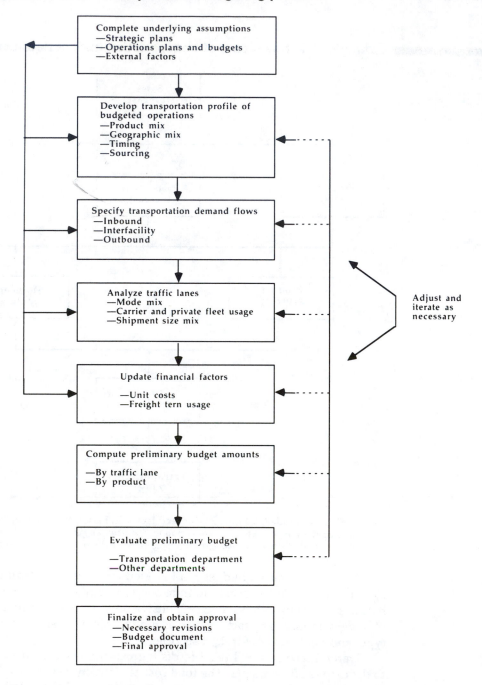

Exhibit 17.5 Transportation budgeting process

Complete underlying assumptions
—Strategic plans
—Operations plans and budgets
—External factors

Develop transportation profile of
budgeted operations
—Product mix
—Geographic mix
—Timing
—Sourcing

Specify transportation demand flows
—Inbound
—Interfacility
—Outbound

Analyze traffic lanes
—Mode mix
—Carrier and private fleet usage
—Shipment size mix

Update financial factors
—Unit costs
—Freight tern usage

Compute preliminary budget amounts
—By traffic lane
—By product

Evaluate preliminary budget
—Transportation department
—Other departments

Finalize and obtain approval
—Necessary revisions
—Budget document
—Final approval

Adjust and
iterate as
necessary

Source: Ernst and Whinney, *Transportation Accounting & Control: Guidelines for Distribution and Financial Management* (Oak Brook, Illinois: National Council of Physical Distribution Management, 1983), p. 41. Reprinted by permission.

In addition, the budget process requires judgments about how to allocate costs that are simultaneously associated with two or more different units of output. Before making such judgments, it is necessary to identify suitable bases for allocation such as product line, customer, and shipment size. A traffic manager then must try to allocate relevant costs as accurately as possible. Cost allocation decisions are inherently judgmental, yet the effectiveness of the overall budgeting process depends on a consistent and effective approach to this task.

Exhibit 17.6 is an example of a flexible budget report used to analyze cost variances by traffic lane. It can be seen that total cost figures for the transportation of products A, B, and C are separated into their variable and fixed components. A comparison of budgeted versus actual costs includes a breakdown of the differences according to the type of variance (that is, volume, rate, mode mix, freight absorption, or fixed budget). This breakdown helps to explain the reasons for the overall cost variance of $54,710.

Freight Payment and Auditing

The effective control of freight payments involves four key tasks. The first is to achieve accuracy in the preparation of bills of lading and freight bills. The second task is to ensure that the lowest legally applicable rate is paid at all times. The third is to establish procedures for the timely processing and payment of freight bills. These procedures will need to be sensitive to the flexible credit arrangements that carriers can now offer. The fourth, and by far the most important, task is to analyze historical freight payment information and to develop meaningful feedback to those involved in the purchase and use of transportation services.

To carry out these responsibilities, a traffic manager has to decide whether to commit the use of internal resources (people as well as financial capabilities), or to purchase these services externally. This issue actually involves five options (discussed in Chapter 7): (1) manual, in house; (2) manual, consulting firm; (3) computer, in house; (4) computer, consulting firm; and (5) bank payment plan.

Besides the previously discussed design considerations for a good transportation information system, the traffic function needs to consider the capability of a freight payment system to perform the following five important functions:

1. Preauditing. Checking freight bills for accuracy before payment by the shipper is known as preauditing. Although preauditing is useful for reducing the incidence of both overcharge and undercharge errors, its success primarily depends on the system's ability to update rate files and to retrieve rates accurately. The preaudit capability allows the shipper to reduce the incidence of overcharges by paying the correct amount the first time. Another important benefit is the improved ability to invoice customers accurately for freight charges incurred.

Exhibit 17.6 Flexible budget report — Variance analysis by traffic lane

| Cost Type | Product | Budget | Actual | Variances* | | | | | | |
				Volume	Rate	Mode mix	Freight absorption	Fixed budget	Total
Variable	A	$114,250	$102,825	$5,710	$4,420	$(12,560)	$13,855		$11,425
	B	150,750	135,675	7,540	5,830	(16,580)	18,285		15,075
	C	267,000	240,540	13,350	10,260	(29,070)	31,920		26,460
		532,000	479,040	26,600	20,510	(58,210)	64,060		52,960
Fixed	A	17,750	15,875					$1,875	1,875
	B	12,375	11,750					625	625
	C	34,500	35,250					(750)	(750)
		64,625	62,875					1,750	1,750
Total	A	132,000	118,700	5,710	4,420	(12,560)	13,855	1,875	13,300
	B	163,125	147,425	7,540	5,830	(16,580)	18,285	625	15,700
	C	301,500	275,790	13,350	10,260	(29,070)	31,920	(750)	25,710
		$596,625	$541,915	$26,600	$20,510	$(58,210)	$64,060	$1,750	$54,710

*( ) indicates unfavorable variances

Source: Ernst and Whinney, *Transportation Accounting & Control: Guidelines for Distribution and Financial Management* (Oak Brook, Illinois: National Council of Physical Distribution Management, 1983), p. 72. Reprinted by permission.

2. *Postauditing.* Checking freight bills for accuracy after payment is referred to as postauditing. Since overcharges are essentially subject to a three-year statute of limitations (see Chapter 13), the need for postauditing is not as time sensitive as preauditing. From a cash flow perspective, however, overcharge amounts are not realized until the claim is actually paid by the carrier to which the claim is filed. Postaudits also are beneficial for reducing the impacts from duplicate or unidentified billings.

3. *Rate-analysis capability.* This capability supports the analysis of transportation alternatives. Typical analyses focus on opportunities for assembly and distribution, options for shipment routing, and impacts of packaging on rates. An effective freight rate retrieval system is generally regarded as an important cornerstone of a responsive rate-analysis capability.

4. *Report generation.* Many companies have found that information relating to past shipments is useful for a wide range of applications in traffic, purchasing, and marketing. A comprehensive freight payments system should have the capability to generate required reports automatically.

5. *Rate retrieval.* In an effort to reduce paper-handling activities and expenditures for clerical and supporting staff, firms are turning to automated rate-retrieval systems. These systems may require a large investment and a large volume of transactions to achieve low unit costs for processing individual freight bills. Microcomputers, though, are beginning to make rate retrieval feasible for low-volume shippers.

Evaluating alternative freight payment systems requires traffic managers to consider two important issues. The first is the extent to which each available option is capable of performing the basic system functions of preaudit, postaudit, bill payment, rate analysis, report generation, and rate retrieval. The second issue is the effect the alternative might have on the overall management of the traffic function itself. The most significant consideration here is the desired degree of internal control over all payments made by vendors serving the company. In some companies, the prevailing culture (and policy) may effectively preclude the opportunity for external involvement in any activity related to accounts payable. Perhaps the most frustrating situation is when company policy prohibits involving outside firms in matters related to freight payments, even though internal capabilities are clearly deficient. This situation might occur, for example, when the responsibility for freight payments is actually a low priority for (and probably a nuisance to) a centralized accounts payable function. In instances like this, the traffic manager needs to make company personnel aware of alternative approaches. Other elements that may have an impact on the traffic function include (1) the financial resource needs in the traffic area, (2) the data processing and computer capabilities of traffic staff, (3) personnel and staffing requirements, and (4) cash flow implications.

General Electric Corporation (GE) has developed an innovative freight payment arrangement with a bank in Connecticut. GE writes checks on an account that has no cash balance. When the carrier submits the checks for payment, the bank notifies GE, and funds are transmitted to the bank to cover the total amount of checks payable on a particular day. GE has complete control over its funds and can receive earnings from them right up to the time of payment to the bank. The bank charges a nominal monthly fee plus a per-check charge for this service. This type of arrangement is likely to become more popular as small- and medium-size banks pursue growth strategies in the newly deregulated banking environment.

Exhibits 17.7 and **17.8** are examples of the reports commonly made available to firms involved in bank payment plans. In addition to assuring the timely and accurate payment of freight charges, services like these are valuable to the traffic function's report-generating capabilities. Because obtaining suitable reports of expenditures from internal sources is sometimes difficult for the traffic area, this attribute is often a factor in deciding to contract with outside sources for the payment and auditing of freight bills.

Borden Chemical paid $22.5 million in freight charges in 1980, then later filed claims for $240,000 in duplicate payments and $318,000 in overcharges. The overcharges represented unnecessary payouts of nearly 2.5 percent. Further, some of these funds were not returned directly to Borden; one half were paid to audit firms involved in identifying these overcharges.

This situation spurred the development of Borden Chemical's Freight Rating, Auditing, and Payment System (FRAPS), which is illustrated in **Exhibit 17.9**[9]. Borden receives a carrier freight bill for payment along with a copy of the bill of lading prepared by Borden for the shipment. On approval, freight bills are sent to Cass Bank and Trust in St. Louis, where they are first audited for duplicate payments and then paid by check to the carrier. Once each week, the bank collects the information in summary form and reports back to Borden. The freight bill and bill of lading data for each shipment are then sent back to the Borden computer where each is audited against a rate data file containing about 300,000 individual rates. Any freight bills that do not equal the computer-based rate are highlighted on an exceptions report.

The agreement with Cass is that Borden will be reimbursed for any differences. The transportation systems group at Cass then files a claim with the carrier for the amount of the overcharge. This process relieves Borden of the lost cash-flow problem while the claim is outstanding. If the bank's transportation group cannot resolve the issue with the carrier within a reasonable time, the problem reverts to Borden.

Overall, this system permits Borden to operate with a smaller staff and a

Exhibit 17.7

FIRST FREIGHT PAYMENT PLAN —
MANAGEMENT INFORMATION SYSTEM:
CARRIER TRANSACTION SUMMARY REPORT

FROM 3/29/87 to 5/02/87

CARRIER NUMBER	CARRIER NAME	CURRENT			YEAR TO DATE BEGINNING 1/01/87		
		NUMBER OF SHIPMENTS	TOTAL AMOUNT	TOTAL WEIGHT	NUMBER OF SHIPMENTS	TOTAL AMOUNT	TOTAL WEIGHT
18-3	ANDREWS & PIERCE INC. P O	1	20.40	500	1	20.40	500
21-7	SPECTOR FREIGHT SYSTEM I	0	.00	0	2	288.80	2,117
23-9	NEW PENN MOTOR EXPRESS I	1	450.82	5,874	2	519.30	6,563
26-3	MCLEAN TRUCKING COMPANY	2	298.95	1,391	15	1,576.40	13,263
30-7	ST. JOHNSBURY TRUCKING CO.	15	605.21	7,336	59	3,486.90	42,319
36-5	CAROLINA FREIGHT CARRIER	0	.00	0	1	50.33	252
41-2	HERMANN FORWARDING COMPANY	0	.00	0	2	140.27	804
43-4	INTERSTATE MOTOR FREIGHT	3	324.41	4,755	2	623.48	11,227
47-8	HEMINGWAY TRANSPORT INC.	0	.00	0	1	186.43	645
48-9	SANBORNS MOTOR EXPRESS I	42	3,450.00	48,372	112	7,670.77	110,282
82-2	SEVERANCE MOVING & TRUCK	1	25.83	450	2	7,670.77	119,282
118-8	HELMS EXPRESS DIV. RYDER	31	4,739.21	49,262	42	6,542.87	67,142
164-5	ROADWAY EXPRESS, INC. A	1	50.62	343	6	314.53	2,084
173-6	SPRINGMEYER SHIPPING CO.	1	358.32	2,672	1	358.32	2,672
233-2	INTERCITY TRANSPORTATION	1	26.94	130	2	55.00	280
296-8	THURSTON MOTOR LINES INC.	45	4,104.11	121,263	133	17,147.37	446,512

Source: Adapted from The First National Bank of Boston, "First Freight Payment Plan—Management Information System." Reprinted by permission.

Exhibit 17.8

FIRST FREIGHT PAYMENT PLAN—MANAGEMENT INFORMATION SYSTEM
WEIGHT BREAK ANALYSIS REPORT

FROM 8/01/87 TO 8/31/87

	SHIPMENTS OF MINIMUM WEIGHT	**SHIPMENTS EXCLUDING MINIMUM WEIGHT**								
		UP TO 99	100 TO 499	500 TO 999	1,000 TO 1,999	2,000 TO 5,999	6,000 TO 19,999	20,000 TO 31,999	OVER 31,999	TOTALS
HOLMES TRANS INC. ACC CARRIER NUMBER 521-1										
NUMBER OF SHIPMENTS	3	1	2	1	0	1	0	0	0	5
WEIGHT		60	911	500	0	2,000	0	0	0	3,471
AMOUNT	35.01	20.80	77.52	52.30	.00	77.44	.00	.00	.00	228.14
COST PER 100 LBS		34.66	8.50	10.47	.00	3.87	.00	.00	.00	6.57
COST PER SHIPMENT	11.67	20.80	38.76	52.38	.00	77.44	.00	.00	.00	45.62
RED STAR EXPRESS LINES 0 CARRIER NUMBER 541-5										
NUMBER OF SHIPMENTS	1	0	0	1	0	0	0	0	0	1
WEIGHT		0	0	600	0	0	0	0	0	600
AMOUNT	21.41	.00	.00	34.26	.00	.00	.00	.00	.00	34.26
COST PER 100 LBS		.00	.00	5.71	.00	.00	.00	.00	.00	5.71
COST PER SHIPMENT	21.41	.00	.00	34.26	.00	.00	.00	.00	.00	34.26

Source: Adapted from The First National Bank of Boston, "First Freight Payment Plan—Management Information System." Reprinted by permission.

Exhibit 17.9 **Rate control and bank freight payment system**

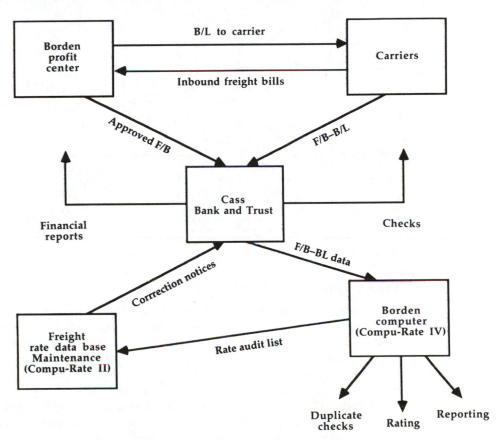

Source: "FRAPS: Borden Chemical's Marriage of Rate Control and Bank Freight Payment," *Computer Systems Report* 1 no. 3 (March, 1982). Reprinted with permission of TCS Inc. and Computer Systems Report, Inc.

lesser commitment to overhead than would be needed otherwise. Rating, billing, and claims processing thus reverts from an internal capability to an external service for a fee directly related to the number of freight bills processed.

Since the early 1980s, when the FRAPS system was first installed at Borden Chemical, new generations of freight payment systems have evolved. One of these, which reflects the current state of the art, was recently installed at the A. O. Smith Company's Consumer Products Division. **Exhibit 17.10** illustrates this system.

■■■■■■ *Exhibit 17.10* **Rate retrieval and freight payment system**

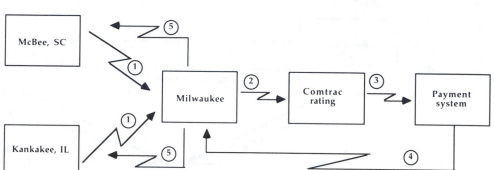

1) B/Ls transmitted to Milwaukee data center
2) Milwaukee transmits B/Ls to Comtrac for rating.
3) Comtrac transmits rated B/Ls to payment service.
4) Payment service transmits paid detail to Milwaukee.
5) Milwaukee transmits plant detail back to shipping location.

Source: A. O. Smith Company, Consumer Products Division. Reprinted by permission.

Productivity and Performance Measurement

Although the terms ''productivity,'' ''utilization,'' and ''performance'' (sometimes called ''efficiency'') are often used interchangeably by distribution managers, each term has its own unique meaning.[10] *Productivity* is the ratio of real output produced to real resources consumed. *Utilization* is the ratio of used capacity to available capacity. *Performance* is the ratio of actual output to standard output (or standard hours earned to actual hours).

Exhibit 17.11 summarizes recommended methods of measuring productivity, utilization, and performance in traffic management. Such measurements, of course, did not offer much prescriptive information about how to achieve productivity.

Exhibit 17.12 identifies a number of additional performance measures that may be used to evaluate activities and performance levels in the traffic area. To be most useful, these measures must be tracked carefully over time. Although broad aggregate measures are helpful, measures broken down by product groups, geographic regions, origin points, or individual modes and carriers will add useful detail to the information available.

Appendix 17.1 offers practitioners a format that can assist in the task of measuring and reporting the benefits of transportation cost-control management. Since each of the recommendations in the appendix is an opportunity for cost reduction, they also may be interpreted as ways to improve productivity.

Statistical Methods for Control

The use of statistical methods offers an exceptionally powerful alternative to the conventional ways in which the management process of control takes place.[11]

■■■■■■■ *Exhibit 17.11* **Control measures for traffic management**

Productivity	Utilization	Performance
Ton-miles transported/ total actual transportation cost	Total transportation capacity used/total transportation capacity paid for	Actual transportation cost/budgeted transportation cost
Stops served/total actual transportation cost		Actual transportation cost/standard transportation cost*
Volume of goods transported to destination/ total actual transportation cost		Actual transit times/ standard transit times
Shipments transported to destination/total actual transportation cost		
Standard ton-miles/total actual transportation cost		

*Standard based on cost if work performed by private fleet transportation, or based on engineered standard costs.

Source: A. T. Kearney, Inc., *Measuring and Improving Productivity in Physical Distribution* (Oak Brook, Illinois: National Council of Physical Distribution Management, 1984), p. 170. Reprinted by permission.

Conventional methods require an analyst to measure the difference between what was observed and what was expected, and to take the necessary corrective action. By contrast, statistical process control (SPC) requires an understanding of the variability of the process itself before making such management decisions.[12] For example, to analyze delivery time experience from several vendors, it is necessary to know not only the average time elapsed from the issuance of a purchase order to the receipt of a shipment but also the likely variation of delivery times.

Exhibit 17.13 is a flowchart that describes statistical process control. Like the initial steps in the traditional approach to control, the first three steps of SPC are (1) design system, (2) establish standards, and (3) perform process. Once these steps are taken, however, SPC poses three questions. First, is the activity measured "in statistical control"? In other words, do all the observed measurements fall reasonably close to the average? Second, is the activity's performance "capable"? Capability here refers to how far performance variations stray from the levels that ordinarily would be expected. Third, do the activity measurements meet standards? That is, is the average performance on target?

The control chart, perhaps the most widely used statistical approach to gaining insight into these questions, permits an examination of activity measurements in relation to both upper and lower "control limits." These limits, which

■■■■■■■ *Exhibit 17.12* **Suggested performance measures**

Key Indicator	Comments
Transportation costs	
Per total sales	Useful in discounting the effects of
Per unit sold	product price cutting or price increases. Indicate rising costs relative to sales volume, but not the reason why
Per ton-mile	Excellent measures of the effectiveness
Per cube-mile	of transportation purchases. Circumstances determine which measure is most suitable.
Per hundredweight	Hundredweight measure is widely used
Per cube shipped	and easy to capture; cube measure may be a more accurate measure for some companies.
Per mile	Primarily used to measure performance of private or dedicated contract fleets
Per order	Indicates the impact of transportation on the landed cost of an order. Often of interest to marketing.
Transportation service	
On-time shipments per total shipments	Important indicator that is difficult to obtain
Actual transit time per standard transit time	Especially useful for private or dedicated fleets
Tonnage shipped per tonnage capacity purchased	Appropriate for truckload and bulk shipments' effective utilization of transportation dollars
Cube shipped per cubic capacity purchased	
Weight/order, cube/order, units/order	Indicate the customer's changing purchasing patterns or changing physical characteristics of products
Miles shipped/order	Highlight emerging geographic concentration or dispersion
Weight/cube	Indicates a changing product mix or new physical characteristics

Source: Based on Masao Nishi, ''Measuring the Transportation Manager's Contribution to Company Profits,'' *Handling and Shipping* (Presidential Issue, 1983–84), pp. 83–88.

are statistically derived, help the analyst identify the instances in which observed behavior departs significantly from what was expected. Such departures often indicate that a problem exists. The search for underlying explanations and causes may then proceed in an organized, efficient manner.

Exhibit 17.13 Statistical process control

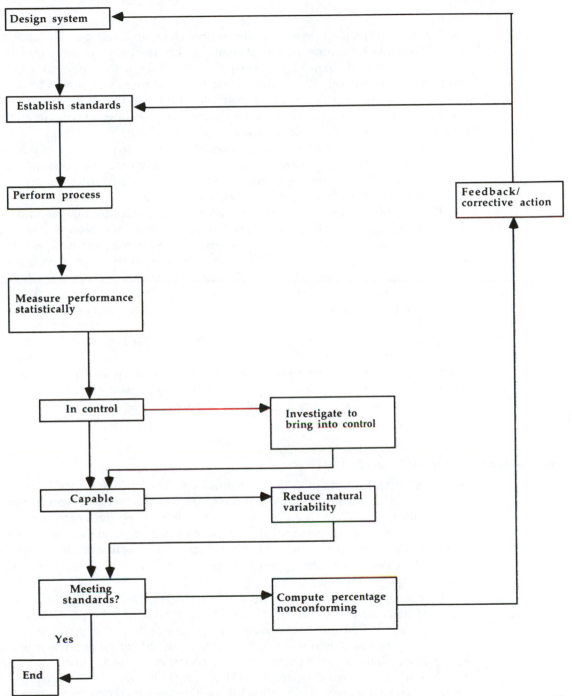

Source: C. John Langley, Jr., "Information-Based Decision Making in Logistics Management," *International Journal of Physical Distribution and Materials Management* 15 no. 7 (1985), p. 50. Reprinted by permission.

Exhibit 17.14 shows two example control chart applications. The first resulted from an examination of transit times (in minutes) of shipments traveling over a 260-mile traffic lane. The three points that exceed the upper and lower limits are out of control. The analyst needs to investigate these points to determine the causes and prevent their repetition. The second chart shows the percentages of carrier freight bills found by a particular shipper to contain errors. Only one point is high enough to be out of control.

Although these two examples are relatively simple, any one of a number of activities in traffic management could have been illustrated. Potential applications might include claims experience, on-time delivery performance, and private truck operations (fuel mileage, vehicle breakdowns, and so on).

These statistical methods offer new and exciting ways to analyze and improve on the quality of the traffic function and its activities. Nonetheless, the successful implementation of an SPC program has prerequisites. First, it is critical that senior logistics and corporate managers support the implementation of such a program. The adoption of SPC methods means management has to change some of the traditional ways of doing things. The best place to gain initial confidence is at the highest level possible in the firm. A common denominator among firms that have realized success is the direct involvement of top executives, including the chief executive officer. Second, for best results, statistical methods should be viewed as just one component of an overall quality management program. Other program elements may include goal formulation and identification, quality policies, supervisory training, quality awareness, error cause identification and removal, and performance management.

Computer Usage Patterns

Just as other parts of the firm have benefited from the increased speed, storage, and convenience of computers, so has the traffic function become more efficient as a result of using them. A recent survey of logistics executives provides some insights into how companies use computers advantageously in the traffic area.[13] From the results of this survey (see **Exhibit 17.15**), it is clear that many firms have already developed computer support for routine tasks as well as for traffic planning.

Computers are used most commonly in activities relating to freight bill payment (81 percent), freight bill auditing (65 percent), bill of lading generation (62 percent), and freight rate maintenance (60 percent). As a group, all these are routine, day-to-day activities that generally involve little or no decision making. Fewer firms show computer support of management activities such as freight cost budgeting (55 percent), routing (51 percent), shipment consolidation (36 percent), and carrier selection and evaluation (30 percent).[14] Nonetheless, recent advancements in hardware and software development, and in decision support

Exhibit 17.14 **Control chart examples**

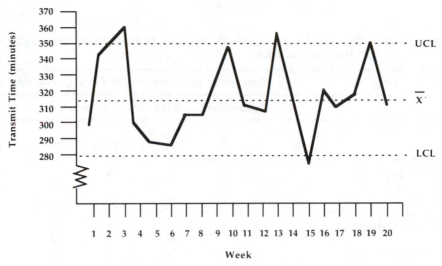

(a) Transit Time Performance

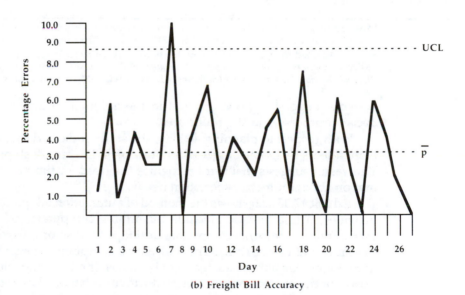

(b) Freight Bill Accuracy

UCL = Upper control limit
LCL = Lower control limit
$\overline{X}$ = Mean (of measurements)
$\overline{p}$ = **Mean (of percentages or proportions)**

Source: C. John Langley, Jr., "Information-Based Decision Making in Logistics Management," *International Journal of Physical Distribution and Materials Management* 15 no. 7 (1985), p. 51. Reprinted by permission.

■■■■■ *Exhibit 17.15* **Automation and computerization strategies in the traffic area**

Activity	*Percent of Respondents in Each Category*					
	Automated	**Methods of automation**				
		Main-frame	Mini	Micro	Service bureau	Carrier
Freight bill payment	81	39	0	4	36	1
Freight bill auditing	65	17	1	5	41	1
Bill of lading generation	62	54	4	2	1	1
Freight rate maintenance	60	27	2	15	12	3
Freight cost budgeting	55	31	8	17	2	0
Routing	51	43	5	4	0	0
Shipment forecasting	49	36	3	8	0	0
Shipment tracing	47	22	10	0	2	14
Shipment consolidation	36	22	3	9	1	2
Carrier selection and evaluation	30	17	3	9	2	0
Private fleet management	28	11	3	13	0	0

*The results in this table are from a telephone survey of 100 logistics and traffic executives taken in 1985. Totals across rows do not add to 100 because multiple responses were permitted.

Source: Adapted from C. John Langley, Jr., Robert J. Quinn, and Stephen I. Levine, "Microcomputers in Logistics: 1985," *Proceedings of the Annual Conference of the National Council of Physical Distribution Management* (St. Louis, Missouri: NCPDM, 1985). Reprinted by permission.

systems, suggest a growing trend in the automation and computerization of decision processes in the traffic area.

With respect to planning activity, the study indicated that 49 percent of the respondents had some degree of automation in the area of shipment forecasting. This result suggests that firms recognize the need for timely and accurate information about future transportation needs.

Exhibit 17.15 also shows the method of automation adopted for specific traffic activities. These methods include mainframe computers, minicomputers, microcomputers (personal computers, desktops, and so on), third-party "service" bureaus, and carrier firms. The mainframe computer is generally the source of automation for almost all the activity areas cited. Nonetheless, the growing power of the microcomputer, coupled with its relatively low cost, has resulted in an increasing acceptance of its use in the traffic area. Third-party service bureaus are being used widely in activities relating to freight bill rating, auditing, and payment. Also, it is apparent that traffic managers are relying heavily on the information-processing capabilities of carrier firms for matters relating to shipment tracing.

Several additional findings of the survey are noteworthy. First, a preferred strategy for automation is to link microcomputers to a mainframe or a minicomputer or to one another through some form of local area network (LAN). Second, there is an observable trend toward using computers dedicated to logistics and traffic support. This arrangement is often preferable to having to rely on the capabilities of a centralized computer department that shares its facilities with other departments and functions within the firm. Third, there appears to be a growing trend toward linking computers in the logistics and traffic areas of the firm with those of customers, vendors, service bureaus, and company-owned plants, warehouses, and distribution centers. **Exhibit 17.16** illustrates the variety of computer links now in use.

Exhibit 17.16 **Computer data links**

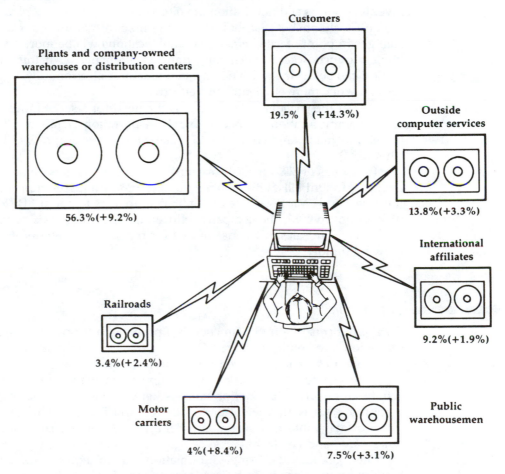

(Total exceeds 100% because some users have multiple links. Figures in parenthesis indicate the percentages of respondents who plan to establish links.)

Source: Dave Russ and Christine Fehlner, "Computers: Who's Got 'Em and What They're Doing with Them," *Handling and Shipping Management* (May, 1984) p. 42. Reprinted by permission.

Computers are only as effective as the software they use. Software suitable for conducting the various traffic management activities is apparently being developed rapidly.[15] Traffic managers should therefore expect a wide range of alternatives to be available.

Electronic Data Interchange

Electronic data interchange (EDI) among shippers, affiliates, customers, vendors, transportation companies, and warehouse operators is gaining popularity in the traffic field. Managers now see ordering, billing, payment, and continual inventory status and notification of volume projections taking place throughout the entire channel of distribution. For example, a manufacturer aware of daily changes in point-of-sale demand at retail locations will be more adept at scheduling production and anticipating transportation needs. If this kind of information can be made available to the manufacturer electronically, the entire process can work more efficiently and effectively.

One of the challenges facing EDI is the need for a set of industry standards and a communications protocol. The Transportation Data Coordinating Committee has made, and continues to make, significant progress in this area (see Chapter 4).

Two areas of EDI especially exciting for traffic managers involve (1) the processing of freight bills, bills of lading, and claims and (2) the transmission of rate information. K Mart was one of the first companies to adopt EDI for freight bill processing. In 1983, the company reduced its operating costs by 0.8 percent of $18.6 billion in sales.[16] The use of EDI for transmitting rate information is also growing rapidly.

Summary

For many firms, the traffic function is a prime area for computer-supported information and control systems. There is a wealth of opportunity to see meaningful advances and improved control in the traffic area.

Control refers to the task of measuring performance and taking corrective actions to ensure actual performance is kept in line with objectives. An information system is an integral part of the control process. The framework for a transportation information system identified two major components: key requirements and key techniques.

The most important single information system requirement is to determine key information needs, including information about (1) mode cost and service; (2) alternative distribution channels, customer service levels, and inventory tradeoffs; (3) fleet assets; (4) organizational units; and (5) integrated distribution opportunities. This information should be comprehensive, consistent across divisions, reliable, and detailed.

Key requirements also include the elements of good system design, such as (1) paperwork, (2) confidentiality, (3) back-up capability, (4) cash flow, (5) speed and reliability, (6) financial commitment, (7) compatibility, (8) accuracy, and (9) quality of information.

There are a number of key techniques for management control. One of these techniques, a transportation accounting system, should accumulate, organize and report financial and operating information to the traffic manager. A recent study, however, indicates that systems currently in use do not provide traffic managers with reliable or sufficient information to make effective decisions. Standard costing procedures and individual shipment costing methods are two important costing techniques.

Flexible budgeting, one of the most valuable tools to control transportation expenditures, establishes cost and revenue projections for a range of activity levels.

Freight payment systems accomplish four key tasks. They (1) achieve accuracy in preparing shipping documents, (2) ensure that the lowest legal rates are paid, (3) establish procedures for timely payment, and (4) permit comprehensive analyses of historical freight payment data. Important capabilities of such systems include preauditing, postauditing, rate analysis, report generation, and rate retrieval.

Productivity and performance measurement is at the heart of control. Productivity is the ratio of real output produced to real resources consumed. Utilization is the ratio of used capacity to available capacity. Performance is the ratio of actual output to standard output. To be effective, these measures need to be broken down by product groups, geographic regions, origin points, modes, and carriers and tracked over time.

Statistical process control (SPC), a relatively new technique to the traffic field, focuses on the variability of performance as well as on the average performance. Control charts, which comprise an important part of SPC, help analysts isolate any unusual results that require further investigation for causes and cures.

There is a clear trend toward using state-of-the-art hardware, software, and electronic data interchange technology to assist decision making and operations related to traffic management. Technological progress has already been made in the traffic function, but the true potential has only begun to be tapped.

Selected Topical Questions from Past AST&L Examinations

1. Until a few years ago, the performance of distribution managers was measured on the criteria of cost reduction and profit improvement. Deregulation, the cost of energy, and inflation have changed that, and today's distribution manager has an expanded, more complex role within the company. Discuss at least five

other performance standards that can be used for measuring distribution managers. (Fall 1982)

2. You've been asked to give a talk at a local AST&L college chapter. The topic is "Small Computer Applications in the Shipper-Carrier Relationship." What points would you include in this talk? Specifically, what applications and examples would you present? (Spring 1985)

3. The concept of responsibility accounting usually defines three kinds of centers: investment centers, profit centers, and cost centers. Define each type of center, and give examples of where they might be applied in a transportation context. (Spring 1985)

Notes

1. Ernst and Whinney, *Transportation Accounting & Control: Guidelines for Distribution and Financial Management* (Oak Brook, Illinois: National Council of Physical Distribution Management, 1983), p. xxv.

2. Ernst and Whinney, *Transportation Accounting & Control,* p. xxv.

3. This paragraph has been adapted from C. John Langley, Jr., "Information-based decision making in logistics management," *International Journal of Physical Distribution and Materials Management* 15 no. 7 (1985), pp. 41–55.

4. Douglas M. Lambert, "Distribution cost, productivity, and performance analysis," in *The Distribution Handbook,* James F. Robeson and Robert G. House, eds. (New York: The Free Press, 1985), p. 277.

5. Ernst and Whinney, *Transportation Accounting & Control.*

6. Financial accounting systems tend to develop natural cost accounts rather than functional cost accounts. This approach makes it difficult to trace costs to specific traffic or transportation functions. In addition, these systems generally post revenues and expenses on a "cash" basis rather than on an "accrual" basis. Since cash systems post revenues and expenses when they are incurred, it is sometimes difficult to "match" revenue items with expense items accurately.

7. Ernst and Whinney, *Transportation Accounting & Control,* p. 57.

8. Gene R. Tyndall, John R. Busher, and James D. Blaser, "Accounting and control in physical distribution— Phase I: Transportation," *Annual Conference Proceedings of the National Council of Physical Distribution Management,* 1984, pp. 83–84.

9. "FRAPS: Borden Chemical's marriage of rate control and bank freight payment," *Computer Systems Report* 1 no. 3 (March 1982), p. 1.

10. A.T. Kearney, Inc., *Measuring Productivity in Physical Distribution: A $40 Billion Dollar Goldmine* (Oak Brook, Illinois: National Council of Physical Distribution Management, 1978), p. 9.

11. This discussion has been adapted from C. John Langley, Jr., "Information-based decision making in logistics management," *International Journal of Physical Distribution and Materials Management* 15 no. 7 (1985), pp. 41–55.

12. C. John Langley, Jr. and J. L. Hartzell, "Statistical process control applications in materials management," *Annual Conference Proceedings of the National Council of Physical Distribution Management* (1983); J. H. Foggin, "Improving motor carrier productivity with statistical process control techniques," *Transportation Journal* 24 no. 1 (Fall 1984), pp. 58–74.

13. See C. John Langley, Jr., Robert J. Quinn, and Stephen I. Levine, "Microcomputers in logistics: 1985," *Annual Conference Proceedings of the National Council of Physical Distribution Management* (1985); see also Dave Russ and Christine Fehlner, "Computers: Who's got 'em and what they're doing with them," *Handling and Shipping Management* (May 1984), pp. 40–46.

14. The low figure (28 percent) for private fleet applications is an indication that only a small percentage of the respondents were involved in private fleet operations.

15. Knowledge of the kinds of software appropriate for logistics and traffic management applications may be gained by consulting a comprehensive reference source such as *Physical Distribution Software*, published annually by Arthur Anderson & Co., Management Information Consulting Division, Stamford, Connecticut. The most recent version of this document is reprinted in *Annual Conference Proceedings of the Council of Logistics Management* (1986).

16. Joan Feldman, ed., ''The bytes that bind: Electronic data interchange,'' *Handling and Shipping Management* (May 1984), p. 63.

Appendix 17.1 Measuring and Reporting Specific Benefits Resulting from Transportation Cost-Control Management

COST SAVINGS
A. Rate Negotiations on Traffic Previously Moving
   Measure of savings is difference between new rate and old rate.
B. Classification Rating Changes
   Measure of savings is the effect of the new rating on rates on which traffic is moving or will move.
C. Distribution or Materials Management Cost Savings
   Measure of savings is the net savings in such costs that occur from implementing transportation department's recommendation.
D. Consolidation of Small Lot Shipments
   Measure of cost savings is the difference between total cost of shipping via the consolidation (or distribution) point and the cost of shipping each small lot separately.
E. Contracts for Specific Traffic Movements
   Measure of cost savings is the difference in cost from the previous arrangement. Only report contracts for traffic previously moving but not under contract or a contract replacing a contract at a lower cost. All other benefits from contracts on specific movements are reported as "Cost Efficiencies."
F. Legal and Regulatory Proceedings
   Measure of cost savings is the amount recovered.
G. Negotiation of Favorable Commodity Description Interpretations on Traffic Previously Moving
   Measure of cost savings is difference in rates under the negotiated description compared to the previously used description.
H. National Contracts Provided by Corporate
   Measure of cost savings is difference in cost via lowest cost method otherwise available.
I. Discount Programs
   Measure of savings is actual refunds from carriers or credits from corporate.
J. Management of Private Trucks or Ships
   Measure cost savings by comparing private transportation costs to

Source: James E. Isbell, Jr., "Measuring and Reporting Benefits Resulting from Transportation Cost Control Management," in *Logistics: Concepts and Applications—Proceedings of the Logistics Research Forum* (1983), pp. 227–230. Reprinted by permission.

negotiated for-hire carrier costs or if none exist, to achievable levels of negotiated rates.

COST EFFICIENCIES

A. Rate Negotiations on New Traffic
   Measure of cost efficiency is difference between negotiated rate and the normal rate level.
B. Correcting Misdescriptions
   Measure of cost efficiency is the difference in rate under new description compared to the previously used description.
C. Changes in Carrier, Freight Forwarder, Port or Mode of Carriage
   Measure of cost efficiency is difference in cost between new and previous arrangements.
D. Cost Avoidance
   Report only if (1) an increased cost to the company is clearly prevented as a direct result of action taken by the reporting unit; and (2) the action required the same degree of research and analysis required in rate negotiations. Measure of cost efficiency is the increased cost that clearly was avoided.
E. Negotiation of Favorable Commodity Description Interpretations on New Traffic
   Measure of cost efficiency is difference in movement cost under negotiated description and description carrier held to be applicable.
F. Management of Private Railroad Cars or Barges
   Measure cost efficiency (or inefficiency) by comparison to previously set and periodically adjusted standards of utilization.
G. Contracts for Specific Traffic Movements

   - New Traffic
     Measure of cost efficiency is comparison to normal rate levels.
   - Second or Subsequent Years of Multi-year Contracts
     Measure of cost efficiency or inefficiency is comparison to contemporaneous rate levels under which skilled transportation management would move such class of traffic absent a contract.
   - Contract Replacing Another Contract at Higher Cost but with Quantifiable Benefit
     Measure of cost efficiency is comparison to contemporaneous rate levels under which skilled transportation management would move such class of traffic absent a contract.

COST RECOVERIES

A. Audit of Freight Charges

   - Rate Errors on Freight Bills
     Measure of cost recovery is actual amount of reduction.
   - Errors in Freight Charged on Vendors Invoices
     Measure of cost recovery is actual amount of credit taken.

- Errors in Freight Charged on Customer Invoices
  Measure of cost recovery is actual amount of debits paid by customers.
- Duplicate Bills Rejected
  Measure of cost recovery is the actual freight cost rejected.

B. Overcharge Claims
  Measure of cost recovery is actual amount paid by carrier less cost (or commission), if any, paid for outside audit.
C. Freight Loss or Damage Claims
  Measure of cost recovery is the full amount paid by the transport carrier and/or the insurance carrier.

Exhibit A17.1 suggests a method for documenting cost savings, cost efficiencies, and cost recoveries.

Exhibit A17.1 **Transportation cost control management benefits**

	Beneficiary					
	The company		**Other**		**Total**	
	Quarter	Ytd	Quarter	Ytd	Quarter	Ytd
Cost savings						
Rate negotiations						
Classification ratings						
Distribution or mt ls mgt						
Consolidation of small lots						
Contracts — Specific movements						
National contracts						
Discount programs						
Mgt of private trucks or ships						
Other						
Subtotal						
Cost efficiencies						
Rate negotiations						
Cost avoidances						
Mgt of private rail cars and barges						
Contracts — Specific movement						
Other						
Subtotal						
Cost recoveries						
Audit of freight charges						
Overcharge claims						
Loss and damage claims						
Subtotal						
Total benefits						
Reporting Unit_____						

chapter | 18

Organizing for Traffic Management

chapter objectives

After reading this chapter, you will understand:
The organizational issues and alternatives relevant to the traffic function.
The circumstances under which each of these alternatives may be appropriate.
The overall effectiveness of a firm's traffic function.
The issues of organizational growth and development as they relate to the traffic function.

Introduction

*Modern Traffic
Management Orientation*

*Elements of Good
Organizational Design*

Major Organizational Influences
Size of firm
Industry
Customer service philosophy
Logistics and transportation costs
Complexity of inbound and
outbound movements
Management orientation

*Organizational Issues
and Alternatives*
Line versus staff
Centralization versus decentralization

Functional versus
product-line orientation
The traffic function as part of a
distribution division
Integration of inbound and
outbound traffic activities
Span of control
Approaches to coordination

*Measuring
Organizational Effectiveness*

*Organizational Growth in the
Traffic Area*

The Issue of Titles

Summary

Questions

Introduction

Traffic management may be thought of as a process by which a firm mobilizes and coordinates the resources necessary to make effective decisions about the purchase and use of transportation services. As discussed in Chapter 1, this process involves the planning, implementation, and control of traffic activities in the context of overall logistical and corporate strategies and priorities. By now, it should be clear that decisions made in the traffic area can have an impact on the competitiveness and profitability of the firm as a whole. Only through a thoughtful and systematic approach to organization of the traffic management area will progress be made in this direction.

This chapter focuses on the people involved in traffic management and addresses a number of important organizational issues. Besides offering some helpful advice about the organizational alternatives that need to be considered, this chapter presents the premises, principles, and characteristics of good organization design. The application of these guidelines will facilitate decision making in the traffic area and will help make effective use of people's talents.

Although a certain degree of organizational formality clearly is necessary, it must be viewed only as a means to an end. This chapter emphasizes a number of specific issues about organizational structures. Nevertheless, the real goal is to find the most effective means of mobilizing the necessary resources to accomplish the objectives of the traffic function.

Modern Traffic Management Orientation

For many years, the traffic function in many firms was one that top management had little familiarity or understanding of. This was due largely to the complex ways in which carriers published their rate and service offerings in their tariffs, as well as to the myriad procedures that needed to be followed to determine the specific price for any particular transportation service. Few top executives felt they had a good grasp on what activities actually took place in the traffic area, and most traffic managers seemed to find this relationship a comfortable one. The goals for the traffic function were stated simply: to make sure that the products and materials were at the right place, at the right time, and in the proper condition, and also to minimize transportation cost whenever possible. In the regulated environment, traffic managers generally had a limited number of price-service options to choose from and, as a result, needed far fewer analytical skills than they need in today's largely deregulated environment.

In progressive companies, the traffic function is now viewed as a key component of the firm's overall logistical capabilities, and the traffic manager has a good deal of control over a wide range of activities related to the movement of materials, product, and supporting information. Elements contributing to this trend include deregulation and a greater awareness by buyers of all aspects of the purchase decision, including transportation and logistical responsibilities.

Traffic managers are increasingly regarded as key participants in the development of overall logistical strategies for the movement of products and materials at individual firms. Thus, today's successful traffic managers must be good general managers. **Exhibit 18.1** profiles the most common traffic management activities.

A number of dramatic changes have taken place in recent years in the traffic management area of the firm, as well as in the broader area of business logistics management. That the word ''modern'' is now used in the context of these two areas of management is largely the result of the significant and methodical evolutionary process discussed in Chapter 1. Several observations are worth reviewing.

First, the traffic management area has experienced a phased evolution similar to that of the logistical process itself. Second, the once fragmented activities within the traffic area have become integrated, so that today, it is common for most of the activities listed in Exhibit 18.1 to fall within the overall responsibility

■■■■■■ *Exhibit 18.1* **Representative traffic management activities**

1. *Identification of transportation needs*
   Working closely with people in the marketing and operating areas of the business to identify needs for transportation services as far in advance as possible, and to develop sound business strategies for meeting those needs.
2. *Carrier selection*
   Recognize that transportation companies are actually vendors selling a service rather than a product and that the universally applicable principles of good purchasing management should apply as well to the purchase of transportation services. Many companies have developed formal processes for screening and selecting carriers they do business with.
3. *Rate/service negotiations and contracting for carrier services*
   Given the pricing flexibility in today's market for transportation services, formal efforts at rate and service level negotiations can produce remarkable results. Deep discounts and acceptable levels of service are available from many providers of service in exchange for commitments to have certain tonnage, volumes, and so on. Regarding carriers as strategic elements of the firm's distribution capability is essential.
4. *Shipment planning*
   Developing a set of policies that should guide the process of selecting carriers to handle individual shipments, and planning for the movements of individual shipments as far in advance as possible. Effectiveness will depend on the ability to forecast demand for finished products and the certainty of production schedules and materials needed.
5. *Consolidation and load planning*
   To strategically take advantage of opportunities to consolidate shipments from one or more facilities and to use pool or distribution services as appropriate. Benefits should include economies of scale, service improvements, and simplification of the administrative effort in the traffic management area.
6. *Ordering service*
   Making daily arrangements with individual carriers for the pickup and delivery of individual shipments. Goals are to arrange for, monitor, and control the response of individual carriers to requests for service.
7. *Prerating and preparation of bills of lading*
   Before the actual transportation movement, to insert the commodity description, class rating, and anticipated freight charges on the bills of lading for the shipments.
8. *Staging of loads/carrier pickup*
   Being sure that shipments are available at the time promised to carriers, and that they have been staged and prepared accordingly for convenient carrier pickup. May also include developing appointment schedules for carrier pickups.
9. *Tracing/expediting*
   To determine the specific whereabouts of particular loads as needed, and to take steps to expedite their transport when necessary.
10. *Freight bill audit and payment*
    Includes the audit of freight bills for overcharges by carriers; the processes of freight bill monitoring, payment, and control; and the monitoring and control of any discount or allowance programs with individual carriers.

(continued)

Exhibit 18.1 (continued)

11. *Claims*
    Filing of claims for freight loss or damage, and the monitoring and control of freight claims experience by carrier.
12. *Private carriage*
    To take full responsibility for the management, operation, and control of a firm's private fleet or other available form of proprietary transportation capability.
13. *International transportation*
    Responsibility for all activities relating to import or export of products and materials as well as to finished-goods output of the firm.
14. *Legal activities*
    To make certain that all traffic activities are consistent with local, state, and federal laws and regulations, and to monitor the activities of carriers to see that their management, operating, and pricing practices are in observance of the applicable laws and regulations.
15. *Service-level monitoring, evaluation, and control*
    To regularly evaluate the quality of transportation service being provided by carriers. This requires devising a formal system that will feed back directly into the carrier selection decision process.
16. *Transportation activity, carrier, and cost analysis*
    To regularly analyze the different types of transportation activities (by volume and expenditure levels) taking place in the company, and to segment these results by carrier, freight lane, customer, and so on. This requires devising a formal system that will feed back directly into the carrier selection decision process.
17. *Management information system*
    To see that traffic activities and management needs are supported by the availability of accurate, timely information. The management information system should be responsive to the principal information needs in the traffic area, and should be accessible by those people having a need for such information.

of a single manager. Even though many traffic organizations have retained a strong functional orientation, the various activities are usually coordinated by a single manager. Third, more than ever before, traffic managers now anticipate problems and customer needs rather than simply react to them.

Fourth, the traffic area has become a more meaningful component of an effective logistics organization through the need to consider "tradeoff" opportunities with other areas such as inventory, warehousing, administration and systems planning, production planning, purchasing, and customer service. Fifth, there has been a trend toward recognizing the strategic value of effective traffic decision making to the firm as a whole. Whereas previously the traffic management function had a narrowly defined focus, recent experience has shown that decisions made in this area can have an impact on how a firm relates to its customers. Finally, although the traffic function is also operating in a changing business environment, it appears that effective strategic planning in the traffic area has been successful in diluting some of the adverse effects of this uncertainty.

Elements of Good Organizational Design

Before discussing organizational alternatives, it is helpful to review a number of underlying premises, principles, and characteristics of good organizational design that should be in evidence in any successful traffic organization. These elements are as follows:

1. *Organizational structure has strategic importance.* A firm may pursue various business strategies to achieve its goals. The selection of an organizational framework for the traffic area is itself a decision of strategic importance.

2. *Organizational structure should be consistent with strategy.*[1] The organizational structure chosen for the area of traffic management should be in response to the overall goals, objectives, and strategies selected for traffic and logistics management, not vice versa. Although a common temptation is to mold other strategies to fit within the existing organizational structure, the organization should be designed around, and correspond to, the strategic plan. For example, if the strategic plan is structured around geographic regions or product lines, the organizational structure should be consistent with that priority.

3. *Specialization should not limit coordination.* Specialization within any organizational structure generally promotes greater expertise and efficiency in performing individual activities. It also places a greater demand on the need for effective communications and coordination to facilitate the conduct of activities. Thus, functional specialization within the traffic area should not adversely affect the ability of all functions to relate and effectively coordinate with one another.

4. *Authority and control need to be commensurate with responsibility.* A traffic manager with responsibility for making cost-effective purchases of transportation services must be able to influence the choice of carriers. Otherwise, the financial-performance goal is less likely to be met.

5. *Organizational structure must be flexible.* Whatever the degree of formality or informality in the traffic organization, it should be flexible enough to respond effectively to changing business conditions. The organization must be able to identify the need for change when it exists and to respond to it appropriately.

6. *Responsibilities should be clearly defined.* Each person with management responsibilities in the traffic organization needs to understand the cor-

porate, logistics, and traffic missions and objectives, as well as how specific job responsibilities fit within that framework. Traffic managers should develop a concise statement of the strategic and operational objectives that need to be met regularly. Through conscientiously adhering to this process, each person in the traffic area will make the most meaningful contribution to the goals and objectives of the traffic function.

7. Leadership people in the traffic area must have access to upper-level management. Whether through the business logistics management structure or through some form of coordinative mechanism, traffic management should help formulate higher-level business strategies. Key traffic decisions and tradeoff opportunities must be considered at the most appropriate level in the firm. For example, when it becomes apparent that delivery-time promises made by a company's sales force mean using high-priced transportation services but little in the way of additional sales, the organization must be able to facilitate resolution of the matter.

Major Organizational Influences

The enhanced role of the traffic function is a direct result of the well-documented trend toward more, rather than less, sophistication and formality in the management of all logistical activities, including traffic. Although the degree of formality and sophistication will vary from firm to firm, several elements help explain whether a firm will benefit much from a more formal approach to management. These elements are discussed next.

Size of Firm

There is no question that there is a greater likelihood of organizational formality in the traffic area among larger firms. Larger firms must regularly deal with traffic matters on a far greater scale than smaller firms, and the degree of complexity and the need for coordination become more critical as the size of the firm itself increases. Large firms are more likely to have separate divisions, operate more facilities, and create a clearer distinction between line and staff activities. Large organizations also employ more people, which raises questions about appropriate span of control and organizational structure. Thus, large firms generally will need organizational formality and sophistication.

Industry

The nature of the trade will influence the amount of management attention given to inbound or outbound movements. Industries having a greater concern for inbound movements of product and materials include construction, finance, insurance, real estate, services, and public administration. The construction industry, for example, is characterized by a significant materials management responsibility, but there is virtually no outbound physical distribution management. Further, examining data on commodity flows into an area such as Washington, D.C., will show a tremendous imbalance between inbound and outbound tonnages. Little freight moves out of Washington, D.C., and most of what moves in is in the category of paper and paper products for the public administration activities so prevalent there.

Many companies in these industries, of course, may need to consider logistical responsibilities on the outbound side as well. Service firms of all types place a great deal of importance on managing the distribution of their offerings to the marketplace. However, this concern is defined principally in the context of overall logistical responsibilities and does not result in many purchases or much use of freight transportation services. For this reason, the service sector, broadly defined, is predominantly involved in inbound, not outbound, transportation services.

Firms in the extractive and agricultural industries are more intensive consumers of outbound transportation and distribution services. The tonnages that must be moved require that attention be directed to the traffic management area.

Most companies engaged in manufacturing and the retail or wholesale trades emphasize both inbound and outbound transportation and logistical responsibilities. Companies that provide transportation, communications, electricity, gas, and sanitary services also need effective management of both outbound and inbound movements. Outbound traffic flows in these industries, however, are often somewhat unconventional. For example, outbound logistical needs for a power company would include using a grid to distribute stored electrical power, and satellites would be used to distribute the output of communications companies.

Customer Service Philosophy

Many firms have recognized that the traffic function can influence the level of service provided to customers. Good customer service can translate into a competitive advantage for a firm. Thus, it is easy to understand why many companies view their logistical and traffic capabilities as among their chief strengths in the marketplace.

Although there are numerous ways of measuring customer service, some of the more popular ones are product availability, order cycle consistency and length, service flexibility, service information, service malfunction, and postsale product support. These measures were identified in a study conducted for the Council of Logistics Management.[2] This study also examined the importance of these measures by industry; the results of this investigation are shown in **Exhibit 18.2.**

Product availability and order cycle time were chosen as the two most im-

Exhibit 18.2 **Importance of customer service elements by industry**

Elements	All Industries	All Manufacturing	Manuf. Chem. and Plastics	Manuf. Food	Manuf. Pharm.	Manuf. Elect.	Manuf. Paper	Manuf. Machine	Manuf. All Others	All Merchandising	Merch. Consumer	Merch. Industrial
Product availability	42.4*	42.7	44.5	37.1	39.7	32.7	41.3	56.3	50.5	43.1	40.5	43.9
Order cycle	20.7	19.4	17.4	21.4	28.0	17.4	12.3	10.7	18.0	25.5	26.2	20.2
Distribution system flexibility	11.5	11.6	10.6	12.9	10.6	12.9	18.5	17.3	12.4	10.1	9.0	12.9
Distribution system information	12.6	12.4	11.7	14.8	9.0	16.7	20.1	1.0	9.5	11.8	14.0	8.0
Distribution system malfunction	7.7	8.0	9.1	10.3	7.8	7.9	4.5	4.0	5.4	7.2	8.2	10.0
Postsale product support	4.5	5.1	6.2	2.3	2.9	11.7	1.8	10.0	4.1	2.3	2.1	5.0

*More points indicate greater importance. Relative rankings represent average distribution of 100 points to elements.

Source: Bernard J. LaLonde and Paul H. Zinszer, *Customer Service: Meaning and Measurement* (Oak Brook, Illinois: National Council of Physical Distribution Management, 1976). p. 117. Reprinted by permission.

portant customer service variables among nearly all industries surveyed. That "consistency" and "dependability" are the principal management concerns regarding these variables is certainly understandable. Decisions made in the traffic and logistics areas can have a significant influence on whether acceptable levels of service are provided to customers, so the strategic importance of decision making in these areas of management cannot be ignored.

Since the traffic function is an integral part of the firm's more comprehensive and complex logistical system, it should be expected that customer service capability will be greatest in firms having a more formal, and therefore more responsive, organizational structure.

Logistics and Transportation Costs

The firms that experience high costs in the traffic and logistics areas are most likely to benefit from a greater degree of organizational formality and sophistication. Although this is certainly logical, it is necessary to recognize that there is more than one way of expressing cost, and it is important to distinguish between the alternatives. Specifically, the term "cost" can refer to one or more of the following:

- Total cost (as in absolute dollars)
- Cost as a percent of sales
- Cost per unit (such as cost per hundredweight)

Each of these is computed differently, yet they all refer to the same concept. Companies involved in the manufacture or resale of high-valued products typically have the lowest traffic and logistics costs when expressed as a percent of sales, but they also have a relatively high total cost. It is also not uncommon for these products to have high distribution costs when costs are expressed in terms of cost per unit.

Exhibit 18.3 shows the results of a recent study that profiled distribution costs in a number of key industries in the United States. Although distribution cost for the average of all companies surveyed was found to be 7.5 percent of sales, and $33.35 per hundredweight, substantial variation from these averages is apparent. It should be noted that there is some bias regarding the use of Exhibit 18.3, in that it focuses on broadly defined distribution cost, not on traffic costs alone.

A more meaningful perspective of cost can be gained if it is viewed in terms of (1) its relation to other functional costs that are incurred; and (2) the extent to which it is controllable, in contrast to noncontrollable. It is important to view the issue of traffic management costs in comparison to the magnitude of other activity-area costs incurred by the firm. For example, if traffic costs are high even

■■■■■■■ *Exhibit 18.3* **Average distribution cost by industry**

	Percent of Sales	*$/cwt*
All companies	7.50	33.35
Industrial nondurable	8.10	16.08
Plastics	8.65	4.90
Chemicals	6.40	3.50
Nonchemical	8.82	12.50
Hospital and medical supply	8.15	58.55
Industrial durable	7.40	37.64
Industrial durable < $10/lb	8.07	27.49
Consumer nondurable	7.14	36.58
Food products		
Dry and packaged food	8.66	6.88
Canned and processed food	8.58	4.79
Temperature-controlled food	7.94	7.03
Nonfood in grocery channel	9.39	9.48
Pharmaceuticals > $10/lb	3.77	119.35
Pharmaceuticals < $10/lb	5.59	22.63
Consumer durable	8.86	44.48
Retail stores	8.22	29.13

Source: Herbert W. Davis, ''Physical Distribution Costs 1984: In Some Companies, a Profit Contribution in Others, Cost-Price Margins Continue to Dwindle,'' *Proceedings of the Twenty-Second Annual Conference of the National Council of Physical Distribution Management* (Dallas, Texas, September, 1984), p. 37. Reprinted by permission.

in comparison to warehousing, inventory, packaging, advertising, or product development, a higher degree of organizational formality would be appropriate.

Regarding the issue of control, because there are sometimes few alternatives for the traffic manager to consider, a more formal and sophisticated organizational structure would probably do little to enhance performance in this area. For example, a large volume shipment of grain is likely to be captive to the rail or water modes, thus limiting the choices for the traffic manager. Alternatively, the overall cost of distributing consumer nondurable goods is relatively high, as seen from Exhibit 18.3, yet there is a good deal of choice and potential control of the cost and service aspects of such a movement.

Complexity of Inbound and Outbound Movements

Traffic management becomes a more formidable task when the complexity of a firm's inbound and outbound transportation movements increases. Organizational formality is helpful to a firm that has numerous suppliers and customers, uses a variety of modes and carriers, and has shipments of varying sizes and physical characteristics.

Companies having a more sophisticated procurement function will experience a greater need for expert coordination with the traffic area. For example, the use of contemporary approaches to inventory management such as materials requirements planning (MRP) and just-in-time deliveries have increased the dependence on the traffic area for precise timing of inbound transportation movements. Offshore sourcing and overseas subassembly will also lead to a need for greater organization formality in the traffic area.

Regarding outbound transportation decisions, the degree of complexity will increase as more sophisticated and elaborate channels of distribution are in evidence. For example, the outbound distribution of grocery products from manufacturer to ultimate consumer is typically complex, emphasizing the need for more effective management of the traffic area. Furthermore, companies that have adopted the use of distribution requirements planning (DRP) have recognized the need for a high level of coordination and precision in the timing of outbound product movements.

Management Orientation

Generally speaking, it is helpful to adopt the prevailing management style of the firm. For example, a high level of formality in the way the firm is organized strongly signals that the traffic function would benefit internally from a similarly formal organizational structure.

Another way to address this issue is to assess whether the firm and the logistics area have been able to develop effective management systems that facilitate strategy formulation and decision making. The traffic area of the firm does lend itself well to systems management, and evidence of the systems approach in the firm should speak well for its potential applicability in the traffic area.

A study conducted by A.T. Kearney and *Traffic Management* magazine contains an interesting observation about the development of the logistics function.[3] In investigating whether the apparent level of sophistication was related to the type of industry, the answer was yes and no. The study did identify electrical and electronic products, food and drugs, and chemicals and allied products industries as the most progressive and innovative, but it also cautioned readers that these three industries achieved an earlier start than others in the area of logistics management. Their conclusion is that other industries are moving steadily toward a broader physical distribution management, and that the rate of progress is not necessarily related to the type of industry. This statement similarly holds true also for the traffic area of the firm.

Organizational Issues and Alternatives

A number of key issues must be addressed when evaluating alternative organizational designs for the traffic function.[4] These issues are examined next.

Line versus Staff

The issue here is not whether the traffic function should be a staff function only or a line function only; rather, the issue is how the traffic area can be organized so that these two important functions can take place and achieve the desired results. A line operation will eventually prove ineffective unless supported by a competent staff capability. On the other hand, staff recommendations will prove useless unless there is an effective mechanism for implementation.

The principal distinction between line and staff is that the line function is responsible for making and implementing decisions, whereas the staff function provides support and guidance to the line function. In many companies, particularly divisional, decentralized organizations, the staff function also is responsible for overall policy formulation. However, the implementation responsibility is left solely to the line function.

Exhibit 18.4 places traffic management activities into line and staff categories. This exhibit also makes an important distinction between the strategic and operational nature of various line activities. The line-strategic activities have far-reaching effects for the traffic area as well as for the logistics area. The line-operational activities are oriented more toward the conduct of day-to-day matters. Although a single principal orientation is identified for most activities, responsibility for private carriage is regarded as a line-strategic and a line-operational activity. Activities related to international transportation, management information systems, and service-level monitoring, evaluation, and control are oriented closely to the staff function as well as to both types of line functions. In areas such as these there is a reasonable expectation of dispute among line and staff managers about "territorial rights."

Centralization versus Decentralization

The distinction between centralization and decentralization in organization structure is based upon the degree of authority and profit responsibility delegated to specific operating units. Within an enterprise, units or divisions are considered highly decentralized if each is able to function on an almost autonomous basis. In a fully decentralized organizational structure, each division would be responsible for providing its own logistical requirements.[5]

Rather than argue whether centralization or decentralization of traffic activities is more appropriate in any given situation, it is more helpful to identify the

███████████ *Exhibit 18.4* **Line versus staff nature of traffic management activities**

Activity	Principal Orientation
Identification of transportation needs	Staff
Carrier selection	Line-strategic
Rate/service negotiations and contracting for carrier services	Line-strategic
Shipment planning	Staff
Consolidation and load planning	Staff
Ordering service	Line-operational
Prerating and preparation of bills of lading	Line-operational
Staging of loads/carrier pickup	Line-operational
Tracing/expediting	Line-operational
Freight bill audit and payment	Line-operational
Claims	Line-operational
Private carriage	Line-strategic and line-operational
International transportation	All
Legal activities	Staff
Service-level monitoring, evaluation, and control	All
Transportation activity, carrier, and cost analysis	Staff
Management information system	All

key conditions under which each would be more suitable.[6] In particular, centralization would be preferable when (1) there is a high priority on a consistent, firmwide implementation of overall policies and procedures; (2) there is a noticeable lack of depth of managerial talent in the traffic area; (3) highly capable management information systems serve as an effective link between operating units; or (4) there is a high cost of decentralization or, conversely, a high opportunity cost in terms of lost economies of scale from not centralizing traffic responsibility. In addition, the centralization of a firm's overall logistical responsibilities would be a strong signal to similarly centralize the traffic activities.

Some conditions under which decentralization would be better include when

- The types of products handled differ from unit to unit.
- Different markets are served and different vendor networks are chosen for purchased materials.
- It is necessary to maintain close contact with transportation firms in individual freight lanes and market areas.

- There is a high priority on providing meaningful opportunities for people to develop and refine their traffic management skills before assuming responsibility on a larger scale.
- There is an emphasis on providing a high level of customer service (although many firms are centralized in the traffic area and manage to provide a highly competitive level of customer service).

Exhibits 18.5 and **18.6** show typical organizational structures that might exist under the conditions of centralized and decentralized traffic management responsibilities.

Exhibit 18.5 **Example of centralized traffic function**

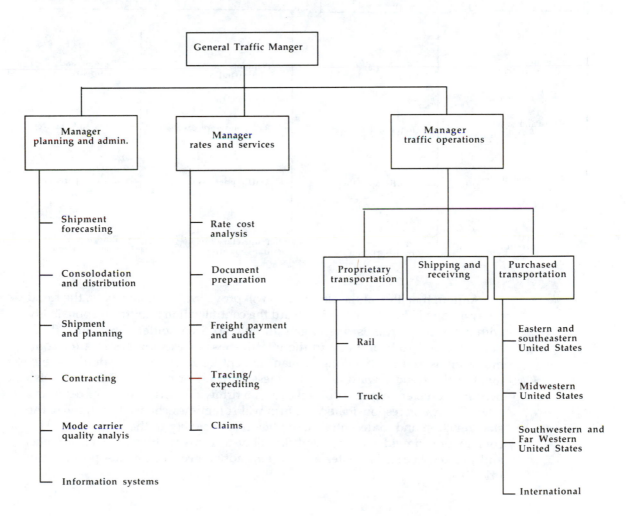

■■■■ *Exhibit 18.6* **Example of decentralized traffic function**

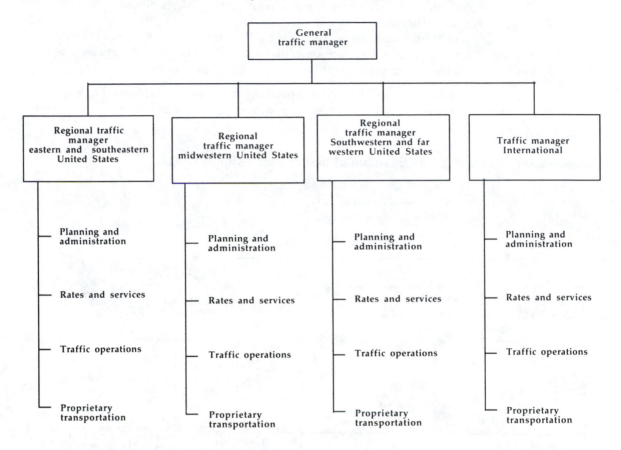

Although neither of the alternatives will prove best in all instances, the trend today among U.S. businesses is toward the centralization of traffic responsibility within the firm. This is due largely to enhanced computer capabilities (both hardware and software) and to the effectiveness of existing systems for communications and the handling of management information. An additional element is the trend toward doing business with fewer carriers. This trend also permits a greater degree of control over the firmwide consumption of resources related to traffic responsibilities. A firm will sometimes choose to centralize the staff function and to decentralize the line responsibility in the traffic area. This combination provides the needed flexibility to firms with individual products and markets, yet it facilitates a consistent adherence to firmwide policies and priorities.

Functional versus Product-Line Orientation

Conceptually, this topic is quite similar to the discussion of centralization versus decentralization. A functionally oriented traffic department is one having a well-defined set of centrally available resources for making and implementing decisions. These resources and capabilities are available to assist in the purchase and use of transportation service across all product lines. A traffic department that has a product-line orientation has separate areas of traffic activity, one for each area of product-line responsibility.

The principal benefit of functional orientation is the wealth of expertise that results from being able to concentrate resources in such a strong, central capability. Alternatively, the most significant misgiving is that it is difficult to customize decision making to the peculiarities or special needs of individual product lines. Although progress is being made in this direction, a functionally oriented traffic department is generally not very sensitive to such needs.

The traffic department oriented toward product lines, however, benefits from traffic decisions being tailored to the needs of individual product lines. This type of organization would be appropriate when the customer service priorities differ widely among individual product lines. A major drawback to this organizational strategy is that economic considerations frequently preclude the opportunity to have sufficient depth of management talent in all areas where traffic decisions are made. In such cases, the best recommendation is to rely to some extent on centralized capabilities that may be able to provide benefit to more than one product line at a time.

The Traffic Function as Part of a Distribution Division

Recent studies have indicated a trend toward using the distribution division as an acceptable organizational alternative for the logistics function.[7] For example, a survey of logistics executives showed that the number of people with management responsibilities within a distribution division of a larger company increased from about 14 percent in 1980 to 20 percent in 1984. This figure decreased to approximately 18 percent during 1985, largely because of more frequent use of the centralized logistics department.

When the traffic organization exists within a distribution division, traffic management generally assumes a supporting role, in that it actually provides a service to the various operating units of the business. The traffic and logistics functions serve the rest of the firm just as would a centralized accounting area. A major advantage to this type of arrangement is the strong, consistent application of centralized policies in the traffic area. Alternatively, sometimes this leads to an insensitivity to the needs of particular operating units (or products or markets) and has potentially adverse effects on customer service.

Integration of Inbound and Outbound Traffic Activities

This issue becomes significant when economies can be gained by having a single traffic department that assumes responsibility for the purchase and use of both inbound and outbound transportation services. Though this situation would occur most commonly in a firm having an integrated logistics system, it also appears frequently when the direction of inbound and outbound movements are complementary and where the types of transportation equipment needed for both moves have a basic similarity.

Closely related to the coordination of inbound and outbound is the use of a firm's private fleet capability. Since there is a definite trend toward managing private fleets as separate business units, the opportunity to haul a company's inbound and outbound shipments in opposite directions is certainly viewed as a key to productivity.

Span of Control

The term ''span of control'' refers to the extent of authority either held by any individual person or ascribed to any specific position within an organization. The span of control is typically measured by how many people report to a given person. For example, a general traffic manager at corporate headquarters may have a span of control that extends to eight regional traffic managers.

It is easy to confuse the concept of span of control with the issue of how much authority a manager really has. Here the number of reporting people becomes less meaningful. A firm's general traffic manager may have few people reporting directly but still has total responsibility and authority to see that all traffic activities are managed toward the achievement of the firm's logistical and corporate goals.

What the appropriate span of control should be for each person in the traffic area is an important issue. Because some managers are more capable than others, they can assume a greater span of control. Other managers work effectively when job responsibilities have been defined more narrowly. In addition, the talents of subordinates frequently affect the perception of a manager's capabilities, especially when the manager is near the top of the organizational hierarchy. This relationship implies that many high-level managers are effective because they delegate decision making to subordinates and do not attempt to do everything themselves.

The scope of span of control is closely related to the diversity of the business itself. Companies with multiple product lines, for example, may find it advantageous to decentralize traffic activities to reduce the span of control of certain persons. Alternatively, single-product companies, or companies having only a few product lines with similar transportation needs, may find that a broad span of control is appropriate.

Exhibit 18.7 illustrates the relationship between span of control and the basic nature of organizational structure. Each of the examples shown includes sixty-four operatives, but each illustrates the result from incorporating a different span of control. In Exhibit 18.7(a), where the span of control is sixty-four, each operative reports to a single manager. The organizational structure is obviously flat and wide, a natural result when the span of control is so large. In Exhibit 18.7(c), where the span of control is two, the organizational structure is characteristically narrow and tall. Organizations having a narrow span of control generally have more managers. On the other hand, the narrow span of control supports more direct supervision of people in the organization.

Approaches to Coordination

The effectiveness of decision making in the traffic area relates to the ability of a firm to coordinate its various resources toward the achievement of preestablished goals and objectives. In the general field of management, there are three conceptual approaches to effective coordination:[8]

1. Use basic management techniques. These techniques include the rules and programs, as well as the managerial hierarchy and the joint planning used for coordination. **Exhibit 18.8** characterizes these alternative mechanisms in terms of complexity, cost, and information processing capability.

2. Increase coordination potential. The potential for coordination may be enhanced through two general mechanisms. The first is the effective use of informatior systems to achieve an appropriate level of vertical coordination throughout the managerial hierarchy. Second is the use of several techniques that facilitate lateral coordination across the chain of command. These techniques include direct contact between key people; liaison roles to become involved in interdepartmental issues; committees and task forces; integrator roles (for example, project manager); managerial linking roles; and a matrix organization approach.

3. Reduce the need for coordination. Two of the principal approaches to simply reducing the need for coordination are (1) the creation of "slack resources" (for example, extra people, longer lead times, excess budget) so that individual departments do not sense as great a need for coordination; and (2) the creation of "self-contained" tasks so that an individual department does not have to work closely or share resources with other units within the company.

Exhibit 18.7 Relationship between span of control and organizational structure

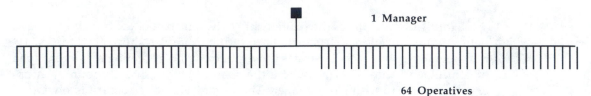

1 Manager

64 Operatives

Span of management = 64

(a) Very wide span of management* and very flat operational structure

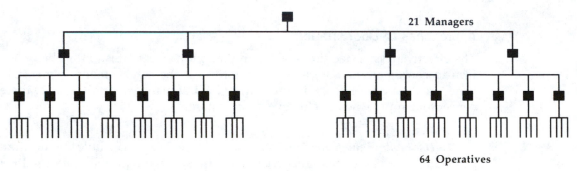

21 Managers

64 Operatives

Span of management = 4

(b) Narrower span of management and taller operational structure

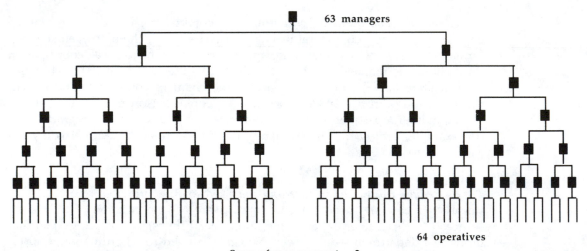

63 managers

64 operatives

Span of management = 2

(c) Very narrow span of management and very tall organizational structure

■■■■■ *Exhibit 18.8* **Mechanisms for coordination and control**

Mechanism	Complexity	Cost	Information Processing Capacity
1. Rules and Programs	Simple	Cheap	Low
2. Hierarchy			
3. Joint Planning			
4. Either, or Both			
Formal Information Systems (e.g. MIS) Lateral Relations (e.g. teams)	Complex	Costly	High

Source: Adapted from Michael L. Tushman and David A. Nadler, "Information Processing as an Integrating Concept in Organizational Design," *Academy of Management Review* 3, no. 3 (July, 1978), Fig. 2, p. 618.

■■■■■ *Measuring Organizational Effectiveness*

Although there are many different ways in which the effectiveness of the traffic organization can be measured, a number of elements (such as flexibility, response time, coordination, communication, satisfaction, productivity, cost efficiency, and goal orientation) regularly appear in effective traffic organizations. The priorities attached to each of these elements should be decided by both the traffic manager and higher level logistics and corporate management, and a workable scheme should be developed to measure and regularly report on the status of each of these concerns. The key performance indicators identified in Chapter 17 (see Exhibit 17.12) can be helpful in measuring the effectiveness of a firm's traffic function, as well as in providing an early warning of changes that may have an impact on the company. Although these indicators are far more specific than the elements mentioned above, they represent a useful set of measurements for evaluating the overall effectiveness of the traffic function. For best results, any general observations concerning the traffic function should be made on the basis of several available measures, not on the basis of a single indicator.

---

*Professor Stoner uses the term "span of management" rather than "span of control" because "management" more accurately indicates the breadth and scope of the supervisor's function. As managers, we do much more than merely control our subordinates' work. Stoner discusses span of management in pages 290–299 of his text.

Source: James F. Stoner, *Management*, (Englewood Cliffs, New Jersey: Prentice-Hall, 1982), p. 293. Reprinted by permission.

480 Chapter 18 Organizing for Traffic Management

Organizational Growth in the Traffic Area

Organizational structure should follow strategy, in that the organization should be structured to help achieve predetermined goals. Given this premise, and the knowledge that goals and strategies do change over time (or at least evolve), it can be expected that the recommended organizational structure will change similarly over time.

Exhibit 18.9 illustrates an interesting thesis concerning five phases of growth

Exhibit 18.9 The five phases of growth

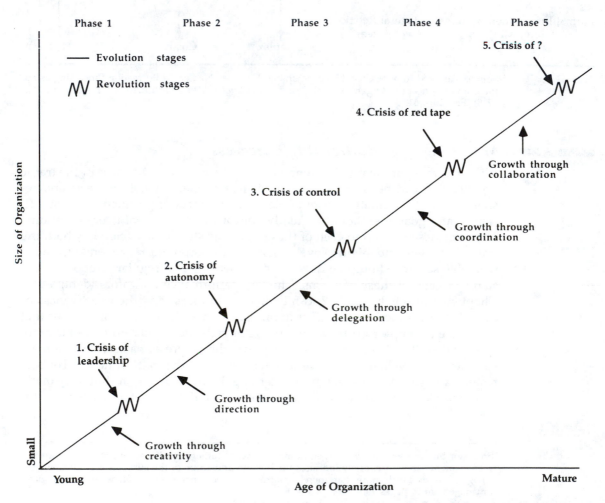

Source: Larry E. Greiner, "Evolution and Revolution as Organizations Grow," *Harvard Business Review* 52, no. 4 (July–August, 1972), p. 41. Copyright 1972 by the president and fellows of Harvard College; all rights reserved. Reprinted by permission.

■■■■ *Exhibit 18.10* **Organization practices during evolution in the five phases of growth**

Category	Phase 1	Phase 2	Phase 3	Phase 4	Phase 5
Management focus	Make and sell	Efficiency of operations	Expansion of market	Consolidation of organization	Problem solving and innovation
Organization structure	Informal	Centralized and functional	Decentralized and geographical	Line-staff and product groups	Matrix of teams
Top management	Individualistic and entrepreneurial	Directive	Delegative	Watchdog	Participative
Control system	Market results	Standards and cost centers	Reports and profit centers	Plans and investment centers	Mutual goal setting
Management reward emphasis	Ownership	Salary and merit increases	Individual bonus	Profit sharing and stock options	Team bonus

Source: Larry E. Greiner, "Evolution and Revolution as Organizations Grow," *Harvard Business Review* 52, no. 4 (July–August, 1972), p. 45. Copyright 1972 by the president and fellows of Harvard College; all rights reserved.

that an organization (such as a traffic organization) experiences as it evolves from a young to a mature organization. The five major stages of organizational evolution are seen to be creativity, direction, delegation, coordination, and collaboration. At some time late in the phases associated with each of these stages, however, some sort of crisis is likely to occur. Respectively, the first four of these have been termed leadership, autonomy, control, and red tape. Since there are a number of possibilities for the fifth, Exhibit 18.9 shows a question mark for that particular crisis.[9] **Exhibit 18.10** profiles the specific organization practices that typically occur during each of these five phases.

■■■■ *The Issue of Titles*

The Official Directory of Industrial and Commercial Traffic Executives (the "Blue Book") is published annually by the Traffic Service Corporation. Along with a variety of other useful material, it contains key items of information relating to traffic executives in many of the commercial and industrial firms located in the United States.[10] The directory includes the identities and titles of each executive and manager, listed by company and operating division. Although the highest ranking traffic executives carry a wide range of titles, the most common are general traffic (transportation) manager; director (manager), corporate traffic (transportation); national traffic manager; and traffic manager.

Summary

A number of organizational issues and alternatives are relevant to the traffic function. The principal areas of concern are (1) the underlying principles and characteristics of good organizational design that should be in evidence in the traffic area, (2) the major influences that will determine the most appropriate organization for any individual traffic function, (3) the organizational design issues confronting management, (4) the measurement of organizational effectiveness, and (5) the topic of organizational growth.

Organizational influences include (1) the size of the firm, (2) the industry, (3) the customer service philosophy, (4) the logistics and transportation costs, (4) the complexity of inbound and outbound movements, and (5) the overall management orientation. Although there is sometimes a temptation to make a major organization decision based only on one or two of these factors, the best advice is to analyze each of these carefully before committing resources to any particular alternative.

Organizational issues and alternatives that need to be addressed include (1) line versus staff, (2) centralization versus decentralization, (3) functional versus product-line orientation, (4) the traffic function as part of the distribution division, (5) span of control, (6) and the coordination of various activities. The organizational decision in the traffic area can be quite complex and must be made in concert with the overall goals and objectives of the organization.

Selected Topical Questions from Past AST&L Examinations

1. Discuss briefly the major differences between the responsibilities and objectives of the traffic manager as opposed to the director or vice-president of physical distribution. (Spring 1980)
2. Logistics consultant John F. Magee has stated, ''Logistical system management poses some puzzling organizational problems to the typical functionally organized firm.''
   (a) What functional areas of the firm are generally considered to be part of the logistics functions? (*Hint:* One is packaging.)
   (b) Discuss at least two approaches or organizational alternatives to achieving a high degree of coordination between the functional areas of a logistics system. (Spring 1982)

Notes

1. For an understanding of the fundamental work done on this topic, see Alfred D. Chandler, *Strategy and Structure* (Cambridge, Massachusetts: MIT Press, 1962).
2. Bernard J. LaLonde and Paul H. Zinszer, *Customer Service: Meaning and Measurement* (Oak Brook, Illinois: National Council of Physical Distribution Management, 1976).
3. Jack W. Farrell, ed., "Organization study: Distribution departments gain ground," *Traffic Management* 20 no. 9 (September 1981), p. 47.
4. An interesting discussion of organizational alternatives as they generally apply to the distribution function may be found in James P. Falk, "Organizing for effective distribution," *Proceedings of the Annual Conference of the National Council of Physical Distribution Management* (Oak Brook, Illinois: NCPDM, 1980), pp. 181–189. An excellent academic treatment of the topic of distribution organization may be referenced in David H. Maister, "Organizing for physical distribution," monograph series of the *International Journal of Physical Distribution and Materials Management* 8 no. 3, pp. 134–178.
5. Donald J. Bowersox, *Logistical Management*, 2d ed. (New York: Macmillan, 1978), pp. 449–450.
6. Two articles that lend significant insight into these issues are John R. Harold, "A case for the centralized traffic manager," *Handling and Shipping Management* (July 1982), pp. 61–68; Patrick Gallagher, "Corporate transportation finds its role at FMC," *Handling and Shipping Management* (May 1982), pp. 85–92.
7. Data supporting the statements made in this paragraph may be located in the results of yearly studies concerning career patterns of logistics executives conducted by Professor Bernard J. LaLonde of the Ohio State University. These results appear annually in the Annual Conference Proceedings of the National Council of Physical Distribution Management (now Council of Logistics Management). The most recent citation is Bernard J. LaLonde and Larry W. Emmelhainz, "Career patterns of logistics executives: 1985," *Proceedings of the Annual Conference of the Council of Logistics Management* (Oak Brook, Illinois: Council of Logistics Management, 1985).
8. James F. Stoner, *Management*, 2d ed. (Englewood Cliffs, New Jersey: Prentice-Hall, 1982), p. 284.
9. Larry Greiner, "Evolution and revolution as organizations grow," *Harvard Business Review* 52 no. 4 (July–August, 1972), p. 41.
10. *The Official Directory of Industrial and Commerical Traffic Executives* (Washington, D.C.: The Traffic Service Corp., 1982).

Selected Bibliography

General References

Traffic Management

Bryan, Leslie A. *Traffic Management in Industry.* New York: The Dryden Press, 1953.

Cushman, Frank M. *Transportation for Management.* Englewood Cliffs, N.J.: Prentice-Hall, 1953.

Davies, G. J. and R. Gray. *Purchasing International Freight Transport Services.* England: Gower Publishing, 1985.

Flood, Kenneth U., Oliver G. Callson, and Sylvester J. Jablonski. *Transportation Management,* 4th ed. Dubuque, Iowa: Wm. C. Brown, 1984.

McElhiney, Paul T., and Charles L. Hilton. *Introduction to Logistics and Traffic Management.* Dubuque, Iowa: Wm. C. Brown, 1968.

Morse, Leon W. *Practical Handbook of Industrial Traffic Management,* 6th ed. Washington, D.C.: Traffic Service Corp., 1980.

Murr, Alfred. *Export/Import Traffic Management and Forwording,* 5th ed. Cambridge, Md.: Cornell Maritime Press, 1977.

Vanderleest, Henry W. and Michael L. Johnston. *Cases in Transportation Management.* Prospect Heights, Ill.: Waveland Press, 1983.

Van Metre, Thurman W. *Industrial Traffic Management.* New York: McGraw-Hill, 1953.

Way, William Jr. *Elements of Freight Traffic.* Washington, D.C.: Regular Common Carrier Conference, ATA, 1959.

Wilson, G. Lloyd. *The Principles of Freight Traffic.* Washington, D.C.: Traffic Service Corp., 1935.

Logistics

Ballou, Ronald H. *Business Logistics Management: Planning and Control,* 2d ed. Englewood Cliffs, N.J.: Prentice-Hall, 1985.

Blanchard, Benjamin S. *Logistics Engineering and Management,* 2d ed. Englewood Cliffs, N.J.: Prentice-Hall, 1981.

Blanding, Warren. *Practical Handbook of Distribution/Customer Service.* Silver Spring, Md.: Marketing Publications, Inc., 1985.

Bowersox, Donald J., David J. Closs, and Omar K. Helferich. *Logistical Management,* 3d ed. New York: Macmillan, 1986.

Christopher, Martin. *The Strategy of Distribution Management.* England: Gower Publishing, 1985.

Coyle, John J. and Edward J. Bardi. *The Management of Business Logistics,* 3d ed. St. Paul, Minn.: West, 1984.

Davis, Grant M. and Stephen W. Brown. *Logistics Management.* Lexington, Mass.: D. C. Heath, 1974.

Guelzo, Carl M. *Introduction to Logistics Management.* Englewood Cliffs, N.J.: Prentice-Hall, 1986.

Heskett, James L., Nicholas A. Glaskowsky, Jr., and Robert M. Ivie. *Business Logistics,* 2d ed. New York: Ronald Press, 1973.

Hutchinson, Norman E. *An Integrated Approach to Logistics Management.* Englewood Cliffs, N.J.: Prentice-Hall, 1987.

Johnson, James C. and Donald F. Wood. *Contemporary Physical Distribution and Logistics,* 2d ed. Tulsa, Okla.: PennWell, 1982.

Lambert, Douglas M. and James R. Stock. *Strategic Physical Distribution Management.* Homewood, Ill.: Richard D. Irwin, 1982.

Magee, John F., William C. Copacino, and Donald B. Rosenfield. *Modern Logistics Management.* New York: John Wiley and Sons, 1985.

Mossman, Frank H., Paul Bankit, and Omar K. Helferich. *Logistics Systems Analysis.* Washington, D.C.: University Press of America, 1977.

Robeson, James F. and Robert G. House, eds. *The Distribution Handbook.* New York: Free Press, 1985.

Rose, Warren. *Logistics Management: Systems and Components.* Dubuque, Iowa: Wm. C. Brown, 1979.

Schary, Philip B. *Logistics Decisions: Text and Cases.* New York: Dryden, 1984.

Shapiro, Roy D. and James L. Heskett. *Logistics Strategy: Cases and Concepts.* St. Paul, Minn.: West, 1985.

Taff, Charles A. *Management of Physical Distribution and Transportation,* 7th ed. Homewood, Ill.: Richard D. Irwin, 1984.

Transportation

Bowersox, Donald J., Pat J. Calabro, and George Wagenheim. *Introduction to Transportation.* New York: Macmillan, 1982.

Coyle, John J., Edward J. Bardi, and Joseph L. Cavinato. *Transportation.* St. Paul, Minn.: West Publishing Co., 1986.

Fair, Marvin L. and Ernest W. Williams, Jr. *Transportation and Logistics,* rev. ed. Plano, Tex.: Business Publications, 1981.

Harper, Donald W. *Transportation in America,* 2d ed. Englewood Cliffs, N.J.: Prentice-Hall, 1982.

Hazard, John L. *Transportation: Management, Economics, Policy.* Cambridge, Md.: Cornell Maritime Press, 1977.

Kneafsey, James T. *Transportation Economic Analysis.* Lexington, Mass.: D. C. Heath, 1975.

Leib, Robert C. *Transportation: The Domestic System,* 3d ed. Reston, VA.: Reston, 1985.

Locklin, Philip D. *Economics of Transportation,* 7th ed. Homewood, Ill.: Richard D. Irwin, 1972.

McElhiney, Paul T. *Transportation for Marketing and Business Students.* Totowa, N.J.: Littlefield, Adams, 1975.

Mossman, F. H. and Newton Morton. *Principles of Transportation.* New York: Ronald Press, 1957.

Sampson, Roy J., Martin T. Farris, and David L. Shrock. *Domestic Transportation: Practice, Theory, and Policy,* 5th ed. Boston: Houghton Mifflin, 1985.

Sweeney, Daniel J., et al. *Transportation Deregulation: What's Deregulated and What Isn't.* Washington, D.C.: NASSTRAC Publications, 1986.

Talley, Wayne Kenneth. *Introduction to Transportation.* Cincinnati: Ohio South-Western, 1983.

Wood, Donald F. and James C. Johnson. *Contemporary Transportation,* 2d ed. Tulsa, Okla.: PennWell, 1983.

Purchasing and Materials Management

Ammer, Dean S. *Materials Management and Purchasing,* 4th ed. Homewood, Ill.: Richard D. Irwin, 1974.

Burt, David N. *Proactive Procurement: The Key to Increased Profits, Productivity, and Quality.* Englewood Cliffs, N.J.: Prentice-Hall, 1984.

Cavinato, Joseph L. *Purchasing and Materials Management.* St. Paul, Minn.: West, 1984.

Corey, E. Raymond. *Procurement Management: Strategy, Organization and Decision-Making.* Boston: CBI, 1978.

Dobler, Donald W., Lamar Lee, Jr., and David N. Burt. *Purchasing and Materials Management: Text and Cases.* New York: McGraw-Hill, 1984.

Hedrich, Floyd D. *Purchasing Management in the Smaller Company.* New York: American Management Association, 1971.

Heinritz, Stuart F. and Paul V. Farrell. *Purchasing Principles and Applications,* 7th ed. Englewood Cliffs, N.J.: Prentice-Hall, 1986.

Leenders, Michael R., Harold E. Fearon, and Wilbur B. England. *Purchasing and Materials Management,* 7th ed. Homewood, Ill.: Richard D. Irwin, 1980.

Murray, John E. Jr. *Purchasing and the Law.* Pittsburgh, Pa.: Purchasing Management Association, 1980.

Tersine, Richard J. and John M. Campbell. *Modern Materials Management.* New York: North-Holland, 1977.

Zenz, Gary J. *Purchasing and the Management of Materials,* 5th ed. New York: John Wiley and Sons, 1981.

Trade and Professional Journals

Air Cargo, 1156 15th Street, Washington, DC 20005

Aviation Week and Space Technology, McGraw-Hill Publishing Co., 330 West 42nd Street, New York, NY 10036

Canadian Transportation & Distribution Management, 1450 Don Mills, Ontario M3B 2X7, Canada

Computer Systems Report, P.O. Box 453, Exton, PA 19341

Container News, 6255 Barfield Road, Atlanta, GA 30328

Defense Transportation Journal, National Defense Transportation Association, 727 N. Washington St., Alexandria, VA 22314-1976

Distribution, Chilton Co., Inc., Radnor, PA 19098

Fleet Owner, McGraw-Hill Publishing Co., New York, NY

Focus on Physical Distribution Management, The Institute of Physical Distribution Management, Management House, Parker Street, London WC2B 5PT, England

Handling and Shipping, Industrial Publishing Corporation, 614 Superior Ave., W. Cleveland, OH 44113

Inbound Logistics Guide, Thomas Publishing Co., Inc., New York, NY 10119

Industrial Marketing, 740 Rush Street, Chicago, IL 60611

International Journal of Physical Distribution and Materials Management, 198–200 Keighley Road, Bradford DB9 4JQ, Yorkshire, England

Journal of Business Logistics, Council of Logistics Management, 2803 Butterfield Road, Oak Brook, IL 60521

Journal of Purchasing and Materials Management, National Association of Purchasing Agents, Inc., 11 Park Place, New York, NY 10007

Journal of Transport Economics and Policy, London School of Economics and Political Science and the University of Bath, Clover Down, Bath, BA 2 YAY, England

Logistics and Transportation Review, University of British Columbia, Vancouver, Canada V6T 1W5

Modern Packaging, Modern Packaging Corp., 575 Madison Ave., New York, NY

Modern Railroads, 2020 Oakton Street, Park Ridge, IL 60068

Owner Operator, Chilton Co., Inc., Radnor, PA 19089

Purchasing, Cahners Publishing Co., Inc., 270 St.
   Paul St., Denver, CO 80206
Railway Age, Simmons-Boardman Publishing Co.,
   508 Birch St., Bristol, CT 06010
The Private Carrier, Private Carrier Conference
   Incorporated, 2200 Mill Road, Alexandria,
   VA 22314
Traffic Bulletin and Traffic World, Traffic Service
   Corporation, 815 Washington Building,
   Washington, DC 20005
Traffic Management, Cahners Publishing Co., Inc.,
   270 St. Paul St., Denver, CO 80206
Traffic Quarterly, Eno Foundation for
   Transportation, Westport, CT 06880
Transportation in America, Transportation Policy
   Association, P.O. Box 33633, Washington,
   DC 20033
Transportation Journal, American Society of Traffic
   and Logistics Inc., 1816 Norris Place,
   Louisville, KY 40205
Transportation Practitioners Journal, Association of
   Transportation Practitioners, 2218 I.C.C.
   Bldg., Washington, DC 20423
Transportation Science, Operations Research
   Society of America, 428 E. Preston Street,
   Baltimore, MD 21202
Transport Topics, American Trucking
   Associations, Inc., 1616 P. Street,
   Washington, DC
Waterways Journal, 701 Chemical Bldg., St. Louis,
   MO 63101
Weekly Letter, The American Waterways
   Operators, Inc., 1600 Wilson Blvd., Suite
   1101, Arlington, VA 11109
What's Happening in Transportation,
   Transportation Association of America, 1100
   17th Street, N.W., Washington, DC

Other Sources of Information Related to Traffic Management

Annual Catalog of Distribution Directories and
   Guides, *Traffic Management* Magazine, 221
   Columbus Ave., Boston, MA 02116
ATA General Library-Monthly Bulletin, American
   Trucking Associations, Inc., 1616 P. Street,
   N.W., Washington, DC 20036
Bibliography on Logistics and Physical

Distribution Management, Council of
   Logistics Management, 2803 Butterfield
   Road, Oak Brook, IL 60521
Brandon's Shipper & Forwarder, One World
   Trade Center #3169, New York, NY 10048
Current Literature in Traffic and Transportation,
   Northwestern University, The
   Transportation Center, Evanston, IL 60201
Distribution Software from Top to Bottom,
   Arthur D. Little, Inc., Cambridge, MA
Hazardous Materials, Glencoe Publishing Co.,
   Encino, CA
Highway Common Carrier Newsletter, The
   Regular Common Carrier Conference–ATA,
   2200 Mill Road, Alexandria, VA 22314
Information Letter, The Association of American
   Railroads, American Railroad Building,
   Washington, DC 20036
Information Sources in Transportation, Material
   Management, and Physical Distribution,
   Greenwood Press, 51 Riverside Avenue,
   Westport, CN 06880
Motor Carrier Computer News, Regular
   Common Carrier Conference, 2200 Mill
   Road, Alexandria, VA 22314
National Highway and Airway Carriers and
   Routes, National Highway Carriers
   Directory, 936 S. Betty Dr., P.O. Box 6,
   Buffalo Grove, IL 60090
Physical Distribution Software, Arthur Anderson
   & Co., Stamford, CT
TAA Report, Transportation Association of
   America, 1100 17th Street, N.W.,
   Washington, DC 20036
The Official Directory of Industrial and
   Commercial Traffic, The Traffic Service
   Corp., Washington, DC
The Official Railway Equipment Register,
   National Railway Publication Company,
   424 West 33rd Street, New York, NY
   10001
The Official Railway Guide, National Railway
   Publication Company, 424 West 33rd Street,
   New York, NY 10001
Traffic World's Question and Answer Book, The
   Traffic Service Corporation, 815 Washington
   Building, Washington, DC 20005
Transport (De)Regulation Report, Cahners
   Publishing Company, P.O. Box 716, Back
   Bay Annex, Boston, MA 02117

Selected Topical References—Part I

The Changing Nature of Traffic Management

Barrett, Colin. "If regulation's sun sets . . . " *Proceedings—Twenty-Fourth Annual Meeting Transportation Research Forum* 24 (1983): 552–557.

Bowersox, Donald J. "Emerging from the recession: The role of logistical management." *Journal of Business Logistics* 4 no. 1 (1983): 21–34.

Campbell, John H. "From traffic manager to logistician." *MSU Business Topics* (Autumn 1980): 25–30.

Dadzie, Kofi Q. and Wesley J. Johnston. "Skill requirements in physical distribution management career-path development." *Journal of Business Logistics* 5 no. 2 (1984): 65–84.

Early, Rollie W. "Innovative purchasing of transportation to solve distribution problems." *Annual Proceedings of the National Council of Physical Distribution Management* (1984): 571–584.

Eder, Peter F. "Report on joint conference—Part one: Changing concepts of transportation." *Traffic Quarterly* 38 no. 2 (April 1983): 166–192.

Germane, Gayton E. *Transportation Policy Issues for the 1980's.* Reading, Mass.: Addison-Wesley, 1983.

Goodwin, Vern H. Jr., Jack J. Holder, Jr., and Robert W. Baker. "Future role of distribution and transportation." *Annual Proceedings of the National Council of Physical Distribution Management* (1979): 172–198.

Haupt, Richard. "Profile of today's transportation user." *Traffic Quarterly* 40 no. 1 (January 1986): 55–64.

Herron, David P. "The educational needs of physical distribution managers." In *The Distribution Handbook,* James F. Robeson and Robert G. House, eds. New York: Free Press (1985): 849–855.

Heskett, James L. "Challenges and opportunities for logistics executives in the 80's." *Journal of Business Logistics* 4 no. 1 (1983): 13–20.

———. "Sweeping changes in distribution." *Harvard Business Review* (March–April 1973).

Johnson, James C. "Seven transportation megatrends for the late 1980s." *Transportation Practitioners' Journal* 53 no. 1 (Winter 1986): 164–180.

Karrenbauer, Jeffrey. "Distribution: An historical perspective." In *The Distribution Handbook,* James F. Robeson and Robert G. House, eds. New York: Free Press (1985): 3–14.

Koontz, Harold. "Management and transportation." *Transportation Journal* 6 no. 4 (Summer 1966): 20–26.

LaLonde, Bernard J. "A reconfiguration of logistics systems in the 1980s: Strategies and challenges." *Journal of Business Logistics* 4 no. 1 (1983): 1–12.

LaLonde, Bernard J., John R. Grabner, Jr., and James F. Robeson. "Integrated distribution systems: Past, present, and future." In *The Distribution Handbook,* James F. Robeson and Robert G. House, eds. New York: Free Press (1985): 15–27.

Lambert, Douglas M., James F. Robeson, and James R. Stock. "An appraisal of the integrated physical distribution management concept." *International Journal of Physical Distribution and Materials Management* 9 no. 1 (1978): 84.

Metz, Peter J. "Megacarriers & transportation products." *Handling and Shipping Management* (Presidential Issue, 1983–84): 70–78.

Muskin, Jerold B. "The physical distribution infrastructure." *Traffic Quarterly* 37 no. 1 (January 1983): 115–133.

Nishi, Masao and Patrick Gallagher. "A new focus for transportation management: Contribution." *Journal of Business Logistics* 5 no. 2 (1984): 19–29.

Poist, Richard F. "Managing distribution during the remainder of the 1980's: How distribution managers should prepare themselves." *37th Annual Meeting American Society of Traffic and Transportation* (1982): 237–249.

Schonberger, Richard J. and James P. Gilbert. "Just-in-time purchasing: A challenge for U.S. industry." *California Management Review* 26 no. 1 (Fall 1983): 54–68.

Sherwood, Charles S. and William C. Rice. "Computer literacy needs in transportation." *Journal of Business Logistics* 6 no. 1 (1985), 1–12.

Spangler, Earl J. "Full scope activities through logistics." *Handling and Shipping Management* (Presidential Issue 1982–83): 30–38.

Stenger, Alan J. "Organizing and managing tomorrow's traffic management activities." *32nd Annual Meeting American Society of Traffic and Transportation* (1977): 53–67.

Stewart, Wendell. "Traffic and transportation managements' role in the evolving distribution organization." *36th Annual Meeting American Society of Traffic and Transportation* (1981): 11–44.

Stewart, Wendell M., et al. "A look to the future." In *The Distribution Handbook*, James F. Robeson and Robert G. House, eds. New York: Free Press (1985): 44–72.

Walter, C. K. and Norman H. Erb. "Evaluating a new transportation and physical distribution curriculum." *Logistics Spectrum* (Summer 1983): 27–33.

Waters, L. L. "Deregulation—For better or for worse." *Business Horizons* 24 no. 1 (January–February 1981).

Transportation Services and Selection

Auguello, William J. *Transportation Insurance in Plain English*. New York: Thomas, 1986.

Baker, Gwendolyn H. "The carrier elimination decision: Implications for motor carrier marketing." *Transportation Journal* 24 no. 1 (Fall 1984): 20–29.

Ballou, Ronald H. and Mohammed Chowdhury. "MSVS: An extended computer model for transport mode selection." *The Logistics and Transportation Review* 16 no. 4 (1980): 325–338.

Bardi, Edward J. "Carrier selection from one mode." *Transportation Journal* 13 no. 1 (Fall 1973): 23–34.

Baumol, William J. and H. D. Vinod. "An inventory–theoretic model of freight transportation demand." *Management Science* 16 (March 1970): 413–421.

Beilock, Richard and James Freeman. "Deregulated motor carrier service to small communities." *Transportation Journal* 23 no. 4 (Summer 1984): 71–82.

Blumenfeld, Dennis E., Randolph W. Hall, and William C. Jordan. "Trade-off between freight expediting and safety stock inventory costs." *Journal of Business Logistics* 6 no. 1 (1985): 79–99.

Brown, Terence A. "Shippers' agents and the marketing of rail intermodal service." *Transportation Journal* 23 no. 3 (Spring 1984): 44–52.

——. "Shippers' associations: Operations, trends, and comparative prices." *Transportation Journal* 21 no. 1 (Fall 1981): 54–66.

Bruning, Edward R. and Peter M. Lynagh. "Carrier evaluation in physical distribution management." *Journal of Business Logistics* 5 no. 2 (1984): 30–47.

Buffa, Frank P. and John I. Reynolds. "A graphical total cost model for inventory-transport decisions." *Journal of Business Logistics* 1 no. 2 (1979): 120–143.

Chow, Garland and Richard Poist. "Rate bureau cost and benefits: The carrier perspective." *Proceedings—Twentieth Annual Meeting Transportation Research Forum* 20 (1979): 432–439.

Cobert, Ronald N. and Robert L. Cope. "How to avoid paying twice for the same shipment when using a broker." *Traffic World* (January 1985): 72–75.

Corsi, Thomas M., Curtis M. Grimm, and Robert Lundy. "ICC exemptions of rail services: Summary and evaluation." *Proceedings—Twenty-Sixth Annual Meeting Transportation Research Forum* 26 (1985): 86–92.

Crum, Michael R. "The expanded role of motor freight brokers in the wake of regulatory reform." *Transportation Journal* 24 no. 4 (Summer 1985): 5–15.

Cunningham, Lawrence F. "Transportation and distribution management: The new options." In *The Distribution Handbook*, James

F. Robeson and Robert G. House, eds. New York: Free Press (1985): 497–509.

Delaney, Robert V. "Managerial and financial challenges facing transport leaders." *Traffic Quarterly* XL no. 1 (January 1986): 29–53.

Eigen, Charles. "Choosing a property broker." *The Private Carrier* (September 1985): 25–28.

Evans, R. E. and W. R. Southard. "Motor carriers' and shippers' perceptions of carrier choice decision." *The Logistics and Transportation Review* 10 no. 4 (1974): 145–150.

Harrington, Lisa. "Carrier selection criteria change with the times." *Traffic Management* (September 1983): 59–62.

Jerman, Roger E., Ronald D. Anderson, and James A. Constantin. "Shipper versus carrier perceptions of carrier selection variables." *International Journal of Physical Distribution and Materials Management* 9 no. 1 (1978): 29–38.

Kogon, Gary R. "Shippers' associations: Operations, trends, and comparative prices—Comment." *Transportation Journal* 21 no. 4 (Summer 1982): 76–77.

Kotzen, Herb. "Working with a property broker: Some common-sense rules." *The Private Carrier* (September 1985): 9–11, 52.

McGinnis, Michael A., Thomas M. Corsi, and Merrill J. Roberts. "A multiple criteria analysis of modal choice." *Journal of Business Logistics* 2 no. 2 (1981): 48–68.

McGinnis, Michael A. and Thomas M. Corsi. "Are the modes really competitive?" *Distribution Worldwide* (September 1979): 40.

Morash, Edward A. "Customer service, channel separation, and transportation intermediaries." *Journal of Business Logistics* 7 no. 1 (1986): 89–107.

Piercy, James E. and Ronald H. Ballou. "A performance evaluation of freight transport modes." *The Logistics and Transportation Review* 14 no. 2 (1978): 99–115.

Schary, Philip B. and Miguel Linares. "Transportation and small business: Practices, problems and attitudes." *Proceedings—Twenty-First Annual Meeting Transportation Research Forum* 21 (1980): 534–542.

Speh, Thomas and George D. Wagenheim.

"Demand and lead-time uncertainty: The impacts on physical distribution performance and management." *Journal of Business Logistics* 1 no. 1 (1980): 95–113.

Statter, Bradley. "A model of industrial buyer behavior: Physical distribution management and the motor carrier selection—A task approach." *Proceedings—Twenty-Fourth Annual Meeting Transportation Research Forum* 24 (1983): 490–497.

Stock, James R. and Bernard J. LaLonde. "The transportation mode decision revisited." *Transportation Journal* 17 no. 2 (Winter 1977): 51–59.

Urycki, Hank. "The pitfalls of working with property brokers." *The Private Carrier* (September 1985): 33–35.

Williams, Ernest W. "A critique of the Staggers Rail Act of 1980." *Transportation Journal* 21 no. 3 (Spring 1982): 5–15.

Williamson, Kenneth C., Marc G. Singer, and Roger A. Peterson. "The impact of regulatory reform on U.S. for-hire freight transportation: The users' perspective." *Transportation Journal* 22 no. 4 (Summer 1983): 64–70.

Transportation Pricing

Antitrust Concerns of Rail Shippers Operating Under the Staggers Rail Act. St. Louis, Mo.: National Grain and Feed Association, 1981.

Antitrust Issues in Transportation: Questions and Answers. Washington, D.C.: National Industrial Traffic League, 1981.

Bohman, Ray Jr. *Guide to Cutting Your Freight Transportation Costs Under Trucking Deregulation,* 2d ed. Gardner, Mass.: Bohman Industrial Traffic Consultants, Inc., 1982.

Boske, Leigh B. "An analysis of recent developments in railroad maximum rate regulation." *ICC Practitioners' Journal* 48 no. 3 (March–April 1981): 294–311.

Bradley, John C. "Antitrust compliance programs in the motor carrier industry—A primer on why and how." *ICC Practitioners' Journal* 49 no. 4 (May–June 1982): 395–412.

Bunce, Elliot. "Special problems relating to collective ratemaking." *ICC Practitioners' Journal* 51 no. 6 (September 1984): 583–590.

Calderwood, James A. "Antitrust — Private rights of action." *Handling and Shipping Management* (June 1983): 109.

——. "Antitrust and shippers: Problems or opportunities?" Remarks before the Drug and Toilet Preparation Traffic Conference, *The Private Carrier* (April 1984): 27.

Carr, Ronald G. "Railroad-shipper contracts under section 208 of the Staggers Rail Act: An antitrust perspective." *ICC Practitioners' Journal* 50 no. 1 (November–December 1982): 29–41.

Colquitt, J. C. *The Art and Development of Freight Classification.* Washington, D.C.: National Motor Freight Traffic Association, 1956.

Corsi, Thomas M., Curtis M. Grimm, and Robert Lundy. "ICC exemptions of rail services: Summary and evaluation." *Proceedings — Twenty-Sixth Annual Meeting Transportation Research Forum* 26 (1985): 86–92.

Corsi, Thomas M. and Merrill J. Roberts. "Patterns of discrimination in the collective ratemaking system." *Proceedings — Twenty-Third Annual Meeting Transportation Research Forum* 23 (1982): 621–630.

Cunningham, Wayne H. J. "Freight modal choice and competition in transportation: A critique and categorization of analysis techniques." *Transportation Journal* 21 no. 4 (Summer 1982): 66–75.

Davis, Grant M. "The collective ratemaking issue: Circa 1984." *Transportation Practitioners Journal* 52 no. 1 (Fall 1984): 60–68.

Davis, Grant M. and John E. Dillard, Jr. "Collective ratemaking — Does it have a future in the motor carrier industry?" *ICC Practitioners' Journal* 49 no. 6 (September–October, 1982): 619–625.

Eckardt, Robert C. "Market dominance in the Staggers Act." *ICC Practitioners' Journal* 48 no. 6 (September–October, 1982): 662–686.

Emrich, Richard S. M. and Vernon J. Hann. "Implementing Staggers: A free rail transportation market through proportional rates." *Traffic World* (December 3, 1984): 111–120.

Ezard, P. H. B. and R. J. Lande. "Computerization of railway freight tariffs."

Proceedings — Twenty-Fifth Annual Meeting Transportation Research Forum 25 (1984): 415–419.

Feldman, Joan M., ed. "Is there life after collective ratemaking." *Handling and Shipping Management* (April 1984): 63–65.

Flexner, Donald L. "Potential problem areas for shippers and carriers under the antitrust laws. *ICC Practitioners' Journal* 51 no. 6 (September 1984): 571–582.

Flexner, Donald L. and Wm. Randolph Smith. "The Keogh doctrine: Practical implications after square D." *Traffic World* (June 16, 1986): 77–82.

Gallagher, Patrick F. "Transportation pricing in the post-deregulatory period." *Proceedings of the Logistics Research Forum* (1983): 109–119.

Gardiner, Paul S. "Rate bureau functions without antitrust immunity: A suggested strategy for motor freight carriers." *ICC Practitioners' Journal* 46 no. 5 (July–August 1979): 651–668.

Gawlik, Dennis M. and Keven B. Boberg. "The evolution and implications of the market dominance concept in railroad ratemaking." *Transportation Law Journal* 13 no. 2 (1984): 259–286.

Goldstein, Andrew P. "The single product rule and rail freight transportation — A contrasting view." *Transportation Practitioners Journal* 53 no. 1 (Winter 1986): 151–163.

Hagan, Paul. "Freight forwarding roles change at dizzying pace." *Air Cargo World* (March 1985): 38–41.

Heaver, T. D. and James C. Nelson. *Railway Pricing Under Commercial Freedom: The Canadian Experience.* Vancouver, B.C.: University of British Columbia, 1977.

Heisley, E. Stephen. "Antitrust implications for the future." *Regional-National Educational Conference — American Society of Traffic and Transportation.* Louisville, Ky., May, 1983.

Hoover, Harward Jr. "Pricing behavior of deregulated motor common carriers." *Transportation Journal* (Winter 1985): 55–61.

Horn, Kevin and John E. Tyworth. "The impact of railroad boxcar deregulation: A case study of transcontinental lumber." *Proceedings — Twenty-Fifth Annual Meeting Transportation Research Forum* 25 (1984): 280–289.

Kalish, Steven J. "Antitrust considerations for shippers in a changing environment." *Transportation Practitioners Journal* 52 no. 2 (Winter 1985): 185–195.

Lindeman, Eric D. "DSI rail links shippers to rails in new version of rate data system." *Traffic World* (November 19, 1984): 19–22.

Linzer, Joel. "Antitrust year in review: Recent developments in antitrust defenses for the transportation industry." *ICC Practitioners' Journal* 50 no. 1 (November–December 1982): 42–50.

Lynagh, Peter M. "Physical distribution: Seventies style." *Business Perspectives* (Summer 1973): 28–31.

Lynagh, Peter M. and Richard F. Poist. "Women's perceptions regarding careers in transportation and distribution." *Proceedings—Twenty-Fifth Annual Meeting Transportation Research Forum* 25 (1984): 182–186.

Jerman, Roger E., Ronald D. Anderson, and James A. Constantin. "How traffic managers select carriers." *Distribution Worldwide* (September 1978): 21–24.

Keenan, William Q. "Shipper antitrust immunity when dealing with interstate common carriers." *ICC Practitioners' Journal* 47 no. 5 (July–August 1980): 522–537.

Marien, Edward J. "Formula rates: Their time has come again." *Distribution* 81 (January 1982): 54–58.

———. "Rate basing systems: Simplify, simplify." *Distribution* 82 (February 1983): 38–43.

Mattox, Elmer A. and Edward J. Marien. "Formula rates: An idea whose time has returned." *Traffic World* (October 18, 1976): 77–81.

McGibbon, James R. and Douglas E. Rosenthal. *Antitrust Primer for Motor Carrier Executives.* Washington, D.C.: American Trucking Association, 1982.

Ongman, John Will. "U.S. antitrust ramifications for the Canadian transport industry." *Proceedings—Twenty-Third Annual Meeting Transportation Research Forum* 23 (1982): 1–6.

Popper, Andrew F. *Shipper Antitrust Liability in a Rate-Regulated Market: Fundamental Inquiries and Analysis.* Washington, D.C.: Washington University, 1979.

Roberts, Merrill J. "Railroad maximum rate and discrimination control." *Transportation Journal* 22 no. 3 (Spring 1983): 23–33.

Selby, George W. Jr. "The nine commandments of lawful cooperation among competitors." *Traffic World* (March 31, 1986): 87–91.

Spychalski, John C. "Paradoxes of the revolution in railway intramodal relationships." *Transportation and Public Utilities Group, American Economic Association* (December 29, 1983).

———. "Progress, inconsistencies, and neglect in the social control of railway freight transport." *Journal of Economic Issues* 17 no. 2 (June 1983).

Stone, James H. and Linda M. Yunashko. "Single factor rate structures can reduce administrative costs for both shipper and carrier." *Traffic World* (December 2, 1985): 102–105.

Thompson, Michael. "The relevance of revenue/variable cost ratios to market dominance determinations." *Proceedings—Twenty-Third Annual Meeting Transportation Research Forum* 23 (1982): 362–368.

Tye, William B. "Balancing the ratemaking goals of the Staggers Act." *Transportation Journal* 22 no. 4 (Summer 1983): 17–26.

———. "On the effectiveness of product and geographic competition in determining rail market dominance." *Transportation Journal* 24 no. 1 (Fall 1984): 5–19.

———. "Revenue/variable cost ratios and market dominance proceedings." *Transportation Journal* 24 no. 2 (Winter 1984): 15–30.

———. "The role of revenue/variable cost ratios: Determinations of rail rate reasonableness." *Proceedings—Twenty-Fifth Annual Meeting Transportation Research Forum* 25 (1984): 214–221.

Tyworth, John E. "Regulatory reform and railroad freight rate structures for lumber." *Proceedings—Twenty-Fourth Annual Meeting Transportation Research Forum* 24 (1983): 659–666.

Wheeler, Porter K. and Ronald L. Freeland. "Joint rate cancellations since the Staggers Rail Act of 1980." *Proceedings—Twenty-Seventh Annual Meeting Transportation Research Forum* 27 (1986): 122–130.

Whitten Herb and Greg Whitten. "The E3

transport pricing system.'' *Traffic World* (January 15, 1979): 30.

———. ''The E3 pricing system: Corrections and amplifications.'' *Traffic World* (April 9, 1979): 104–105.

———. *The Railroad and Motor Carrier Freight Rate Complex and Its Development.* Washington,

D.C.: U. S. Department of Transportation, Office of Policy Review, 1972.

Wooldridge, William C. ''The single product rule and rail freight transportation.'' *Transportation Practitioners Journal* 4 (Summer 1985): 512.

Selected Topical References—Part II

Transportation Planning Framework

Biggs, Dee and Bill Coyne. ''Precluding crises through practical contingency planning.'' *Annual Conference Proceedings of the National Council of Physical Distribution Management* (1983): 213–231.

Constantin, James. ''Planning perceptions and concepts.'' In *The Distribution Handbook*, James F. Robeson and Robert G. House, eds. New York: Free Press (1985): 73–92.

———. ''A framework for logistics planning.'' In *The Distribution Handbook*, James F. Robeson and Robert G. House, eds. New York: Free Press (1985): 93–109.

Fulchino, Paul E. and George T. Mauro. ''Reform impacts and strategic shipper responses.'' *Annual Conference Proceedings of the National Council of Physical Distribution Management* (1980): 372.

Hale, Bernard J. ''Contingency planning for distribution managers.'' In *The Distribution Handbook*, James F. Robeson and Robert G. House, eds. New York: Free Press (1985): 118–142.

''How to write a transportation contract.'' *Inbound Traffic Guide* (April 1985): 32–34.

Harper, Donald V. *Basic Planning and the Transportation Function in Small Manufacturing Firms.* Minneapolis, Minn.: University of Minnesota, 1961.

Jerman, Roger E. and Ronald D. Anderson. ''Physical distribution: A contingency approach.'' *Transportation Journal* 16 no. 2 (Winter 1976): 13–19.

Lambert, Douglas M. and James R. Stock. ''Strategic planning for physical distribution.'' *Journal of Business Logistics* 3 no. 2 (1982): 26–46.

Langley, C. John Jr. ''Strategic management in

transportation and physical distribution.'' *Transportation Journal* 22 no. 3 (Spring 1983): 27–54.

———. ''A framework for applying the strategic management process to logistics.'' *Proceedings of the Logistics Research Forum* (1982): 243–260.

Langley, C. John Jr. and William D. Morice. ''Strategies for logistics management: Reaction to a changing environment.'' *Journal of Business Logistics* 3 no. 1 (1982): 1–15.

Lynagh, Peter M. and Richard F. Poist. ''Managing physical distribution/marketing interface activities: Cooperation or conflict?'' *Transportation Journal* 23 no. 3 (Spring 1984): 36–43.

McGinnis, Michael A. ''Strategic planning— Dodging the traps.'' *Canadian Transportation and Distribution Management* (May 1985): 41–42.

Potter, Ronald S. ''Transportation planning: An overview.'' *Annual Conference Proceedings of the National Council of Physical Distribution Management* (1980): 228–233.

Ringbakk, Kjell A. ''Why planning failed.'' *European Business* (Spring 1971).

Shapiro, Roy D. ''Get leverage from logistics.'' *Harvard Business Review* (May–June 1984): 119–126.

Steiner, George A. *Strategic Planning.* New York: Free Press, 1979.

Temple, Barker and Sloane, Inc. *Transportation Strategies for the Eighties.* Oak Brook, Ill.: National Council of Physical Distribution Management, 1982.

Zinzser, Paul H. ''Operational planning.'' In *The Distribution Handbook*, James F. Robeson and

Robert G. House, eds. New York: Free Press
(1985): 110–117.

Contracting, Costing, and Negotiating

Adams, Aden C. and Carl W. Hoeberling.
"Future of contract rates in rail
transportation." *ICC Practitioners' Journal* 47
no. 6 (September–October 1980): 661–664.

Allen, Benjamin J. "The potential for
discrimination with rail contracts—One
point of view." *The Logistics and
Transportation Review* 17 no. 4 (1981): 371–
385.

Altrogge, Phyllis D. "Railroad contracts and
competitive conditions." *Transportation
Journal* 21 no. 2 (Winter 1981): 37–43.

Anderson, David L. and Dean H. Wise. "Rail
contract rate escalators: The hidden costs of
shipper-carrier agreements." *Traffic World*
(July 8, 1985): 64–68.

Armstrong, Richard D. "Relative importance of
linehaul and terminal costs for individual
motor carrier shipments." *Proceedings—
Twenty-Sixth Annual Meeting Transportation
Research Forum* 26 (1985): 459–463.

Bagby, John W., James R. Evans, and Wallace R.
Wood. "Contracting for transportation."
Transportation Journal 22 no. 2 (Winter 1982):
63–73.

Barks, Joe. "URCS works: But what's in it for
you?" *Distribution* 81 (March 1982): 52–55.

Barrett, Colin. "Antitrust law and the frumious
bandersnatch." *Distribution* 82 (April 1983):
45–51.

——. "Is the fix out for ratemaking (rail
contract secrecy)?" *Distribution* 83 (January
1984): 17.

Barrett, Colin, ed. "Conflicts over contracts: A
modern-day fable." *Distribution* 84 (February
1985): 38–46.

Barry, John J. "Transportation: Purchasing a
service." *Inbound Traffic Guide* (January
1983): 72–81.

Bernstein, Robert S. "Railroad contract rates."
Traffic World (June 7, 1982): 101.

Bielenberg, Judith M. and Terrell J. Harris.
"Exploitation of rail contract rates." *ICC
Practitioners' Journal* 47 no. 6 (September–
October 1980): 665–672.

Blackwell, Richard B. "Pitfalls in rail contract
rate escalation." *ICC Practitioners' Journal* 49
no. 5 (July–August 1982): 486–502.

Borts, George H. "Long-term rail contracts—
Handle with care." *Transportation Journal* 25
no. 3 (Spring 1986): 4–11.

Corber, Robert J. "An experts advice: The well-
drafted contract: Basic principles,
guidelines." *Transport Topics* 2529 (January
30, 1984): 2, 21.

Crichton, Jean. "RCA recipe for transportation
purchasing." *Inbound Traffic Guide* (January
1985): 19–23.

Dart, Robert C. Jr. *Contracting for Coal
Transportation: Rail Service Agreements Under
the Staggers Act.* Washington, D.C.:
McGraw-Hill, 1982.

Davis, Bob J. and C. K. Walter. *Contract Railroad
Rates.* Macomb, Ill.: Center for Business and
Economic Research, Western Illinois
University, 1984.

Domonkos, Gloria. "Opportunities and
challenges of rail contract rates."
*Proceedings—Twenty-Second Annual Meeting,
Transportation Research Forum* 22 (1981): 19–
24.

Farris, Martin. "Purchasing reciprocity and
materials management." *Journal of
Purchasing and Materials Management*
(Summer 1981): 27–35.

Fisher, Roger and William Ury. *Getting to Yes.*
Boston: Houghton-Mifflin, 1981.

Foster, Thomas A. "Negotiating with carriers—
Railroads." *Distribution* 81 (November 1982):
40–45.

——. "Negotiating with carriers—Railroads,
part two." *Distribution* 82 (January 1983): 54–
57.

Gilbert, Ernist. "Problems in purchasing
transportation." *National Association of
Purchasing Managers 67th International
Conference* (May 10, 1982).

Hill, Donald M. and Kenneth O. Nilsen. "New
needs for transportation cost information in
logistics planning." *Annual Conference
Proceedings of the National Council of Physical
Distribution Management* (1981): 646–688.

Hill, Stephen G. "Contract rates: Increasing rail
profitability." *ICC Practitioners' Journal* 46 no.
2 (January–February 1979): 222–232.

Hoffman, Kurt, ed. "Freight payments: 'Special

report': Are your freight bills coming back to haunt you?'' *Distribution* 84 (March 1985): 45–52.

Hoffman, Kurt, ed. "Freight payments: 'Special report': Of bills, bankruptcies and balance dues." *Distribution* 84 (May 1985): 28–36.

Hoffman, Stanley. "Checklist for contracts between shippers and railroads." *Eleventh AICCP Eastern Transportation Law Seminar.* Washington, D.C., 1981.

Illich, John. *The Art and Skill of Successful Negotiation.* Englewood Cliffs, N.J.: Prentice-Hall, 1973.

Illich, John and Barbar Schindler Jones. *Successful Negotiating Skills for Women.* Reading, Mass.: Addison-Wesley, 1981.

Johnson, Mark A. and Stephen C. Yevich. "An overview of the ICC's uniform rail costing system." *Proceedings — Twenty-First Annual Meeting Transportation Research Forum* 21 (1980): 379–387.

Karass, Chester L. *Give and Take: The Complete Guide to Negotiating Strategies and Tactics.* New York: Thomas V. Crowell, 1974.

Kline, Jeffrey C. "Transport costing in practice at General Mills." *Proceedings — Seminar on Transport Pricing, Costing, and User Charges Transportation Research Forum, Washington, D.C. Chapter* (April 5–7, 1982): 29–31.

Maltz, Arnold B. "Know your carrier's costs? Factors influencing LTL motor carriers." *Proceedings — Twenty-Sixth Annual Meeting Transportation Reseach Forum* 26 (1985): 483–492.

Marien, Edward J. "Measuring the financial condition of motor carriers." *The Private Carrier* (April 1984): 16–25.

Marien, Edward J. and C. Joseph Miller. "Rail contracts — Is it the right way to go?" *Annual Conference Proceedings of the National Council of Physical Distribution Management* (1981): 390–407.

McBride, Mark E. "An evaluation of various methods of estimating railway costs." *The Logistics and Transportation Review* 19 no. 1 (1983): 45–66.

———. "Economic costs, railway costs, and the uniform rail costing system." *Proceedings — Twenty-Third Annual Meeting Transportation Research Forum* 23 (1982): 375–382.

Miller, C. J. "Railroad contract rates: A license to innovate." *ICC Practitioners' Journal* 49 no. 6 (September–October 1980): 646–660.

Morton, J. Robert. "Contract rates by rail — A tool in ratemaking." *ICC Practitioners' Journal* 49 no. 4 (May–June 1982): 413–419.

O'Conner, Tom, John M. Robinson, and William F. Huneke. "The uniform railroad cost system: Implications for transportation management." *Proceedings — Twenty-Fourth Annual Meeting Transportation Research Forum* 24 (1983): 368–375.

Raiffa, Howard. *The Art and Science of Negotiation.* Cambridge, Mass: Belknap Press, 1982.

Regular Common Carrier Conference. *A Carrier's Handbook for Costing Individual Less-Than-Truckload Shipments.* Washington, D.C., 1982.

Robinson, Alan M. "Innovative transportation management: Contracting for uncertainty." *Proceedings — Twenty-Seventh Annual Meeting Transportation Research Forum* 27 (1986): 209–215.

Schneider, Lewis M. et al. "Rail contract services — The new frontier." *Traffic World* (August 3, 1981): 94–98.

———. "Rail contract services — The new frontier." *Traffic World* (August 10, 1981): 98–104.

Schuster, Alan D. "Statistical cost analysis: An engineering approach." *The Logistics and Transportation Review* 14 no. 2 (1979): 151–164.

Schuster, Alan D. and Robert G. House. "An analysis of the determinants of pickup and delivery costs." *Proceedings — Nineteenth Annual Meeting Transportation Research Forum* 19 (1978): 387–397.

Seip, Douglas W. "How IBM's bidding program reduced rates by 20%." *Canadian Transportation and Distribution Management* (February 1983): 41–42.

Shelley, Donald F. "Costing pool shipments: A simple formula for determining the base-delivered prices on goods moving in mixed-load consolidated shipments." *Handling and Shipping Management* 23 (August 1982): 67–68.

Shrock, David L. "Motor carrier cost analysis — The next step." *ICC Practitioners' Journal* 40 no. 5 (July–August 1975): 577–582.

———. "The functional approach to motor carrier costing: Application and limitations."

Proceedings—Twenty-Seventh Annual Meeting Transportation Research Forum 27 (1986): 181–188.

Smith, Jay A. Jr. and Gary R. Fane. *Railroad and Transportation Costing Annotated Bibliography.* Washington, D.C.: Federal Railroad Administration, Office of Policy and Program Development, FRA-OPPD-79-16, April, 1979.

Souza, Donald R. "Shipper-carrier rate negotiation: A new world for the traffic manager." *Traffic World* (March 19, 1984): 34–38.

Taff, Charles A. *Commercial Motor Transportation,* 7th ed. Centerville, Md.: Cornell Maritime Press, 1986.

Talley, Wayne K. "Motor carrier platform costing." *ICC Practitioners' Journal* 50 no. 2 (January–February 1983): 176–195.

——. "Methodologies for transportation cost analysis: A survey." *Transportation Research Record* no. 828 (1981): 1–3.

Trunick, Perry A. "Contracting for transportation." *Handling and Shipping Management* 22 (November 1981): 54–58.

Uggens, Michael W. "Railroad contract rates: A working analysis of section 10713." *ICC Practitioners' Journal* 48 no. 5 (July–August 1981): 526–542.

——. "Negotiating with carriers, rates." *Distribution* 82 (March 1983): 66–72.

Walter, C. K. "Analyses of railroad contract provisions after the 1980 Staggers Act." *Journal of Business Logistics* 5 no. 1 (1984): 81–91.

Warschaw, Tessa Albert. *Winning by Negotiation.* New York: McGraw-Hill, 1980.

Waters, W. G. II. "Statistical costing in transportation." *Transportation Journal* 15 no. 3 (Spring 1976): 49–62.

Wilner, Frank N. "Creating a linkage between rates and costs: A survival tactic for small carriers under deregulation." *ICC Practitioners' Journal* 48 no. 1 (November–December 1980): 65–69.

Wood, Wallace R. "A robust model for railroad costing." *Transportation Journal* 23 no. 2 (Winter 1983): 47–60.

——. "Meaning and measurement of transportation." *International Journal of Physical Distribution and Materials*

Management, Logistics: Interfaces with Marketing and Finance 14 no. 6 (1984): 5–16.

——. "The risk of reliance on the uniform railroad costing system." *Journal of Business Logistics* 6 no. 2 (1985): 142–162.

Yarusavage, George A. "Carrier negotiation and selection after reregulation: One shipper's experience." Unpublished paper submitted for partial fulfillment of the requirements for certified membership American Society of Traffic and Logistics. Louisville, Ky., Rpt. R-5648, 1983.

Shipment Consolidation

Anderson, David L. and Robert J. Quinn. "The role of transportation in long supply line just-in-time logistics channels." *Journal of Business Logistics* 7 no. 1 (1986): 68–87.

Anderson, David L. and Stephen B. Probst, "Innovative shipper transportation options: The post-deregulation experience." *Proceedings—Twenty-Seventh Annual Meeting Transportation Research Forum* 27 (1986): 216–223.

Cooper, Martha C. "Freight consolidation and warehouse location strategies in physical distribution systems." *Journal of Business Logistics* 4 no. 2 (1983): 53–73.

——. "Cost and delivery time implications of freight consolidation and warehousing strategies." *International Journal of Physical Distribution and Materials Management, Logistics: Interfaces with Marketing and Finance* 14 no. 6 (1984): 47–67.

Fulchino, Paul E., David L. Anderson, and Robert J. Quinn. "Motor carriers storm into era of A&D service." *Presidential Issue of Handling & Shipping Management* 26 no. 10 (September 1985): 78–82.

Jackson, George C. "Evaluating order consolidation strategies using simulation." *Journal of Business Logistics* 2 no. 2 (1981): 110–138.

——. "A survey of freight consolidation practices." *Journal of Business Logistics* 6 no. 1 (1985): 13–34.

Lai, Andrew W. and Bernard J. LaLonde. "A computer simulation study of freight consolidation alternatives at the distribution center level." *Proceedings, ORSA/TIMS Joint National Meeting* (November 1976).

Masters, James M. "The effects of freight consolidation on customer service." *Journal of Business Logistics* 2 no. 1 (1980): 55–74.

Newbourne, Malcolm J. A. *Guide to Freight Consolidation for Shippers.* Washington, D.C.: Traffic Service Corp., 1976.

Schuster, Alan D. "The economics of shipment consolidation." *Journal of Business Logistics* 1 no. 2 (1979): 22–35.

Sheahan, Drake. "Know thy carrier: All kinds of cost-saving opportunities exist in LTL consolidation" *Handling and Shipping Management* 23 (June 1982): 44–46.

Shelley, Donald F. "Costing pool shipments: A simple formula for determining the base-delivered prices on goods moving in mixed-load consolidated shipments." *Handling and Shipping Management* 23 (August 1982): 67–68.

Weart, Walter L. "The techniques of freight consolidation." *Traffic World* (March 19, 1984).

———. "Consolidation: Everybody out of the pool." *Distribution* 84 (May 1985): 50–62.

Other Elements of Traffic Management

Auguello, William J. *Freight Claims in Plain English*, rev. ed. Huntington, N.Y.: Shippers National Freight Claim Council, 1982.

Barrett, Colin. *Practical Handbook of Private Trucking.* Washington, D.C.: Traffic Service Corp., 1983.

Basedow, Jurgen. "Common carrier continuity and disintegration in U. S. transportation law — Part I." *Transportation Law Journal* 13 no. 1 (1984): 1–42.

———. "Common carrier continuity and disintegration in U.S. transportation law — Part II." *Transportation Law Journal* 13 no. 2 (1984): 159–188.

Bender, Paul S. "The international dimension of physical distribution management." In *The Distribution Handbook*, James F. Robeson and Robert G. House, eds. New York: Free Press (1985): 777–816.

Bierlein, L. W. *Red Book on Transportation of Hazardous Materials.* Boston: Cahners, 1977.

Calabro, Pat J. and Thomas W. Speh. "Historical perspectives on the freight car supply problem: The role of demurrage." *ICC Practitioners' Journal* 43 no. 4 (May–June 1976): 470–481.

Callahan, Timothy P. "Coping with hazardous material shipping." *Defense Transportation Journal* 41 no. 5 (October 1985): 28–32.

Cavinato, Joseph L. "Pricing strategies for private trucking." *Journal of Business Logistics* 3 no. 2 (1982): 72–84.

Cobert, Ronald N. and Robert L. Cope. "Transportation in depth: How to avoid paying twice for the same shipment when using a broker." *Traffic World* (January 14, 1985): 72–75.

Dulaney, Tom. "The lease vs. buy question." *Distribution* 79 (February 1980): 32–35.

Fair, Marvin L. and John Guandolo. *Transportation Regulation*, 9th ed. Dubuque, Iowa: Wm. C. Brown, 1983.

Feldman, Joan M. "Liability: Are shippers paying too high a price for low rates?" *Handling and Shipping Management* (June 1984): 45–48.

Gill, Lynn Edward. "Delivery terms — Important element of physical distribution." *Journal of Business Logistics* 1 no. 2 (1980): 60–82.

Guandolo, John. *Transportation Law*, 4th ed. Dubuque, Iowa: Wm. C. Brown, 1983.

Henke, T. R. *Managing Your Private Trucking Operation.* Washington, D.C.: Traffic Service Corp., 1976.

House, Robert G. and C. Charles Kimm. "Computer-assisted freight bill rating." In *The Distribution Handbook*, James F. Robeson and Robert G. House, eds. New York: Free Press (1985): 428–440.

Kahn, Fritz. "Railroad circulars: Exemptions are now the rule." *Distribution* 84 (June 1985): 60–62.

Loftin, Fred. "Proposed ocean rules boost liability coverage." *Canadian Transportation and Distribution Management* (October 1983): 46–48.

Mulcahy, Francis J. "Motor carrier cargo liability — An overview." *ICC Practitioners' Journal* 49 no. 3 (March–April 1982): 263–270.

Piercy, John E. "Lost, damaged and astray freight shipments: Some explanatory factors." *Transportation Journal* 20 no. 4 (Summer 1980): 33–37.

Probst, Lester A. *The "Freight Payment Problem": An Analysis of Industry Applied Solutions in the Nineteen Eighties,* 1st rev. Metuchen, N.J.: Transportation Concepts and Services, Inc., 1980.

Roberts, Joseph Michael. "Freight claims liability—Where are we today and where are we heading?" *Traffic World* (August 4, 1986): 77–81.

Safer, Laurie A. "Looser liability looms for freight shippers." *Handling and Shipping Management* (May 1981): 64.

Sigmon, Richard R. *Miller's Law of Freight Loss and Damage Claims,* 4th ed. Dubuque, Iowa: Wm. C. Brown, 1974.

Sorkin, Saul. *How to Recover for Loss and Damage to Goods in Transit, Release No. 1.* New York: Matthew Bender, 1977.

Southern, R. Neil. "Structure of the private motor carrier industry." *The Private Carrier* (November 1982): 11–17.

Stephenson, Fredrick J. and John W. Vann. "Air cargo liability deregulation: Shipper's perspective." *Transportation Journal* 20 no. 3 (Spring 1981): 45–58.

Stern, George. "Innovative pricing to improve utilization of rail plant and equipment." *Annual Conference Proceedings of the National Council of Physical Distribution Management* (1980): 323–330.

Tyworth, John E. "Pricing private freight car service: An analysis of transportation use and the problems of standby car capacity." *ICC Practitioners' Journal* 46 no. 6 (September–October 1979): 789–800.

Voorhees, Roy Dale, Benjamin J. Allen, and Dale J. Pinnekamp. "An examination and analysis of the 'invisible' full-service truck leasing industry." *Transportation Journal* 22 no. 3 (Spring 1983): 64–70.

Wilson, Lawrence B., Paul O. Roberts, and James T. Kneasfsey. "Models of freight loss and damage." *Traffic World* (April 9, 1979): 59–65.

Wolfe, E. Eric. "An examination of risk costs associated with the movement of hazardous materials." *Proceedings—Twenty-Fifth Annual Meeting Transportation Research Forum* 25 (1984): 228–240.

Selected Topical References—Part III

System Support

Allen, Mary K. and Margaret A. Emmelhainz. "Decision support systems: An innovative aid to managers." *Journal of Business Logistics* 5 no. 2 (1984): 128–142.

Berry, Thomas D. and Walters M. Tad. "Criteria of selection of transportation rating systems." *Proceedings of the National Council of Physical Distribution Management* (1982): 32–39.

Buys, Clifford. *Motor Carrier/Shipper Electronic Data Exchange.* Washington, D.C.: American Trucking Association, Management System's Department, 1985.

Cavinato, Joseph L., Alan J. Stenger, and Paul Novoshielski. "A decision model for freight rate retrieval and payment system selection." *Transportation Journal* 21 no. 2 (Winter 1981): 5–15.

Closs, David J. and Omar Keith Helferich. "Logistics decision support system: An integration of information data base and modeling systems to aid the logistics practitioner." *Journal of Business Logistics* 3 no. 2 (1982): 1–14.

Kaminski, Peter F. and David R. Rink. "Industrial transportation management in a systems perspective." *Transportation Journal* 21 no. 1 (Fall 1981): 67–76.

Lancioni, Richard A. and John Grashof. "Physical distribution organization and information systems development." *International Journal of Physical Distribution and Materials Management* 15 no. 1 (Spring 1973): 183–190.

Langley, C. John Jr. "Information-based decision making in logistics management." *International Journal of Physical Distribution and Materials Management* 15 no. 7 (1985): 41–55.

Levy, Michael, William Cron, and Robert Novack. "A decision support system for determining a quantity discount pricing policy." *Journal of Business Logistics* 6 no. 2 (1985): 110–141.

Magee, John F. "SMR forum: What information technology has in store for managers." *Sloan Management Review* (Winter 1985): 45–49.

McFarlan, E. Warren. "Information technology changes the way you compete." *Harvard Business Review* (May–June 1984): 98–103.

Parsons, Gregory L. "Information technology: A new competitive weapon." *Sloan Management Review* 25 no. 1 (Fall 1983): 3–14.

Tyndall, Gene R. and John R. Busher. "Improving the management of distribution with cost and financial information." *Journal of Business Logistics* 6 no. 2 (1985): 1–18.

Voorhees, Roy Dale, John C. Coppett, and Eileen M. Kelley. "Telelogistics: A management tool for the logistics problems of the 1980s." *Transportation Journal* 23 no. 4 (Summer 1984): 62–70.

Warner, Kendrick V. "The challenge of working with a transportation database." *Annual Conference Proceedings of the National Council of Physical Distribution Management* (1983): 290–316.

Control Concepts and Techniques

A. T. Kearney, Inc. *Measuring and Improving Productivity in Physical Distribution.* Oak Brook, Ill.: National Council of Physical Distribution Management, 1983.

Cavinato, Joseph L. *Finance for Transportation and Logistics Managers.* Washington, D.C.: Traffic Service Corp., 1977.

Chow, Garland and Richard F. Poist. "The measurement of quality of service and the transportation purchase decision." *The Logistics and Transportation Review* 20 no. 1 (1984): 25–43.

Dowdle, John R. "Translating the physical distribution strategy into a financial plan." In *The Distribution Handbook*, James F. Robeson and Robert G. House, eds. New York: Free Press (1985): 259–274.

Drugan, Cheryl Grazulis. "Improving productivity in transportation." *Handling and Shipping Management* (September 1982): 93–96.

Ernst and Whinney. *Transportation Accounting & Control: Guidelines for Distribution and Financial Management.* Oak Brook, Ill.: National Council of Physical Distribution Management, 1983.

Foggin, J. H. "Improving motor carrier productivity with statistical process control techniques." *Transportation Journal* 24 no. 1 (Fall 1984): 58–74.

Heschel, Michael. "Developing cost-effective transportation." *Transportation and Distribution Management* 15 (November–December 1975): 31.

Heskett, James L. "Organizing for effective distribution management." In *The Distribution Handbook*, James F. Robeson and Robert G. House, eds. New York: Free Press (1985): 817–836.

Jacobson, Ronald J. "Control chart techniques applied to the analysis of major shipping lanes." *Proceedings of the Logistics Research Forum* (1982): 233–241.

Kreitner, John. "Managing transportation productivity." In *The Distribution Handbook*, James F. Robeson and Robert G. House, eds. New York: Free Press (1985): 511–539.

LaLonde, Bernard J. and Paul Zinzser. *Customer Service: Meaning and Measurement.* Oak Brook, Ill.: National Council of Physical Distribution Management, 1976.

Lambert, Douglas M. "Distribution cost, productivity, and performance analysis." In *The Distribution Handbook*, James F. Robeson and Robert G. House, eds. New York: Free Press (1985): 275–319.

Langley, C. John Jr. and J. L. Hartzell. "Statistical process control applications in materials management." *Annual Conference Proceedings of the National Council of Physical Distribution Management* (1983).

Langley, C. John Jr., Robert J. Quinn, and Stephen I. Levine. "Microcomputers in logistics: 1985." *Annual Conference Proceedings of the Council of Logistics Management* (1985).

Murray, Robert E. "Matrix management and distribution." *Handling and Shipping Management* (Presidential Issue 1980–81).

Nishi, Masao. "Measuring the transportation manager's contribution to company profits." *Handling and Shipping Management* (May 1982): 85–92.

Ray, David, John Gattoma, and Mike Allen. "Handbook of distribution costing and control." *International Journal of Physical Distribution and Materials Management* 10 no. 5/6 (1980): 212–428.

Russ, Dave and Christine Fehlner. "Computers: Who's got 'em and what they're doing with them." *Handling and Shipping Management* (May 1984): 40–46.

Schiff, Michael. *Accounting and Control in Physical Distribution*. Chicago: National Council of Physical Distribution Management, 1972.

Sterling, Jay U. and Douglas M. Lambert. "A methodology for identifying potential cost reductions in transportation and warehousing." *The Logistics and Transportation Review* 5 no. 2 (1969): 1–18.

Todd, Arthur W. "Impact, knowledge and choice—One company's management of inbound transportation." *Proceedings—Twentieth Annual Meeting Transportation Research Forum* 20 (1979): 269–275.

Tyndall, Gene R., John R. Busher, and James D. Blaser. "Accounting and control in physical distribution—Phase I: Transportation." *Annual Proceedings of the National Council of Physical Distribution Management* (1982): 83–84.

"FRAPS: Borden Chemical's marriage of rate control and bank freight payment." *Computer Systems Report* 1 no. 3 (March 1982): 1–3.

Organization

Chandler, Alfred D. *Strategy and Structure*. Cambridge, Mass: MIT Press, 1962.

DeHayes, Daniel W., Jr. and Robert L. Taylor. "Moving beyond the physical distribution organization." *Transportation Journal* 14 no. 3 (Spring 1974): 30–41.

Falk, James P. "Organizing for effective distribution." *Annual Conference Proceedings of the National Council of Physical Distribution Management* (1980): 181–200.

Farrell, Jack W., ed. "Organization study: Distribution departments gain ground." *Traffic Management* 20 (September 1981): 42–50.

Gallagher, Patrick, ed. "Corporate transportation finds its role at FMC." *Handling and Shipping Management* (May 1982): 85–92.

Gill, Lynn Edward. "Organization for effective physical distribution." *Annual Conference Proceedings of the National Council of Physical Distribution Management* (1977): 103–120.

Granzin, Kent L., George C. Jackson, and Clifford E. Young. "The influence of organizational and personal factors on the transportation purchase decision process." *Journal of Business Logistics* 7 no. 1 (1986): 50–67.

Hale, Bernard J. "Organization for effective physical distribution management." *Annual Conference Proceedings of the National Council of Physical Distribution Management* (1977): 121–124.

Harold, John R. "A case for the centralized traffic manager." *Handling and Shipping Management* (July 1982): 61–68.

LaLonde, Bernard J. and Larry W. Emmelhainz. "Career patterns of logistics executives: 1985." *Proceedings of the Annual Conference of the Council of Logistics Management*, 1985.

Lambert, Douglas M. and John T. Mentzer. "Is integrated physical distribution management a reality?" *Journal of Business Logistics* 2 no. 1 (1980): 18–34.

Lynagh, Peter M. and Richard F. Poist. "Assigning organizational responsibility for interface activities: An analysis of PD and marketing manager interfaces." *International Journal of Physical Distribution and Materials Management, Logistics: Interfaces with Marketing and Finance* 14 no. 6 (1984): 34–43.

Maister, David H. "Organizing for physical distribution." *International Journal of Physical Distribution and Materials Management*, 8 no. 3 (1966).

Morehouse, James E. "Developing a competitive

edge: Linking logistics to the rest of your company." *Handling and Shipping Management* (Presidential Issue 1983–84): 6–14.

Stewart, Wendell. "The distribution organization." *36th Annual Meeting American Society of Traffic and Transportation* (1981): 11–45.

Stewart, Wendell and William J. Markham. "The role of the physical distribution manager."

In *The Distribution Handbook*, James F. Robeson and Robert G. House, eds. New York: Free Press (1985): 28–43.

Summers, Gary J. "Intercorporate communications—Management focus for the 80s." *34th Annual Meeting American Society of Traffic and Transportation* (1979): 78–85.

Wolf, Joel C. "Looking at operations." *Handling and Shipping Management* (September 1982): 81–88.

Author Index

Adams, Aden C., 214
Allen, Benjamin J., 214
Anderson, David L., 214, 291
Anderson, Ronald D., 55
Armstrong, Richard D., 260
Arntzen, Bruce C., 105
Augello, William J., 139, 333, 352

Bagby, John W., 214
Baker, Gwendolyn H., 55
Barks, Joseph V., 395
Barrett, Colin, 55, 70, 189, 291, 334
Barry, John J., 214
Bernstein, Robert S., 214
Betz, John, 329, 334
Blackwell, Richard B., 214
Blaser, James D., 435, 454
Bohman, Ray Jr., 139
Borts, George H., 214
Bowersox, Donald J., 23, 291, 483

Bradley, John C., 78
Brodeur, Leon R., 13
Brown, Terrence A., 55
Bruning, Edward R., 55, 214
Bunce, Ellott., 77
Burt, David N., 260
Busher, John R., 435, 454
Butler, Robert M., 139
Buys, Clifford, 105

Callson, Olliver G., 77, 138–139
Carlson, Gene P., 414
Carr, Ronald G., 78
Cavinato, Joseph L., 377
Chandler, Alfred D., 483
Chow, Garland, 77
Cobert, Ronald N., 55
Constantin, James A., 559
Cooper, Martha C., 291
Cope, Robert L., 55

Corsi, Thomas M., 55
Crichton, Jean, 260
Crum, Michael R., 55
Cushman, Frank M., 105

Dart Robert C. Jr., 105, 260
Davis, Grant M., 77
Davis, Herbert W., 469
Dobler, Donald W., 260
Dolce, John, 366
Dulaney, Thomas C., 291

Emmelhainz, Larry W., 483
Emrich, Richard S. M., 139
Evans, James R., 214

Fair, Marvin L., 334, 353
Falk, James P., 483
Farrell, Jack W., 22, 23, 483
Farris, Martin T., 22, 214
Fehlner, Christine, 451, 454
Feldman, Joan M., 189, 352, 455
Fisher, Roger, 261
Flexner, Donald L., 78
Flood, Kenneth U., 77, 138–139
Foggin, James H., 427, 454
Freeland, Ronald L., 139

Gallagher, Patrick 23, 214, 483
Gawlick, Dennis M., 54
Gilbert, Ernest, 189
Gill, Lynn Edward, 189
Granzin, Kent L., 214
Greiner, Larry, 480–481, 483
Grimm, Curtis M., 55
Guandolo, John, 334, 352–353
Guzzardi, Walter Jr., 214

Haan, Vernon J., 139
Hale, Bernard J., 162
Harold, John R., 483
Harrington, Lisa H., 55
Hartzell, J. L., 454
Haupt, Richard, 214, 260
Heisley, Stephen E., 78
Henke, Thomas R., 363
Hoeberling, Carl W., 214
Hoffman, Kurt, 261
Hoffman, Stanley K., 215
House, Robert G., 161–162, 454

Isman, Warren E., 414

Jablonski, Sylvester J., 77, 138–139
Jackson, George C., 214, 291
Jerman, Roger E., 55
Jessen, Jessie C., 397

Kalish, Steven J., 78
Karass, Chester L., 260
Kline, Jeffery C., 252
Kneasfsey, James T., 352
Koontz, Harold, 23
Kursar, Robert J., 105

Lai, Andrew W., 29
LaLonde, Bernard J., 55, 291, 467, 483
Lambert, Douglas M., 22, 454
Langley, C. John Jr., 161, 447–448, 454
Lee, Lamar Jr., 260
Leilich, Robert H., 434
Levine, Stephen I., 454
Levitt, Theodore, 78
Loftin, Fred, 334
Lundy, Robert, 55
Lynagh, Peter M., 55, 214

Maister, David H., 483
Maltz, Arnold B., 260
Marien, Edward J., 105, 214, 275
Martell, John, 55
Masters, James M., 291
McBride, Mark E., 260
McCreary, Edward Jr., 214
McGinnis, Michael A., 162
Michalson, Jeff, 109
Mulcahy, Francis J., 334
Murr, Alfred, 402

Nadler, David A., 479
Newbourne, Malcolm J., 291
Nishi, Masao, 23

Petracek, S. J., 249, 260
Poist, Richard, 77
Probst, Lester A., 189, 352

Quinn, Francis J., 78
Quinn, Robert J., 291, 454

Raiffa, Howard, 261
Ringbakk, Kjell A., 159
Roberts, Joseph Michael
Roberts, Paul O., 352
Robeson, James F., 22, 161–162, 454
Ross, Warren R., 377
Russ, Dave, 451, 454

Safer, Laurie A., 352
Sampson, Roy J., 22
Schrock, David L., 260
Seiden, Elliot M., 78
Seip, Douglas W., 196, 213
Selby, George W. Jr., 78
Sigmon, Richard R., 333

Smith, Wm. Randolph., 78
Solomon, Mark B., 105, 189
Sorkin, Saul, 334
Spangler, Earl J., 11
Spychalski, John C., 139
Steiner, George A., 162
Stenger, Alan J., 23
Stephenson, Frederick J., 352
Stern, George, 315
Stewart, Wendell, 23
Stock, James R., 22, 55
Stone, James H. 139
Stoner, James F., 478, 483
Sutton, Robert M., 260

Taff, Charles A., 105, 291
Talley, Wayne K., 260
Tushman, Michael L. 479
Tye, William B., 77
Tyndall, Gene R., 435, 454
Tyworth, John E., 378

Uggen, Michael W., 214, 233

Ury, William, 261
Urycki, Hank, 55

Vann, John W., 352
Vick, J. B., 98

Wastler, Allen., 261
Waters, L. L., 105, 214
Waters, W. G. II, 260
Weart, Walter L., 105, 266–267, 275, 281
Wheeler, Porter K., 139
Whitten, Greg, 105
Whitten, Herbert O., 105
Wilson, G. Lloyd, 23, 77, 352
Wilson, Lawrence B., 352
Wise, Dean H., 214
Wood, Wallace R., 214

Yarusavage, George A., 207, 214
Young, Clifford E., 214
Yunashko, Linda M., 139

Zinser, Paul H., 161, 467, 483

Subject Index

A. O. Smith Company, 443
Abandonments, 40, 122
Accessorial services, 294–313
Acquisition methods, 2, 17, 153–157
Act to Regulate Commerce, 33
Actual placement, 245, 248
Actual-value ratings, 325–327
Affirmative routing, 108
Agency tariff, 63
Aggregate tender, 132, 289–290
Aggregate-of-intermediate, 122
Agreed rate, 131
Air bill, 170
Air Cargo Guide, 106
Airline Tariff Publishing Company (ATPCO),
    64
Alternating rates, 129, 277
American Arbitration Board, 342
American Delivery System, 41
American Society of Transportation and
    Logistics, 338

American Trucking Association, 337, 364
Antitrust, 72–75, 211
    business review procedure, 74
    and joint rates, 120–121
    penalties, 72
    per se violations, 73–74
    and proportional rates, 124
    and rate bureaus, 64–66
    rule-of-reason violations, 74–75
Any quantity (AQ) rating, 132
Approaches to coordination, 477–479
Arbitraries and differentials, 124–125
Arbitration procedures, 341–342
Assembly and distribution (A&D) carrier, 272
Assembly service, 268, 272, 282
Assigned cars, 20–21
Association of American Railroads, 68, 122,
    210, 337, 342
Auditing freight bills, 20, 45, 171–173,182–185,
    199, 437–443
Autorelease clause, 326–327, 331

Average demurrage agreement, 303
Averitt Express, 29

Backhaul allowances, 374
Bailment, 319, 324
Bank payment plan, 185, 440
Bareboat charter, 386
Base inventory (see Base stock)
Base point, 89
Base stock, 49
Base-excess rate, 130
Basic demurrage agreement, 302
BDC Corp., 374
Bid method, 201–209
Bill of lading, 164–171, App. 7.1
    air, 170
    functions of, 168
    government, 171
    issuance, 168
    liability, 321–322
    livestock, 170
    ocean, 170–171
    order, 170, 180–182
    straight, 170
    through export, 170
    without recourse clause, 173
Blanket rate structure, 127–128
Borden Chemical, 440–443
Boycotts, 73
Breakbulk, 239–241, 265, 267–268, 272, 282
Brokers, 43, 265, 264, 267, 299
Budget process, 19, 159, 433–437
Bumping, 134
Bunching cars, 303
Burlington Northern Industries, 58–59, 126
Business logistics
    evolution, 14–21
    nature, 5
Business review procedure, 74

Capacity Exchange, Inc. (CAPEX), 96
Capacity load, 287
Cargo Data Interchange System (CARDIS), 398
Carload (CL) rate, 128–129
Carmack liability, 319, 327–329, 331, 337
Carriage of Goods by Sea Act (C.O.G.S.A.),
    327–330, 337
Carrier
    dedicated contract, 272–273
    firms and entities, 41–43
    insurance, 20, 47, 51
    legal forms, 29–33
    pricing situations, 254–256
Carrier terminal operations, 239–241, 271–272
Carrying cost, 49
Centralization v. decentralization, 471–474
Charge v. rate, 284

Charters, 386, 391
Chemical Hazards Response Information
    System (CHRIS), 410
Chemical Manufacturers Association,
    122
Chemical Transportation Emergency Center
    (CHEMTREC), 410
Chessie System, 30, 313, 303
Civil Aeronautics Act of 1938, 33
Claims, 20, 336–351
    arbitration, 341–342
    concealed loss and damage, 341
    control of, 342–346
    functional cost category, 243–244
    loss, damage, and delay, 336–346 (see also
        Liability)
    overcharge, 88, 95, 346–349 (see also Freight
        payment)
    prevention program, 346
    role in carrier selection, 342–344
    and terms of sale, 179
    unlawful rates or practices, 66, 349–350
    venue, 341
Class rate, 82–93, App. 4.1
Classes (see Ratings)
Classification
    bureau agreements, 65
    committees, 82–84
    commodity, 82–89
    criteria, 85–89
    rules, 82, 95
    tariffs, 82–85
    yard, 248
Clayton Act of 1914, 72
Clean Water Act of 1977, 412–413
Closed routing (see Negative routing)
Cluster network, 289
Code of Federal Regulations (CFR), 28
Collateral liability, 51
Collect on delivery, 180–182
Coloading, 265
Combination rate, 120, 122–123
Commodity column rate, 94
Commodity rate, 93–94, App. 4.1
Common carriage, 28–29, 59–69
Common cost, 238–239
Common law, 319
Compensated intercorporate hauling, 31, 373
Competitive bidding, 201–209
Complaints and protests
    railway rate, 67–68, 70
    nonrailway rate, 68–71
Computer technology, 6, 12, 14, 151–153, 155,
    270–271, 285, 429, 449–452, 468 (see also
    Electronic data interchange, Information
    system support)
Computer usage patterns 449–452

Concealed loss and damage, 341 (see also Delivery receipt)
Conditional sliding scales, 131–132
Conrail, 29, 30, 122, 298, 313
Consolidated Freightways, 29, 99, 171
Consolidation (see Shipment consolidation)
Constructive placement, 245, 248, 305
Constructively intermediate points, 283
Container rate, 135
Container-on-flat-car (COFC), 36, 40–41, 43, 117

Containerization, 382–384
Contingency planning, 157–158, 210–211
Continuous movement rate, 125–126
Contract carriage, 29–30, 59, 69–70, 272–273 (see also Contracting)
Contract rate, 59, 69–71 (see also, Contracting)
Contracting
    administration, 208–209
    advantages and disadvantages, 197–201
    background research, 195–202, App. 8.2
    bid method, 201–209
    checklist, 215–228
    cycle, 195–209
    definition, 194
    evaluation and award, 206–208
    negotiation method, 202–209
    pitfalls in, 209–211
    preaward survey, 208
    prebid conference, 206
    railway innovations, 199–201
    solicitation, 203–206
Contribution-to-burden, 245
Control
    and contracting, 199
    of cargo claims, 342–346
    and information system support, 429–431
    key indicators, 157
    key techniques of, 431–449
    process 2, 21, 148, 159–160, 423, 426–428
    productivity and performance measurement, 444, App. 17.1
    of routing, 178
    of shipment documents, 171
    statistical methods, 160, 444–449
    system, 21
    tasks, 19
    and terms of sale, 176–178
Coordinated Freight Classification (CFC), 85
Cost analysis, 48–53 (see also, Costing)
Cost recovery index, 68
Cost recovery percentage, 67
Cost sharing, 250–252
Costing
    carload shipments, 245–250
    concept of, 236

contribution-to-burden approach, 245
Highway Forms A and B, 239
individual shipments, 237–239, 433
less-than-truckload shipments, 239–245
Single and Interline Costing Program, 239
standard transportation, 432–433
and traffic management, 250–252
truckload shipments, 245
Council of Logistics Management, 205, 466
Courier service, 41, 126
Credit terms, 50, 185–187
Cube out, 87, 272
Customhouse broker, 393
Customer service, 8, 145, 151–153, 270–271, 273, 285, 429, 466–469, 473
Customs duties, 392

Declared value, 326–327, App. 7.1
Dedicated contract carriage, 272–273
Deductibles, 325–326
Deficit weight, 288 (see also Weight breaks)
Delay claim, 336
Delay cost, 50
Delivered pricing, 178, 273
Delivery receipt, 165, 173–174
Demurrage, 248, 250, 302–309
Density
    in classification, 87
    and linehaul costs, 243
    rates, 132–134
Department of Transportation, 30, 51, 119, 365, 373, 405, 413
Detention, 302, 309–311
Direct and sole cause standard, 321, 324
Direct shipping cost, 48–49
Direct-connector doctrine, 65
Discounts (see also rates)
    phantom, 60
    and terms of sale, 179–180
Distance scales, 89–92
Distribution requirements planning, 470
Distribution Sciences, Inc., 65
Distribution service, 265, 267, 272
Diversion, 298–301
Division of markets, 73
Dock handling, 87, 241–242, 272 (see also Rehandling)
Documentation (see also Electronic data interchange, Information system support)
    contract, exempt, and private carriage, 175
    export-import, 396–398
    hazardous materials, shipping, 164–175
    tasks, 20, 164–167
    terms of sale, 17, 175–180, 393–396
Dow Chemical, 4

Dravo Bargeline, 29
Drawback zone, 392
Driving
    peddle, 241
    stem, 241
DSI RAIL, Inc., 65, 96–97
Dual authority, 30, 35–36
Dunnage, 311
Duplicate payments, 20, 95

Eastern Central Motor Carrier Association, 91
Electronic data interchange, 13, 14, 95–99,173, 199, 452
Electronic filing of tariffs, 96–98
Elmore v. Stahl, 321
Environmental Protection Agency, 405, 410–411
Equivalent-trailer miles, 243
Exception rating, 85
Exclusive deals, 74
Exclusive use of the vehicle, 126
Exempt carriage 32–33, 59
Export trading companies, 400
Export-import
    documentation, 396–398
    facilitators, 386–389
    rate, 125
    transportation, 381–382

Federal Express, 41
Federal Maritime Commission, 391
Federal Register, 27–28
Federal Trade Commission Act of 1914, 72
Federal Trade Commission, 75
Fifteen-day rule, 341
Filed-rate doctrine, 60
Financial Accounting Standards Board (FASB), 367, 369
Firestone Tire & Rubber Co., 13
Fleet
    rail, 375–376
    supervision, 20–21
    truck, 357–375
Flexible budgets, 434, 437
FOB, 175–179 (see also Terms of sale)
Ford Motor Company, 4, 13, 20, 205, 199, 305–309
Foreign trade zones, 392
Formula rate, 99–102
FRAPS, 440, 443
Free time (see Demurrage, Detention)
Freight audit, 154, 437–443
Freight bill, 164, 171–173
Freight Forwarder Act of 1942, 33
Freight forwarders, 40–42, 264, 267, 360
Freight payment, 20, 95, 154, 182–185, 437–443

Freight Rating, Auditing, and Payment System (FRAPS), 440
Freight, all kinds (FAK), 135, 287, 319, 389
Full-cost pricing, 254–255, 370–371
Functional v. product-line orientation, 475

General average, 398
General damages, 325
General Electric Company, 2, 4, 31, 81, 102, 131–132, 371, 440
General Foods Corp., 5
General Mills, 4
General Motors, 21, 151, 305
General rate increases, 65
Going value of the concern, 40, 68
Goodrich, 376
Government (Section 10721) rate, 65, 136
Government bill of lading, 171
Group rate structure, 127–128
GTE, 201, 205–206, 211

Hague Rules, 327
Hammermill Paper, 376
Handling (see Dock handling, Rehandling)
Harter Act, 327
Hazardous materials, 404–412, App. 16.2
Hershey Chocolate Company, 10
Highway Forms A and B, 239
Horizontal relationship, 73

IBM, 195, 201, 205

Illinois Central, 131
In-transit inventory, 49
Incentive rate, 129–130
INCOTERMS, 393
Industrial switching, 245–246
Information system support, 14, 17, 19, 71–72, 183–185, 398, 426–431
Information technology in pricing, 94–102
Insurance
    carrier, 20, 47, 51
    marine, 398–399
Intercoastal Steamship Freight Association, 64
Intermediate point rule, 127
Intermediate switching, 313
Intermediate-point rate, 126
International transportation (see Export-Import)
Interstate commerce, 117–120
Intraplant switching, 313
Intrastate v. interstate rate, 117–120
Inventory
    carrying cost, 49
    in transit, 49
    base, 49
    safety stock, 49–50

Invitation for bids, 205–206

J.C. Penny, 4
Joint action among affiliates, 74–75
Joint agency tariff, 63
Joint bargaining, 74
Joint costs, 238–239, 255
Joint rate, 65, 120–122
Jurisdictional threshold, 67–68
Just-in-time delivery, 12, 272, 470
Justice Department, 74

K Mart, 4, 173
Kaiser, 205
Keogh Doctrine, 73
Knocked down (k.d.) freight, 87

Land-bridge, 384
Lane-load average, 243
Lease v. purchase decision, 366–370
Less-than-carload (LCL) rate, 82, 128–129
Less-than-truckload (LTL) rate, 84, 128–129,135
Letter of credit, 389, 393–396
Liability, 319–332
    air common carriage, 330
    autorelease clause, 326–327, 331
    in classification, 87–89
    collateral, 51
    direct and sole cause standard, 321, 324
    Elmore v.Stahl, 321
    exempt and contract carriage, 331
    general damages, 325
    overland carriage, 319–327
    special damages, 326
    warehouse operator, 324–325
        water common carriage, 327–330
Lighter-aboard ship (LASH), 383, 391
Line versus staff, 471
Liner ship, 382–383
Liner-breakbulk rate, 389–390
Livestock bill of lading, 170
Load planning, 274–290
Loaded to full visible capacity (LFVC), 286
Loading and unloading, 20, 302, 309
Local area network (LAN), 450–451
Local rate, 120
Long ton, 390
Long-and-short-haul clause, 123, 128
Loss and damage claim, 336–342

Making the shipment, 19–20, 164–187
Managerial and operational control, 148
Managerial planning, 197
Manifest, 171
Marine insurance, 392–393, 398–399
Market dominance, 66–67

Materials requirements planning, 470
Materials-logistics organization, 10, 17
Matsui Amendment, 331
Maximum rate, 40
McGraham's Formula, 99
Metric ton, 390
Micro-bridge, 305
Middle Atlantic Conference, 84
Middlewest Motor Freight Bureau, App. 4.1
Mileage allowances, 374
Mileage rate structure, 126–127
Mini-bridge, 384–385
Minimum charge, 40, 136
Minimum weight, 86, 122, 129–130 (see also
        Weight breaks, Weight group)
Minimum-maximum objective position,254
Misrouting, 119–120, 350
Mission statement, 149–150
Mixed shipments, 287–288
Mode selection, 43–44
Motor Carrier Act of 1980, 31, 33–36, 64, 66,
        135, 373–374
Motor Carrier Directory, 106
Multiple tender (see Aggregate tender)
Multiple-car rate, 130

Nabisco Brands, Inc., 253–254
National Classifications Committee (NCC), 83,
        88
National Freight Claims Council (NFCC), 337,
        341
National Highway and Airway Carriers Routes,
        106
National Industrial Traffic League (NITL), 60,
        66, 122, 346
National Motor Freight Classification (NMFC),
        84–85, 288, 326
National Motor Freight Transportation
Association (NMFTA), 84, 89
National Railroad Freight Committee (NRFC),
        82
National Rate Basis Tariff, 89
National Small Shipments Traffic Conference
        (NASSTRAC), 74
National Starch and Chemical Corp., 270
Negative routing, 108
Negotiation, 253–257
    leverage, 16
    method, 202–209
    of rates and servies, 252–257, (see also
        Contracting)
Nested freight, 87
New England Classifications, 85
New England Motor Rate Bureau, 85
No frill rate, 298
Nonvessel operating common carrier
        (NVOCC), 385

Occupational Safety and Health
      Administration (OSHA), 413
Ocean bill of lading, 170–171
Official Guide to Railroads, 106
Open routing, 108
Operating authority (see Legal forms of
      carriage)
Operational planning, 147–148
Operations
      loading and unloading, 20, 274–290, 302,
            309 (see also, Costing, Demurrage,
            Detention, Shipment consolidation)
      claims (see Claims)
      consolidation (see Shipment consolidation)
      documentation (see Documentation)
      fleet supervision, 20–21 (see also, Private
            carriage)
      making the shipment, 164–187
      overview of traffic, 19–21
      scheduling, 19 (see also Shipment
            consolidation, Release policies)
      shipment process, 165
Order bill of lading, 170, 180–182
Order cost, 50
Order cycle, 271, 282
Organization
      approaches to coordination, 477–479
      and customer service, 466–468, 473
      elements of design, 464–465
      evolution, 14–21, 480–481
      influences on structure, 465–470
      integration, 476
      issues and alternatives, 471–479
      materials-logistics, 10, 17
      measuring effectiveness, 479
      phases of business logistics, 14–21
      and private motor carriage, 362
      OT-5 agreement, 375
Over, short, and damage (OS&D), 151 (see also
      Liability, Claims)
Overcharge, 88, 95, 346–349
Overflow shipments, 285–287

Package express, 41, 126
Package testing, 88
Paired and reload rate, 126
Particular average, 398–399
Peddle driving, 241
Per diem, 303–304
Per se violations, 73–74
Performance measurement, 157, 444
Phantom discounts, 60
Pickup and delivery, 132, 241, 245, 272, 281–
      282, 289, 301–302
Piggyback, 36, 40–43, 117
Pipelines, 41
Pitfalls in contracting, 209–211

Planning
      budgets, 19, 159, 433–437
      contingency, 157–158, 210–211
      continuum, 145
      general framework, 145
      goals and traffic, 151
      load, 274–290
      managerial, 147
      negotiation strategy, 252–257
      operational, 147–148
      organizational design, 404–470
      overview, 14–19
      private fleets, 356–377
      shipment consolidation programs, 268–274
      steps in strategic traffic, 148–161
      strategic, 145–148
      and terms of sale (see Terms of Sale)
Point-to-point rate, 94, 126
Pool shipments, 265, 278–282, 301
Port differentials, 125
Post-shipment audit (see Freight
      payment)
Practicably participate, 65
Preaward survey, 208
Prebid conference, 206
Prerate (see Freight payment)
Price-fixing, 73
Primary business test, 30
Private carriage, 356–377
      backhaul allowances, 374
      costs, 362–366, 372
      purchase v. lease decision, 366–370
      rail operations, 375–376
      subsidiary, 373–374
      transfer pricing, 370–372
      trip leasing, 32, 374–375
      truck fleets, 357–375
Private Carrier Conference, 119
Pro number, 171
Producers Price Index, 68
Productivity measurement, 157, 444
Proportional rate structure, 126
Proportional rate, 123–124
Protective services, 301
Protests (see Complaints)
Public notice requirement, 63
Public team tracks, 245
Purchase v. lease decision, 366–370
Railroad Revitalization and Regulatory Reform
      Act of 1976, 72, 76
Rate basis number (RBN), 89–92, 126
Rate bureau, 36, 64–66
Rate computerization, 94–102
Rate escalation clause, 210
Rate pony, 90
Rate publication, 60–66 (see also Tariffs)
Rate regulation, 33, 36, 40, 60–71

Rate retrieval, 94–102, 182–185
Rates
    aggregate tender, 132
    aggregate-of-intermediate, 122
    agreed, 131
    alternating and minimum weights, 129, 277
    any quantity (AQ), 132
    arbitraries and differentials, 124–125
    base-excess, 130
    blanket, 127–128
    carload (CL), 128–129
    class, 82–93, App. 4.1
    combination, 120, 122–123
    commodity column, 94
    commodity, 93–94, App. 4.1
    complaints and protests, 66 (see also, Rate
      regulation)
    conditional sliding scales, 131–132
    container, 135
    continuous, 125–126
    contract, 59, 69–71 (see also, Contracting)
    deductibles, 325–326
    density, 132–134
    discounts, 131, 179–180, 265
    export-import, 125
    formula, 99–102
    freight, all kinds (FAK), 135, 287, 319, 389
    general increases, 65
    government (Section 10721), 65, 136
    group, 127–128
    incentive, 129–130
    intermediate point, 126
    intrastate v. interstate, 117–120
    joint, 65, 120–122
    less-than-carload (LCL), 82, 128–129
    less-than-truckload (LTL), 84, 128–129, 135
    liner breakbulk, 389–390
    local, 120
    long-and-short-haul, 123, 128
    maximum, 40
    mileage, 126–127
    minimum charge, 40, 136
    multiple car, 130
    multitier (base-excess), 130
    no frill, 298
    paired and reload, 126
    point-to-point, 94, 126
    proportional structure, 126
    proportional, 123–124
    public notice, 63
    released value, 134–135, 325–326, 344–346
    Section 22, 65, 136
    single-factor program, 135, 268, 285–286,
      290
    single-factor, 64–65, 120, 124, 300
    single-line, 120
    through, 120
    time-volume, 130–131
    trading center concept, 96
    trainload, 130
    tramp ship, 391
    truckload (TL) 128–129
    uniform structure, 126
    zip-code, 90–92, 99, App. 4.1
Ratings
    AQ, 86, 132
    commodity, 82–89
    exception, 85
    LCL/CL, 82, 128–129
    LTL/TL, 86, 128–129
    released-value, 134–135, 325–326, 344–346
RCA Corporation, 252, 257
Reasonable dispatch, 320, 324–325
Reciprocal switching, 121–122, 313
Reciprocity, 211
Reconsignment, 298–301
Reed-Bulwinkle Act of 1948, 64
Refused v. returned shipment, 174–175
Regular Common Carrier Conference (RCCC),
    239
Regulation, 27–41
    hazardous material transportation, 404–412
    nature of economic, 27, 36
    rate, 33, 36, 40, 60–71
    safety, 27, 365, 412–413
    social, 413–412, App. 16.1
    tariff, 60–63
Regulatory agencies, 33
Rehandling, 241, 242
Released-value rate, 134–135, 325–326, 344–346
Reparation claim, 349–350
Request for proposal, 205–206
Requirements contracts, 74
Revenue adequacy, 68
Roadway Express, 29, 99
Roadway Package Express, 41
Rocky Mountain Motor Tariff Bureau, 99
Roll-on-roll-off (RORO) ship, 382, 391
Roller car, 299
Rule-of-reason violation, 74–74

Safety regulation, 27, 365, 412–413
Safety stock, 49–50
Santa Fe Railway, 29, 30
Satellite terminal, 239–241
Scheduling transportation, 19 (see also,
    Shipment consolidation, Release
    policies)
Sears, 4
Section 22 rate, 65, 136
Selection of transportation services, 16–17, 47–
    52
Sherman Act of 1890, 72
Shipment consolidation, 19, 264–290

cost-service tradeoffs, 271–274
and customer service, 270, 273
general forms, 265–268
opportunities, 274
planning framework, 268–274
release policies, 282
techniques and practices, 274–290
Shipment costing, 237–239, 433
Shipment process, 165
Shippers National Freight Claims Council
(SNFCC), 337–338 341, 346
Shippers' agents, 43, 264, 267, 360
Shippers's associations, 42–43, 264, 267
Shipping Act of 1916, 391
Shipping Act of 1984, 391, 399–400
Short ton, 390
Sight draft, 396
Single and Interline Costing Program (SICP),
239
Single source leasing, 373–374
Single-factor rate program, 135, 268, 285–286,
290
Single-factor rate, 64–65, 120, 124, 300
Single-line rate, 120
Small package express, 41, 126
Solicitation, 203–206
Sourcing, 11, 203–205
Southern Railway System, 102
Span of control, 465, 476–477
Special damages, 326
Special property brokers, 43, 264
Split delivery, 301
Spreadsheet, 14, 344–345
Stagflation, 9, 12
Staggers Rail Act of 1980, 33, 36–40, 64, 66,135,
326, 332, 341
Stand-alone network, 241
Standard & Poor's Register, 204
Standard point location code (SPLC), 268
Standard transportation commodity code
(STCC), 84
Standard transportation costing, 432–433
Statistical process control, 160, 444–449
Stem driving, 241
Stopoff service, 267–268, 282–285, 301–302
Storage charges, 311
Stowability, 87
Straight bill of lading, 170
Strategic business plan, 148–149
Strategic business unit (SBU), 151
Strategic management, 148
Strategic planning, 145–148
Surcharges, 40, 121–122
Surface Freight Forwarder Deregulation Act of
1986, 33, 40
Surface Transportation Act of 1982, 364
Switch order, 305

Switching
industrial, 245–246
intermediate, 313
intraplant, 313
reciprocal, 121–122, 313
System audit, 153, 268

Tariff
class rate, 82–93, App. 4.1
classification, 82–87
commodity rate, 93–94, App. 4.1
computerization, 71–72, 99, 102
electronic filing, 96–98
filing, 60, 71
governing, 60–61, 89, 309
group guide, 88, App. 4.1
pricing systems, 80–102
publication, 63
regulation, 60–63
simplification, 99
supporting, 60–61, 89
types, 60–61
zip-code, 71, 90, 99, App. 4.1
Terminal charges, 391–392
Terms of sale, 17, 175–180, 393–396
Three-way rule, 297
Through export bill of lading, 170
Through rate, 120
Time charter, 386, 391
Time-volume rate, 130–131
Titles, 5, 481–482
Total cost concept, 6
Toto Purchasing and Supply Co., 31, 54, 373
Tracing shipments, 19, 124, 274
Trading center concept, 96
Traffic Bulletin, 89
Traffic management
control (see Control)
decision framework, 14–21
definition, 1, 5
export-import, 380–401
modern orientation, 1–2, 4–6, 12–14, 461–
464
objectives, 2, 16, 149–151
organization, (see Organization)
overview of operations, 19–21 (see also
Operations)
performance measurement, 444
planning, (see Planning)
purchasing management similarities, 18
resources, 17–18
responsibilties, 14–21, 462–464
titles, 5, 481–482
Traffic manager, 5
Trailer-on-flat-car (TOFC), 36, 40–41, 43, 117
plans, 42
retail and wholesale services, 107

Trainload rate, 130
Tramp ship, 385–386, 391
Transfer pricing, 370–372
Transit privileges, 295–298
Transportation
    accounting, 431–432
    selection and acquisition of, 2, 16–17, 47–52,
        153–157
    budget process, 19, 159, 433–437
    carrier firms and entities, 41–43
    cost analysis, 48–53 (see alsoCosting)
    equipment, 43–44, 272
    export-import, 381–389
    hazardous materials, 404–412
    legal influences on, 27–28,33–41
    mode selection, 2, 16–17, 43–47(see also
        Planning)
    pricing, 58–75
    regulation (see Regulation)
    regulations, 27
    system audit, 153, 268
Transportation Act of 1940, 33
Transportation Arbitration Board (TAB), 341
Transportation Brokers Conference of America,
    43
Transportation Data Coordinating Committee
    (TDCC), 96, 452
Transportation Safety Act of 1974, 405–406
Trip leasing, 32, 374–375
Truckload (TL) rate, 128–129
TRUXBUX, 371
Tying agreements, 73

Uniform Classification Committee (UCC), 82,
    88
Uniform Commercial Code (UCC), 28, 168, 175,
    319
Uniform Freight Classification (UFC), 82–84,
    210, 288

Uniform Rail Costing System (URCS), 238–239,
    250
Uniform rate structure, 126
United Airlines, 29
United Nations Conference on Trade and
    Development (UNCTAD), 399–400
United Parcel Service (UPS), 41
United States Code (U.S.C.), 27
Unlawful rates and practices, 66, 349–350
Unnamed points, 126

Value engineering, 200
Variable-cost pricing, 255, 371
Variance analysis, 434
Venue, 341
Vertical relationship, 73
Virtual insurer, 321
Voyage charter, 386

Warsaw Convention, 330, 337
Waterways Freight Bureau, 64
Waybill, 164, 171
Weighing, 311–313
Weight break, 129–130, 274–278
Weight groups, 92–93
Western Trunkline Tariff Bureau, 65
Weyerhaeuser, 4
Without recourse clause, 173

Xerox Corp., 251–252

Yellow Freight Sytem, 29, 99, 173

Zip-code tariff
    rate system, 71, 90–92, 99, App.4.1
    zone, 88–92
Zone of ratemaking freedom (ZORF), 68–69